Mathematics in Mind

The monographs and occasional textbooks published in this series tap directly into the kinds of themes, research findings, and general professional activities of the **Fields Cognitive Science Network**, which brings together mathematicians, philosophers, and cognitive scientists to explore the question of the nature of mathematics and how it is learned from various interdisciplinary angles. Themes and concepts to be explored include connections between mathematical modeling and artificial intelligence research, the historical context of any topic involving the emergence of mathematical thinking, interrelationships between mathematical discovery and cultural processes, and the connection between math cognition and symbolism, annotation, and other semiotic processes. All works are peer-reviewed to meet the highest standards of scientific literature.

Nikolay I. Kazimirov

Mathematics as a Foreign Language

Springer

Nikolay I. Kazimirov
Moscow, Russia

ISSN 2522-5405 ISSN 2522-5413 (electronic)
Mathematics in Mind
ISBN 978-3-032-30720-0 ISBN 978-3-032-30721-7 (eBook)
https://doi.org/10.1007/978-3-032-30721-7

This Springer imprint is published by the registered company Springer Nature Switzerland AG.
The registered company address is: Gewerbestrasse 11, 6330 Cham, Switzerland

Preface

Dear Reader,

This book collects and streamlines materials I have developed for my course on learning mathematics as a foreign language. These ideas, originally conceived as teaching aids, serve as the foundation for this publication. The book includes approximately 190 exercises of varying difficulty, with detailed answers to all of them provided in the Appendix, supporting independent study and self-assessment. Let me highlight several features that distinguish this book from typical mathematics textbooks.

First, the goal is not to teach mathematics or mathematical logic *per se*, although the text abounds with examples from these disciplines. Rather, the aim is to reveal how transparent and comprehensible the *lingua franca* of the modern exact sciences can be once one undergoes a period of adaptation—a journey for which this book is intended to be your guide.

This endeavor is particularly relevant today, given the rapid advancement of artificial intelligence—entities for which mathematical language is native. The algorithms controlling robots, neural networks analyzing data, and large language models generating text are all built upon the strict laws of logic and algebra that constitute the grammar of mathematics. While mastering this language does not grant a complete understanding of the "mind" of a machine, it provides essential insight into the formalized structures that govern automated decision-making.

Second, this work is designed primarily as supplementary reading rather than a traditional textbook. The language of mathematics differs profoundly from any spoken language, and mastery is defined by criteria distinct from standard proficiency scales. Therefore, this book is best used to bridge the gap between intuitive understanding and rigorous formalism, complementing standard courses in logic, set theory, or algebra.

Third, the book does not claim to provide a scientifically precise linguistic description of mathematical phenomena. Passages venturing beyond mathematics and logic should be viewed as illustrative descriptions rather than rigorous scientific linguistic analysis. While this work explores the internal structure of mathematical language from a logico-mathematical perspective, formal linguistic analysis lies beyond its scope and remains an exciting direction for future research.

Fourth, particular attention is paid to the *anatomy of proofs*. The text offers a level of detail often omitted in standard textbooks, where technical nuances are frequently left to the reader. For instance, the proof of ordinal exponentiation via functions with finite support is presented with exhaustive thoroughness. Furthermore, we explicitly analyze the connection between different definitions of ordinal operations—recursive ones versus those based on ordered sums, products, and powers. This duality of approaches is rarely presented so fully in educational literature, yet it is essential for a deep understanding of how mathematical structures are built. Just as fluency in a foreign language requires mastering the subtleties of grammar, fluency in mathematics requires seeing the microscopic details of its logical foundations.

A note on the depth of exposition is also necessary. While we strive for rigor, the primary objective of this book is to demonstrate the *architecture* of mathematical thought. Consequently, for some monumental results—such as Gödel's Incompleteness Theorems, Morley's Categoricity Theorem, or the independence results in Set Theory—we limit ourselves to a conceptual discussion of their significance and consequences rather than formal derivations. Conversely, we pay scrupulous attention to the "microscopic" foundations often glossed over in standard texts: for instance, the rigorous derivation of elementary arithmetic properties from the Peano axioms is presented in full detail. Additionally, we emphasize a model-theoretic approach to proofs in predicate logic, highlighting the semantic link between language and structure.

The book's structure mimics the stages of language acquisition, as reflected in the title. Its content supplements standard mathematical literature by focusing on the descriptive aspect of mathematics—expressed through formalisms and structures.

The text is divided into two parts. Part I is more informal, featuring general discussions and facts presented without rigorous proof. Part II, the "brute force" section, focuses on the direct application of the language through concrete examples, with a heavy emphasis on mathematical logic and proof theory.

Consequently, Part I contains fewer formulas and relies on associative arguments, while Part II is dense with formal derivations, resembling computer code.

Part I is divided into blocks corresponding to proficiency levels: **A1**, **A2**, **B1**, **B2**, and **C1**. Each level is subdivided into sections. Material is often revisited at different levels with increasing rigor, mirroring the spiral approach to language learning. The **C2** level of professional fluency is omitted, as such mastery characterizes "native speakers"—those with a professional mathematical background who, by definition, do not need this guide.

This modular structure enables a non-linear reading approach. Much like one might use a foreign language textbook—selecting the appropriate difficulty level and skipping already mastered basics—the reader is encouraged to bypass the elementary sections if they seem trivial and proceed directly to the more advanced topics. You are free to choose the starting point that matches your current "proficiency," returning to earlier chapters only for specific clarifications.

Part II does not follow the level-based structure. It is organized by key topics—Language, Logic, Arithmetic, and Set Theory—to consolidate necessary statements and definitions referenced in Part I.

We assume the reader is familiar with basic mathematical notation at least at the secondary school level. We will not discuss addition or multiplication tables; these fundamentals belong to the **A0** level (*Beginner*). However, the text quickly advances to abstract concepts often encountered in undergraduate curricula. The advanced mathematical objects mentioned in the early chapters (such as integrals or groups) serve purely as illustrative examples of the language's syntactic universality. Deep comprehension of their properties is not required at this stage; the focus should remain on their structural composition.

We begin with the alphabet, as all further reasoning rests upon this primitive basis. The **A1** level explains how mathematical text is constructed at the basic level—how intricate formulas are formed. Even without fully grasping the meaning (semantics) of a text, one can learn to parse its internal structure (syntax). From my perspective, the absence of explicit "reading rules" for mathematical notation in standard textbooks is a gap this book aims to fill. This level corresponds to the *survival level* (*Elementary*).

At the **A2** level, we touch upon basic concepts of logic. We adhere to the paradigm that there exists a general scientific logic inherent to the human mind, and formal logic, which lies at the heart of mathematical systems. Here, we introduce semantics, giving meaning to symbols. We study reasoning, inference, initial terminology from set theory, and define various types of relations and functions. This level corresponds to *Pre-Intermediate*.

At the **B1** level, we move to formal logic, distinguishing several layers: zeroth-order calculus (propositional logic), first-order calculus (predicate logic), and formal axiomatic theories. We also pay attention to model theory, interpretation, consistency, and completeness. A significant portion is devoted to equality and congruence. Here, the language transforms from a simple ability to "speak" into a tool for competent thought. This level corresponds to *Intermediate*.

With the **B2** level, we cross a threshold in complexity, writing formulas instead of words. We examine the formalism of first-order Peano Arithmetic and its consequences. We also introduce the language of set theory, focusing on finitary mathematics and the representation of sets as trees. The theory of hereditarily finite sets crowns the **B2** level, which corresponds to *Upper-Intermediate*.

The **C1** level offers a thorough analysis of the language of set theory. We provide rigorous definitions for relations and functions, analyze the Axiom of Choice, the Axiom of Regularity, the von Neumann universe, and methods for constructing mathematical structures. It is worth noting that while we focus on the classical Set-Theoretic foundations (ZF/ZFC) as the standard grammar of mathematics, we acknowledge the existence of alternative foundational perspectives, such as Category Theory. Here, we make a phase transition from simple formalism to a high-level language, capable of expressing modern mathematical concepts.

If you successfully travel this road and navigate the thickets of formulas in Part II, I am confident that any apprehension regarding mathematical texts will vanish. You will acquire skills to independently generate such texts and formulate thoughts in the unique language in which, as Galileo Galilei famously noted, the book of nature is written.

Throughout the text, key concepts and designations are highlighted in **bold** upon their first occurrence. Following the spiral approach of language acquisition, initial explanations in Part I often rely on intuition and context, while rigorous mathematical definitions are provided as the exposition advances to higher levels or are consolidated in Part II. To facilitate the transition from passive understanding to active mastery, the exercises range from mechanical calculation tasks to creative proofs. Detailed answers to all of them are provided in the Appendix, making the book fully suitable for independent self-study

Digital Companion

To assist readers in navigating the complex web of mathematical concepts, I have developed an interactive 3D visualization tool, **MathLogic Nexus**. This software allows users to explore the logical dependencies between definitions, theorems, and theories. While the book focuses on the core linguistic and logical foundations, the *MathLogic Nexus* places these concepts into a broader mathematical context, visualizing their connections with Model Theory, Topology, and Computability Theory. It serves as a dynamic map of the "architecture" of mathematical thought. The tool is available online at: https://nexus.mathem.at/

Finally, for ease of navigation, a comprehensive Subject Index and a List of Notations are included at the end of the book.

Acknowledgements This book is the result of a long personal journey through the landscape of mathematical logic, a path I largely navigated as an independent researcher. However, no journey is truly solitary, and I owe a debt of gratitude to those who shaped my mathematical worldview.

The main body of this text was written during my residence in Graz, Austria. I am grateful to this city and its inspiring academic atmosphere for providing the tranquility and focus necessary to bring this work to completion.

I would like to express my deepest respect and gratitude to my scientific advisor, Professor Yuri L. Pavlov, whose guidance during my PhD studies on random forests and graphs instilled in me the discipline of mathematical thought.

Although I did not have the privilege of being his direct student, I am profoundly grateful to Academician Lev D. Beklemishev. His brilliant lectures made available by the Department of Logic at Moscow State University were my primary gateway into modern proof theory. His clarity of exposition served as a benchmark for my own attempts to explain complex concepts.

I wish to thank the editorial team at Springer, and especially Elizabeth Loew, for her patience and support throughout the publication process.

I am grateful to my English tutor, Ekaterina Mozgacheva, for her guidance with the language and for comments that helped improve the clarity of this book.

I am also indebted to the anonymous reviewers for their constructive criticism. Their insistence on expanding the practical components of the book prompted me to add exercises to Part II, significantly enhancing the book's educational value.

Finally, I dedicate this work to my wife, Diana, for her unwavering support and understanding.

Graz, Austria
2026

Competing Interests The author has no competing interests to declare that are relevant to the content of this manuscript.

Contents

Preface v

List of Key Theorems xvii

Part I The Ascent

A1 Survival 3
A1.1 An Illustrative Tale 3
A1.2 The Matryoshka Principle 5
A1.3 Semantics vs. Syntax 13
A1.4 Reflection 17
A1.5 Terminology 19
A1.6 Variables and Bindings 20
A1.7 Features of the Mathematical Language 25
A1.8 Self-check Questions 32
Exercises 33

A2 At the Threshold of Logic 37
A2.1 Concept 41
A2.2 "Truth" and "Falsity" 47
A2.3 Logic and Sets 49
A2.4 Logic and Semantics via Concept Extension 54
A2.5 Logical Reasoning 55
A2.5.1 Understanding Logical Connectives 56
A2.5.2 The Principle of Reduction 59

A2.5.3 Quantifiers ... 60
A2.5.4 On Implication ... 63
A2.6 Constructing Inferences ... 65
A2.6.1 Syllogisms and Abduction ... 66
A2.6.2 Modi ... 71
A2.6.3 Deduction and Induction ... 72
A2.7 On Relations and Functions ... 74
A2.8 Self-check Questions ... 80
Exercises ... 81

B1 Formal Logic ... 87
B1.1 Propositional Logic ... 87
B1.1.1 The Language of Propositional Logic ... 88
B1.1.2 Axioms and Rules of Inference ... 89
B1.1.3 Basic Facts about Propositional Logic ... 93
B1.2 Predicate Logic ... 98
B1.2.1 Terms ... 99
B1.2.2 Formulas ... 100
B1.2.3 Axioms and Inference Rules of Predicate Logic ... 103
B1.2.4 Examples of Theories ... 105
B1.2.5 Congruence of Equality ... 107
B1.2.6 Model of a Language and a Theory ... 109
B1.2.7 Examples of Interpretation ... 110
B1.2.8 Some Concepts of Model Theory ... 115
B1.2.9 Basic Facts about Predicate Logic ... 116
B1.2.10 More on the Theory of Equality ... 121
B1.3 Summary and Self-check Questions ... 123
Exercises ... 125

B2 The Foundations of Mathematics ... 129
B2.1 Peano Arithmetic ... 129
B2.1.1 Language and Axioms ... 131
B2.1.2 Order ... 136
B2.1.3 Variants of Induction ... 138
B2.1.4 Provability and Expressibility ... 141
B2.2 Prolegomena to Set Theory ... 144
B2.3 Hereditarily Finite Sets ... 154
B2.3.1 Language and Axioms ... 154
B2.3.2 The Bracket Model ... 161

B2.4 Self-Check Questions ... 166
Exercises ... 167

C1 A Mathematician's Paradise ... 169
C1.1 Zermelo-Fraenkel Set Theory ... 169
C1.1.1 The ZFC Axiom System ... 169
C1.1.2 Relations, Functions, and Classes ... 174
C1.1.3 Why Functions are Needed ... 178
C1.2 The Axiom of Choice ... 179
C1.3 The von Neumann Universe ... 190
C1.4 Constructions ... 201
C1.4.1 Structures ... 201
C1.4.2 Principles of Structure Building ... 203
C1.4.3 A Concrete Discussion about Models ... 208
C1.5 Summary and Self-check Questions ... 218
Exercises ... 219

Part II Proof-Theoretical Basis

Introduction to Part II ... 227

D1 Language ... 229
D1.1 Elementary Particles of Language ... 229
D1.2 Lexemes ... 233
D1.3 Syntactic Analysis ... 239
D1.4 Simplifications ... 241
D1.5 Graphics ... 243
D1.6 Variables ... 245
Exercises ... 247

D2 Logic ... 253
D2.1 Propositional Logic ... 253
D2.2 Predicate Logic ... 263
D2.2.1 Soundness and Completeness of the Calculus ... 267
D2.2.2 Definability ... 272
D2.2.3 Categoricity ... 274
D2.2.4 Completeness of a Theory ... 278
D2.2.5 Summary of Results ... 282
Exercises ... 283

D3 Arithmetic 289
D3.1 Basic Properties 290
D3.2 Induction 306
D3.3 Elementary Theory of Divisibility 309
D3.4 Sequence Coding 315
D3.5 Toward Gödel's and Tarski's theorems 324
Exercises 328

D4 Set Theory 333
D4.1 Basic Properties 334
D4.2 Finite and Infinite Sets 342
D4.2.1 Comparison of Cardinalities 342
D4.2.2 Finite Sets 346
D4.3 Ordinals and Cardinals 354
D4.3.1 Recursion and Induction 355
D4.3.2 The von Neumann Hierarchy and Rank 361
D4.3.3 Ordinal Arithmetic 362
D4.3.4 Cardinal Arithmetic 377
D4.4 The Axiom of Choice 381
D4.4.1 Countable Choice and its Consequences 382
D4.4.2 The Axiom of Choice and its Equivalents 386
D4.5 On the Interconnection of Formal Theories 391
Exercises 393

Answers to Exercises 399

References 431

List of Notations 437

Index 441

List of Key Theorems

Theorem D2.5 (on Deduction) 264
Theorem D2.8 (on the Soundness of Derivation) 269
Theorem D2.9 (Gödel's Completeness Theorem for Predicate Logic) 270
Theorem D2.11 (Löwenheim-Skolem Downward Theorem) 274
Theorem D2.14 (Löwenheim-Skolem Upward Theorem) 277
Theorem D2.15 (Morley's Categoricity Theorem) 277
Theorem D2.18 (Tarski-Seidenberg) 281

Theorem D3.8 (Main Theorem of Formal Arithmetic) 322
Theorem D3.9 (Fundamental Theorem of Arithmetic) 323
Theorem D3.11 (Gödel's first incompleteness theorem) 326
Theorem D3.12 (Gödel's second incompleteness theorem) 326
Theorem D3.13 (Tarski's undefinability theorem) 327

Theorem D4.2 (Cantor) 343
Theorem D4.3 (Cantor–Bernstein–Schröder) 345
Theorem D4.7 (on Rank) 361
Theorem D4.13 (Equivalents of the Axiom of Choice) 386

Part I
The Ascent

Level A1
Survival

The most incomprehensible thing about the world is that it is comprehensible.

— Albert Einstein

A1.1 An Illustrative Tale

Let us imagine that Mathematics is an exoplanet, one with a rather unconventional structure, inhabited by strange creatures we shall call *Mathematians*. Their means of communication consists of intricate symbols, which they can draw directly in the air, a feat permitted by the composition of their atmosphere. However, unlike the plot of the well-known sci-fi film "Arrival", these "texts", so to speak, appear far more complex while simultaneously possessing a very strict logical structure. An AI-equipped terrestrial bot lands on this planet and attempts to establish contact with the local inhabitants. To do so, it must, of course, find a common language with them, which means, in essence, deciphering their strange language and learning to "speak" it. We use quotation marks for the word "speak" because the process is not accompanied by the air vibrations we would call sound; rather, it is a purely visual phenomenon.

The "intelligent" bot begins to apply its complex structural analysis algorithms to recognize this language based on some pattern known to Earthlings. The first thing it understands is that the language is simultaneously alphabetic and ideographic, as a significant portion of the language elements is repeated many times over, while another part appears relatively infrequently. On the

N. I. Kazimirov, *Mathematics as a Foreign Language*, Mathematics in Mind,
https://doi.org/10.1007/978-3-032-30721-7_1

one hand, this presents a problem, as it is not entirely clear which linguistic methods should be used to parse this language into its constituent parts. On the other hand, it is quickly discovered that almost all ideographic signs correlate with alphabetic sequences—that is, to put it simply, they can be deciphered using a longer but simpler text. This is encouraging.

The next fundamental feature of the Mathematicians' language is that every single graphic image expressing some abstract concept has a specific internal structure, like a Matryoshka doll or a nested tree. That is, it is formed by substituting simple graphemes into certain templates, the variety of which is quite limited. This makes it clear how new "words" and "phrases" in the local language can be generated.

Having figured out the basis and structure of the language, our bot attempts to find the rules that determine which constructions are permissible (i.e., occur frequently in the "conversations" it has observed) and which are strictly impermissible (never occur) or very rarely allowed. Using the method of random `decision trees`, or some other more advanced method of structural analysis, the bot's AI recognizes a set of simple rules for constructing correct texts while, as much as possible, avoiding constructs that might be interpreted as offensive or nonsensical by the local population.

Thus, the syntax of the language becomes more or less clear, and now it is time to recognize some of the language's semantics—that is, to assign meaning to it. To this end, one must try to interpret the correctly constructed texts, either in whole or in part, within some semantic reality already known to us. For example, by finding fragments in terrestrial texts that relate to each other in similar ways. In other words, if in the language of the Mathematians one specific expression is always followed by another no less specific one, then in the terrestrial interpretation, the concrete analogues of these expressions must also follow each other in the same order with a very high degree of probability.

Here, it is important to understand that interpretations can be simple or complex. Simple interpretations are often quite absurd, yet they allow us to assess the very possibility of interpreting a foreign language (and its relative consistency). This brings to mind Isaac Asimov's classic short story "Reason," in which a new-generation robot, servicing an energy-beaming device on a space station, invented a religion that provided a very simple explanation of the need to keep the beam focused on the relay. It began to worship the device, calling it "the Master," dismissed any possibility of the existence of space, Earth, or the energy beam, but instead strove unfailingly to ensure that the instrument needles did not deviate from their required values by a hair's breadth. This completely eliminated the need for humans, declared itself the Master's prophet, and turned the assistant robots into adepts of this new religion. Characteristically, such a simplified metaphysical concept allowed it to perform what was originally prescribed for the robots with great precision and reliability, without any thought for the public good or some distant Earth. It simply selected a simple, self-sufficient fragment of the device's control language and interpreted it in an even simpler way. The interpretation was completely false from the human point of view, but it was consistent and quite functional.

So, our bot begins to devise interpretations or, to put it more scientifically, to build models of the language it has discovered. The construction of various models that have a concrete meaning for Earthlings makes it possible to at least approximately evaluate the semantics of the Mathematians, and also to attempt to build a translator from our language to theirs, and vice versa. After this, one can confidently assert that the terrestrial bot, along with the humans waiting for it in orbit, has found a common language with the natives of the planet Mathematics and can proceed with a cultural exchange. Unless, of course, the standard model of the studied language turns out to be so primitive that our correspondents consider us complete fools and cease all communication.

A1.2 The Matryoshka Principle

So, we have already sketched out the path we are to follow. Let us imagine ourselves as the creators of that very AI bot studying an alien language, and apply our insights to the study of a language that is not so alien, but at times more complex and incomprehensible: the language of mathematics.

First, for brevity, we will denote this language by the sign $\mathfrak{Math}$. It should be noted at once that $\mathfrak{Math}$ actually encompasses a whole family tree of languages and dialects, which are defined more rigorously in each specific context. For now, however, the language $\mathfrak{Math}$ for us is the mathematical language in general, in its broadest sense.

Second, we note that this language possesses many features of natural languages, such as: communicativity (as it enables scientists from different fields and countries to exchange knowledge), the presence of a built-in metalanguage (tools that allow it to describe itself), aesthetic functions (expressed in the inner beauty of the universality and coherence of its concepts), indexicality (indicating its users' affiliation with certain professional groups), unlimited semantic power (since mathematical concepts can work equally well with completely different semantics), and evolvability (the mathematical language develops rapidly and fruitfully, adapting to new realities in science and culture). At the same time, for obvious reasons, the mathematical language cannot be classified as a natural language, since it is not tied to any ethnic group or language family. Moreover, it includes a vast family of formal languages, often resembling programming languages, which, of course, sharply distinguishes it from spoken languages.

Third, the language $\mathfrak{Math}$ is a purely written language. Of course, this language has a "vocalization," i.e., we can read a text aloud in one way or another, but this phonation will be of no use to the listener, as mathematical formulas are extremely difficult to perceive by ear, with the exception of some very simple expressions. This feature of the $\mathfrak{Math}$ language should, at first glance, simplify its study, since we can immediately discard such well-known components of language learning as listening and speaking, leaving only reading and writing. However, following the principles of dialectics, one should not lose sight of the antagonistic component of this simplification—the reduction of perceptual channels, as we now completely reject auditory memory and rely solely on sight and logic (in some cases, muscle memory might help, but in mathematics, it can play a cruel trick).

Like any other written language, our language $\mathfrak{Math}$ begins with an **alphabet**. It must be understood that the alphabet of the $\mathfrak{Math}$ language is much broader than the standard alphabet of any European language (Latin, Greek, or Cyrillic), as in addition to ordinary letters, it includes numerals and a vast number of special signs, somewhat reminiscent of hieroglyphs. However, these symbols, as a rule, have a quite logical origin and precise contextual semantics.

For example, the membership relation symbol ($\in$), proposed by Giuseppe Peano, originates from the first letter of the Greek word *έστί* (estí, "is", i.e., to be a part of, to be within a given concept's scope). A similar origin story applies to the notation for the differential (*d*), proposed by Gottfried von Leibniz, and the square root (from the first letter of the word *radix*), introduced by Christoph Rudolff. A slightly more complex history lies behind the plus sign (+): it is thought to have originated as a shorthand for the Latin conjunction *et* ("and"). We can also recall that the symbols for the universal ($\forall$) and existential ($\exists$) quantifiers are derived from the first letters of the words *Any* and *Exists* by reflecting them (authorship is attributed to G. Gentzen and C. S. Peirce).

All symbols of the alphabet are what we call **graphemes**, i.e., indivisible elementary graphic symbols. It is assumed that a person can easily distinguish these symbols at a glance (with the possible exception of Gothic script).

In addition to the alphabet, we also have a large number of *dependent graphemes*, which serve to construct all sorts of expressions and mean nothing by themselves. These include, for example, various brackets, accents, arrows, operator symbols, and so on. A characteristic feature of dependent graphemes is the presence of *placeholders*, often denoted in practice by variables. These placeholder variables are a kind of assembly point for more complex lexical units of our language $\mathfrak{Math}$. Furthermore, in mathematics and especially in programming, multi-letter notations for functions, operations, or constants are common, such as sin or arctan. These notations can be either standardized or defined "on the fly" within a single article or book. We do not classify such notations as part of the alphabet, as they are composed of alphabetic symbols, but we call them **rigid graphemes** because they act as indivisible symbols. The perception of rigid graphemes as single symbols is context-dependent and can be difficult in some cases. Rigid graphemes can also be either atomic or dependent (i.e., accompanied by placeholders).

A more detailed definition of the alphabet and graphemes is provided in Part II (see sections D1.1 and D1.2).

All graphemes, despite their vast diversity, form a certain finite list of symbols that changes little over time. They are the elementary particles of a text in the language $\mathfrak{Math}$, just as the letters of an ordinary language are the elementary particles from which written words and sentences are composed. And just as in a natural language, the alphabetic symbols and graphemes of the language $\mathfrak{Math}$ are combined to form more complex notations for concepts, relations, and objects.

In addition to their semantic load, which is rooted in various spoken languages and concepts, mathematical notations almost always carry a functional load and obey the "**Matryoshka principle**," which states that all complex words of the mathematical language $\mathfrak{Math}$ are a combination of its graphemes according to predefined rules. More precisely, let's say that a **lexeme** of the language $\mathfrak{Math}$ is, first, any independent grapheme (including, in particular, the symbols of the alphabet), and second, the result of substituting lexemes into a dependent grapheme's placeholders (or in place of its variables).[1] We deliberately do not single out variables here as a special syntactic type of lexeme to simplify the exposition. But later, we will pay close attention to variables.

[1] A more formal and technically precise definition of a lexeme, based on the concept of a *production rule*, will be given in Part II (see section D1.2). However, on an intuitive level, this principle of recursive assembly is key.

Note that in natural language, we have a similar (though not identical) situation: for example, English grammar has a strictly defined word order in a sentence (specific placeholders for the subject, predicate, etc.), while in German and Russian, the verb often determines not only the case of the dependent word denoting the object of the action (a transitive verb) but also the preposition before it. So, we can only assemble sentences from words according to strictly defined rules.

Thus, the Matryoshka principle tells us literally the following: if we have one empty Matryoshka doll and one already assembled Matryoshka doll, we can place the second inside the first, resulting in a more complex assembled Matryoshka. This simple analogy immediately raises the question: what if the dolls are incompatible in size? For this, there is *typing* of lexemes and placeholders. Simply put, every placeholder "accepts" only lexemes of its own type, and this must be taken into account when assembling lexemes. For example, in geometry, you cannot add two points or find their scalar product, as these operations are applicable only to vectors. In linear algebra, you cannot take the determinant of a non-square matrix. In probability theory, you cannot add a probability to an event. But in mathematics, type incompatibility is the exception rather than the rule, as we try to reduce everything to numerical types or use natural type conversion "on the fly." In programming, conversely, the typing of lexemes plays a huge role, and a misunderstanding of it often leads to errors in program code.

Let's look at a simple example of how the process of lexeme-building works. The addition symbol (+) is a dependent grapheme with two placeholders—on the left and on the right—where both the placeholders and the result of the substitution belong to the same type: numbers. We can substitute, for example, the alphabetic symbols a and b into these places, obtaining the lexeme $a + b$. We proceed similarly in the case of multiplication: $c \cdot d$. And now we can take the addition symbol again and substitute the lexemes we have already obtained into its placeholders: $a + b + c \cdot d$. Usually, placeholders are enclosed in parentheses to preserve the structure of the lexeme, so it would be more correct to write it as $(a + b) + (c \cdot d)$. Here we act strictly according to the definition, substituting "old" lexemes into a dependent grapheme, i.e., building upon existing constructions one step "up."

A more familiar method is the reverse procedure—growing the lexeme downward by substituting ready-made lexemes for the variables of another ready-made lexeme. For example, having the lexeme $a+(b\cdot c)$, we can perform the substitution of the lexeme $a + b$ for the variable c, resulting in the lexeme $a+(b\cdot(a+b))$. Such a generation of a lexeme is not by definition, but it can be proven that it also leads to well-formed lexemes, i.e., the lexeme $a+(b\cdot(a+b))$ can, in fact, be assembled "from the bottom up," strictly following the definition. Structural theorems of this kind are usually proven by induction: assuming that the lexeme is correct regardless of the assembly method for lexemes of length less than n, we then prove that this will also be true for a lexeme of length n. Ultimately, it turns out that lexemes can be assembled in almost any way: building them "up" by substituting into dependent graphemes, growing them "down" by replacing variables with more complex lexemes, or combining both approaches at once. Furthermore, the process of assembling lexemes is not limited in complexity, as a result of which one can assemble infinitely many complex mathematical expressions. We are allowed, for example, to build multi-story towers of powers:

$$5^{4^{3^{2^{1}}}}$$

or to depict something like this:[2]

$$x = \sqrt[3]{-\frac{q}{2} + \sqrt{\left(\frac{q}{2}\right)^2 + \left(\frac{p}{2}\right)^3}} + \sqrt[3]{-\frac{q}{2} - \sqrt{\left(\frac{q}{2}\right)^2 + \left(\frac{p}{2}\right)^3}}. \qquad \text{(A1.1)}$$

It should be understood, however, that mathematicians are also human, and they do not want to pile up vast, multi-story expressions with a huge number of symbols. Therefore, additional notations are usually introduced that allow one to locally (within a single article or book, for example) assign part of a complex expression to a simpler expression (e.g., a single letter) and then use it as a shorthand. We adhere to a similar principle in programming languages when we introduce notations for functions and variables. The aforementioned formula (A1.1) can be decomposed into three simpler expressions, for example, like this:

$$x = \sqrt[3]{A + \sqrt{B}} + \sqrt[3]{A - \sqrt{B}},$$

where we have introduced the temporary notations:

$$A \overset{\text{def}}{\rightleftharpoons} -\frac{q}{2}, \quad B \overset{\text{def}}{\rightleftharpoons} \left(\frac{q}{2}\right)^2 + \left(\frac{p}{2}\right)^3,$$

[2] Cardano's formula for the solution of the depressed cubic equation $x^3 + px + q = 0$.

The symbol $\stackrel{\text{def}}{\rightleftharpoons}$ means equality by definition or *assignment*. It is often used to introduce notations and abbreviations in mathematical texts. This symbol should by no means be confused with equality, as it reflects not a semantic but a graphical equality, i.e., a relabeling. The result's invariance with respect to the order of assembly (and decomposition) of a lexeme is illustrated by the following useful construct, which is called a **parse tree**. To make it simpler, we will show the parse tree for Cardano's formula (see Fig. A1.1). Furthermore, we will illustrate the use of the temporary notations A and B in the very same formula with the corresponding parse trees (see Fig. A1.2).

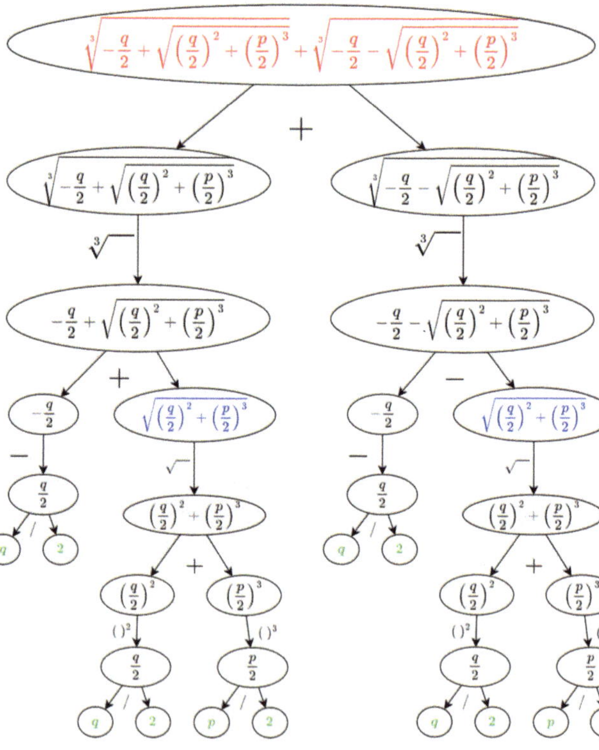

Fig. A1.1. Parse tree for Cardano's formula.

A parse tree is uniquely reconstructed from a formula except in cases where the priority of operations is not specified. Thus, the formula $a + b + c$

can be parsed in two ways depending on the precedence of the two addition operations: $(a + b) + c$ or $a + (b + c)$. Therefore, a parse tree is usually applied only to expressions where all formally required parentheses are in place, thereby unambiguously defining the order of operations. The parse tree is essentially a record of the algorithm for constructing a formula, which specifies how a particular formula is assembled from independent and dependent graphemes. The order of the branches in this tree is important, as the text obtained with its help has directionality. Only when we endow the text with mathematical semantics (for example, understanding the + sign as addition) can we, following the properties of this semantics, disregard the order of the branches, but in that case, the parse tree becomes not *syntactic*, but *semantic*. From the parse tree, it is clear that the construction of the formula can begin from any point, especially since it contains identical branches. We can first construct the expressions under the radicals, then set them aside and construct the formula with the radicals, which will have empty placeholders, and then insert the pre-prepared expressions into these placeholders. The main thing here is not to confuse the order of the placeholders used. While addition might forgive us for such a mix-up, subtraction and division will not; the formula would become incorrect.

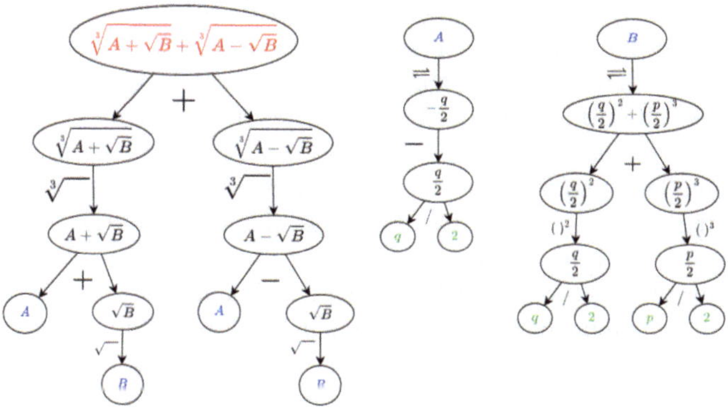

Fig. A1.2. Using temporary notations A and B.

Note also that a parse tree always ends with some elementary graphemes, for example, alphabet symbols or temporary notations (which, as a rule, are also alphabetic). The tree can be modified in three ways: by replacing a branch with some alphabetic symbol (reduction), by attaching a new branch in place of a leaf (a terminal node) of the tree (downward extension), and conversely, by substituting the tree itself in place of a leaf of another tree (upward extension). Through these manipulations, we obtain new lexemes that correspond to the strict definition of lexemes.

The concept of a parse tree, which we use to analyze the structure of formulas, is not merely a convenient visualization. It is a fundamental tool upon which programming languages and artificial intelligence systems are built. When a compiler translates a program written by a human into machine code, the first thing it does is build such a syntactic tree to understand the structure of the commands.

Modern AI systems, such as large language models, do not attempt to recover the grammatical structure of a text by explicitly constructing syntactic trees. Instead, they capture a colossal volume of statistical patterns. What is striking is that the implicit structures they learn correspond so precisely to the rules of the language that classical parse trees have transformed from a means of analysis into a powerful tool for verification, allowing us to confirm that the machine has indeed “understood” the syntax.

Looking at such a beautifully complete idea of lexeme construction, one might assume that any structure built according to the Matryoshka principle is a mathematical expression. This is not actually the case; in each specific formal fragment of the language $\mathfrak{Math}$, certain rules are specified that can be applied to assemble expressions by extending the parse tree. And this is not just about the typing of lexemes, but also about the limited set of symbols used. In other words, a specific formalism operates with specific graphemes. But there are no other restrictions.

It should be noted that the very idea of representing complex expressions through the structured assembly of simpler elements has deep roots. As far back as the 17th century, Gottfried Leibniz dreamed of creating a ’characteristica universalis’—a universal symbolic language that would allow for the formalization of human reasoning. Although his project was not fully realized, the striving for precision and structure in symbolic notation paved the way for modern mathematical notation, where the “Matryoshka principle” and the analysis of expression structure (similar to constructing a parse tree) are an integral part.

A1.3 Semantics vs. Syntax

It might appear that the syntax of a language is primary when constructing a mathematical theory, but this is not entirely accurate. In fact, prior to a rigorous definition of syntax, one must determine the intended richness of the language, the classes of objects to be described, and the relationships within the domain that are deemed essential and require formalization in the notation.

David Hilbert

Therefore, semantics is undoubtedly primary for a language. But once born, a formal language and the mathematical theories built upon it can "forget" their original semantics and be applied in very different fields of knowledge wherever it is "technically" possible. Following David Hilbert, one can note that a formal system works equally well regardless of what we understand by the objects designated in it—be they points, lines, and planes, or tables, chairs, and beer mugs.[3]

However, one must not forget that if we have called points "tables," then we must perform this substitution of concepts totally; that is, from now on, absolutely all formulas in our language will speak of tables where they previously spoke of points. At the same time, one cannot assume tables in places where we previously used lines or planes, i.e., mix initially distinct concepts. In other words, if one is to replace concepts (*choose an interpretation*), it must be done consistently and totally. For while the semantic nature of the described entities themselves does not matter to the formalism, all relations between the entities must be adequately transferred from one domain to another. Roughly speaking, if in geometry we have a formula expressing the "to lie between" relation for three points, then in the domain of tables and chairs, we will also have a working formula signifying the property "to lie between," but this time for three tables.

In the context of modern advances in artificial intelligence, this ability of formal systems to "forget" their original semantics takes on a new dimension. AI systems, by training on vast corpora of data, *distill* knowledge, identifying the most probable connections and patterns. They do not necessarily operate with an understanding of truth or falsity in the human sense, but are able to compute the "correctness" or "admissibility" of constructions based on statistical correlations and patterns extracted from the training data. This

[3] An inexact quote from Constance Reid's book "Hilbert" [50].

allows them, for example, to generate texts or formulas that are formally impeccable and even appear meaningful, even if the internal "semantic reality" on which their generation is based may be completely different from the human one. This underscores how the mathematical language, thanks to its strict structure, allows for the separation of syntax from semantics, creating a universal tool applicable even in cases where its users do not possess a full intuitive understanding of its deeper meanings.

Perhaps the most striking example of the diverse semantic implementations of mathematical theories is *group theory*, which, as is well known, is widely used not only within mathematics itself in more specific theories, but also in physics, chemistry, crystallography, computer science, and even in some humanities disciplines.

Another good example of the varied application of the same theory is the theory of linear operators, which is applied, for example, in quantum mechanics and in algorithms for machine analysis of texts, among many other areas.

Let's try, without delving deep into the art of constructing formalisms, to get a feel for the mechanism that leads us to mathematical formulas, using the example of logical expressions from ordinary (spoken) language. First of all, we need to separate two concepts from each other: objects (subjects) and relations. As in written natural language, we will distinguish the lexemes corresponding to these concepts. **Predicates**[4] are lexemes that denote a relation between objects and/or subjects; they usually have a logical type of semantics. **Terms** are lexemes that denote objects of the subject area under study.

Thus, the aforementioned example of Cardano's formula (A1.1) is a predicate of equality, expressing the identity between the root x and a certain arithmetic expression. And that expression itself, for which we provided a parse tree in Fig. A1.1, is a term, expressing the description of some object (a number). Following this description, starting from specific values of p and q, one can calculate the specific value of x.

Another example of a predicate: the relation $<$ on the integers.

Along with predicates, a more general concept is also used—the **formula**. A formula, like a predicate, is a record of a relationship between objects using predicates and logical symbols. In other words, a formula is a logically constructed description of a relation.

We should emphasize that a predicate can sometimes be defined by a rather complex formula, i.e., not all predicates within a given language are elementary in terms of their definition.

---

[4] From the Latin *praedicatum*—that which is declared, mentioned, or said.

Looking ahead, let us use some logical and mathematical special symbols: $\forall$—"for all," $K \cap M$—the intersection of two sets, $k \in K$—"k is an element of the set K."

Let's suppose our subject area is animals; that is, we are acting as zoologists. Let K be the set of all crocodiles, and M be the set of all animals in Loch Ness. Every animal has a very specific characteristic—color (to avoid overcomplicating the model, we will assume this refers to some single main color). Let the predicate $\mathcal{A}(k, c)$ take a true value if and only if the animal k has the color c. Using these notations, let's write down the following statement: "All crocodiles in Loch Ness are red." We get:

$$\forall k \in K \cap M : \ \mathcal{A}(k, \text{“red”}). \tag{A1.2}$$

Let's break down how this came about. First, we decided to limit the subject area to only animals. Therefore, any variable (in our case, k) automatically denotes an arbitrary animal. Second, we identified two subsets within the general population of animals: "living in Loch Ness" and "crocodiles," without defining these classes explicitly, so it doesn't really matter what words we use to denote them. Third, we introduced another sort of object, which we called colors (again, it is completely irrelevant what is actually hidden behind this word). Finally, we added the notation $\mathcal{A}$ to the theory for a formula that can relate two types of objects: animals and colors. The semantics of this notation is that we understand it as a correspondence between an animal and its color (in logic, this is called a *predicate*, and in natural language, the role of a predicate is usually played by a verb). After that, we wrote formula (A1.2), which can be interpreted as:

> for any k that simultaneously belongs to the set K and the set M, the predicate $\mathcal{A}(k, \text{“red”})$ holds.

Simultaneous membership in two sets is realized by the formula $k \in K \cap M$, which means that k is in the intersection of sets K and M, i.e., it is an element of both sets at the same time. Alternatively, the same formula could be written as: $(k \in K) \wedge (k \in M)$, i.e., using the logical operation of *conjunction* ($\wedge$) instead of the set intersection operation. In passing, let's observe how the "Matryoshka principle" is implemented in the formula $k \in K \cap M$. First, we took the alphabetic intersection symbol $\cap$, which has two placeholders (left and right) where only sets are allowed, and substituted the alphabetic symbols K and M, obtaining the more complex lexeme $K \cap M$. Then we took the membership symbol $\in$, which also has two placeholders, and substituted the variable k and the just-assembled lexeme $K \cap M$, respectively. Thus, the lexeme $k \in K \cap M$ is a two-level Matryoshka doll made of two symbols with placeholders (operator symbols) and three symbols without them (variables). Finally, let's recall the "tables, chairs, and beer mugs." Suppose now we are not dealing with animals, but with all the stars in the Universe. Let K be the set of all neutron stars, M be the set of all stars in our Milky Way galaxy, and the expression $\mathcal{A}(k, c)$ tells us that star k has a spectral class with a predominant color c. In this case, the very same formula (A1.2) will now have the following semantics:

every neutron star in the Milky Way galaxy has a red spectrum.

As we can see, we have completely changed the semantics of our small theory, but the formula (A1.2) has undergone no changes: the formalism remained the same. It is worth noting that a formal theory becomes a theory only when, in addition to the language, we also provide a mechanism for determining which formulas written in that language are true and which are false. For this, a theory is equipped with *axioms*, from which true statements (relative to the given axioms), called *theorems*, are derived according to general logical rules.

The addition of axioms can significantly narrow the range of application of a formal theory, since the theory will be valuable only if, when endowed with a certain semantics from a given subject area, the axioms of the theory are true in that semantics. In other words, if we consider Euclidean geometry about points, lines, and planes with all five postulates, and then replace them with tables, chairs, and beer mugs, we expect that the postulates of Euclid, rewritten in this new language, will turn out to be true. In this case, all the theorems of geometry will also be true for tables, chairs, and mugs, and the theory will be useful in this subject area. Any slight discrepancy between the axioms or theorems and the properties of the real objects to which we are trying to apply the formal theory means that it cannot be used in that case.

To draw an interim conclusion, we note that the language of mathematics is flexible enough to describe the most diverse subject areas. Moreover, it is abstract enough to allow the same language to be used in completely different situations. That is, the mathematical language, once generated by a specific semantics, almost instantly detaches from it, following only its own internal architecture and rules, which are in no way connected to the subject area. This universality of the language guarantees the reliability of all results obtained in it: if the language has been adequately mapped to some domain, and certain axioms are true in this domain, then all theorems proven purely logically from these axioms are also true in it. This property of mathematical theories allows them to be considered on their own, detached from specific implementations, and to obtain universal, absolutely precise theorems.

It should also be noted that within mathematics itself, there are different levels of abstraction and corresponding formal languages and theories. Thus, the language of first-order logic (predicate calculus) is the most abstract and works in almost all of mathematics without restriction. It is the refined core of Aristotelian logic that we call formal logic.

The language of set theory is an "embodiment" of logic in terms of sets and elements; it allows the construction of absolutely all mathematical structures in the form of specific sets (or their classes). However, set theory itself suffers from incompleteness, unlike predicate logic, which leads to various interesting "paradoxes." The next theory in terms of abstractive power should probably be considered the theory of abstract algebraic structures (such as groups and rings), which, as we have already noted, has a multitude of specific implementations in the most diverse areas of research both within mathematics and beyond. On the other hand, the application of theories in specific situations should be done with caution: do not confuse, change, or supplement the introduced concepts; verify the truth of the axioms used in the given subject area; do not introduce new concepts into the language that have any external context without their strict definition within the formal theory itself. Finally, it must be noted that within the power of mathematical theories lies a certain weakness, namely: all results of a theory are impersonal; they cannot "feel" the subtleties that distinguish one implementation of the theory from another. Informally speaking, a mathematical theory "does not see" the difference between its various correct implementations.

A1.4 Reflection

A very important property of the language $\mathfrak{Math}$, as with any sufficiently developed natural language, is the capacity for *reflection*—that is, the ability to describe and analyze its own internal patterns using the language itself.

For this purpose, a special corpus of graphemes is usually set aside in the language, tailored to describe its own elements and constructions. This special part of the language becomes an independent language with its own features, but on the whole, it follows the rules of the more general language it studies. The language used to describe other languages is usually called a **metalanguage**. In a language capable of reflection, the metalanguage is an intrinsic part of it. Note that the encapsulation of the metalanguage within the source language $\mathfrak{Math}$ actually allows the metalanguage to describe itself as well, as part of the language $\mathfrak{Math}$. And here we can provide an illustration from natural language. In any developed language, there are adjectives that qualitatively characterize certain objects and phenomena. In particular, we can consider adjectives that describe the properties of adjectives. For example, any adjective can be characterized as simple or compound depending on its structure (say, "red" is a simple adjective, while "compound" is a compound one). Thus, among all adjectives of a natural language, one can distinguish a class of adjectives that can be used to describe all adjectives. This would be the simplest metalanguage within a natural language.

Various "inconvenient" logical paradoxes are based on reflection. First, there is the well-known liar (or barber) paradox, which we will not repeat here. Second, a paradox can also be devised with adjectives. Since adjectives are capable of saying something about themselves, let us divide them into two classes: *autological* (self-applicable) and *heterological* (non-self-applicable). For example, the adjective "English" is autological, as it is itself an English word. The adjective "polysyllabic" is also autological, as it is itself polysyllabic. In contrast, the adjective "monosyllabic" is heterological, as it is not a monosyllabic word, and "German" is heterological because it is an English, not a German, word. Now let us ask the question: to which class does the adjective "heterological" belong? On the one hand, if it is autological, then it has the property it describes, which means it must be heterological. On the other hand, if it is heterological, it does not have the property it describes, which means it cannot be heterological, so it must be autological. This results in an unsolvable logical paradox.

To avoid falling into such traps, it is usually required to restrict *self-referential* definitions, i.e., definitions that refer to themselves, or even to prohibit them altogether. Thus, the metalanguage is usually separated from the rest of the language and does not say anything about itself. A similar trick was used by Gödel to obtain his most famous theorem on the incompleteness of formal arithmetic. The fact is that arithmetic is capable of analyzing itself if one learns to number arithmetic formulas (i.e., lexemes of the language of arithmetic) in a certain way. In it, too, one can find a statement that will assert something about itself.

But in arithmetic, this paradox is resolved by separating the concept of truth from provability, thereby moving the "bad" statements into a gray area of provability (they become neither provable nor refutable). On the one hand, this saves mathematics from contradictions; on the other hand, it gives reason to doubt its omnipotence. As for the language $\mathfrak{Math}$ that we are concerned with here, as has already been said, we will simply not apply the metalanguage to analyze itself, considering its concepts and lexemes as a superstructure of a higher order. On the one hand, we avoid logical paradoxes; on the other hand, we get our hands on more or less standard linguistic tools that allow us to reason about the language $\mathfrak{Math}$ in general in a formalized way, avoiding as much as possible the means of natural language (in which this book is still written).

In general, it should be noted that such a dualism of language in mathematics is more the rule than the exception. A mathematician always works on two levels of language simultaneously: the concrete and the abstract. The concrete language formally describes the objects and phenomena under study (e.g., linear operators and their properties). The abstract language serves the process of operating with the concrete language and imbuing it with meaning (e.g., for introducing definitions and explaining the goals of formal calculations). This dualism is most clearly observed in mathematical logic, where we investigate some formal system (e.g., Peano arithmetic) using the more abstract tools of logic and set theory. Moreover, above this metatheory of sets, we also have in mind a meta-metatheory that describes the general framework for the study of the given formal system.

On a side note: any work of fiction also has several semantic levels: what the author described literally; what they meant (the hidden meaning, controlled by the author); as well as the involuntary manifestation of the author's own personality in the work.

A1.5 Terminology

The connection between the language $\mathfrak{Math}$ and natural language (in our case, English) remains very strong, primarily through the concept of a **designation**. The fact is that lexemes themselves, as syntactic units, are not endowed with meaning. But in order to give them a certain semantics (like the robot from Asimov's story), we must somehow attach certain concepts to the lexemes. For example, the lexeme sin is used to denote the sine function. There is a certain logic in the chosen notation: first, the graph of a sine wave looks like a curved shoreline, or a bay, and second, the Latin word for bay is *sinus*, from which the notation was taken. This is how most (if not all) verbal notations for concepts are invented in science. But a word from a natural language written on paper ("sinus" or "sine") is a lexeme of that word in that natural language.

From this arises the definition: we call a **designation** a lexeme of a natural language that is assigned to a given concept and has a corresponding lexeme in the language $\mathfrak{Math}$.

So, there is the concept of the sine function, to which the designation "sine" and the lexeme sin (of the language $\mathfrak{Math}$) are attached.

Designations are very convenient when we want to define a concept (locally in a given article/text or globally for all of science) for which there is no established lexeme, and the definition itself is too long to write out each time. For example, the designation "function" singles out a strictly defined concept in mathematics that has a non-trivial definition. Moreover, the natural language word itself carries a certain connotation that (just as in the case of the bay and the sine) builds a bridge between the semantic content of the natural language word and the strictly defined formal mathematical concept. This is practical, economical, and organic.[5]

Thus, the presence of designations taken from natural language that accompany the concepts and lexemes of the language $\mathfrak{Math}$ provides us with a certain basic semantics that allows us to a) better remember the vocabulary of the language $\mathfrak{Math}$, b) read texts in the language $\mathfrak{Math}$ more easily, and c) more quickly assimilate new concepts and the connections between them, which is a significant humanistic factor when working with the language $\mathfrak{Math}$. After all, as the ancient Greeks said, man is the measure of all things! At the same time, we must not forget that the semantics of natural language, which manifests itself in designations, is auxiliary and illustrative (like pictures and graphs); it cannot serve as a basis for proof within the language $\mathfrak{Math}$ itself.

A1.6 Variables and Bindings

Making an analogy with natural language, by **variables** we can mean fairly simple lexemes that are assigned the role of denoting an arbitrary element of a certain class. For example, in natural language, the word "person" refers to

[5] In this text, we distinguish between a mathematical object itself (e.g., a number, a set) and its linguistic representation. We use the word **term** in the strict logical sense (a syntactically correct string denoting an object). For the natural language name of a concept (like "group" or "integral"), we use the word **designation**. This distinction helps avoid the ambiguity often found in standard texts where "term" is used for both.

some person (not a specific person and not all people in general, but simply an arbitrary representative of the set of people). Similarly, in the language $\mathfrak{Math}$, when we want to report something about an arbitrary object of a given kind, we denote it by a certain variable, for example, x or y, and then use this variable in formulas. In essence, a variable is also a temporary notation, but not for a specific formula, but for an arbitrary object from the class of objects we are describing in our language.

It should be noted that despite the rather simple definition of a variable, the concept itself is quite complex from a semantic point of view. If we recall programming languages (which can be considered implementations of the language $\mathfrak{Math}$), variables come in several types, for example, string or numeric. This means that a *type* is assigned to a variable, which limits the possibilities for using that variable. On the one hand, this forces us to perform type checking when we compose any complex lexemes, so as not to, for example, accidentally substitute letters for a numeric variable. On the other hand, it greatly simplifies the syntax, since it is not necessary throughout the entire text (e.g., a math paper or a program listing) to explicitly indicate what values a particular variable can take. It is enough to say something like "let x be an arbitrary natural number" when defining the variable.

The second problem associated with the concept of a variable is the definition of the concept itself. Purely syntactically, it is clear that if there is a variable of a certain type, any lexeme of the same type can be substituted for it. But this at least means that we must distribute all lexemes into types, i.e., impose on them some classification of a non-syntactic property. In formal mathematical theories, this is exactly what is done. When defining the fragment of the language $\mathfrak{Math}$ that will be used in a specific theory, we specify at the outset what the lexemes of a particular type should look like, or we postulate the absence of types. In the latter case, everything is much simpler, since variables can range over all possible objects described by that theory, without any restrictions.

Informally, one can think of a variable as a container of a certain shape, into which only objects of a compatible shape can be placed (triangular prisms into a triangular container, cylinders into a round one, etc.). The objects that fit the shape are the instantiations of that variable within its type. In natural language, the analogues of variables are common nouns. For example, the word "house" in each specific context can denote a specific house or a type of house, but it cannot be applied to a person or an animal.

In mathematics, the identification of classes of objects from the general diversity is achieved by writing a defining *formula* and introducing a *designation*.

At the same time, as already mentioned, the term itself does not belong to the language $\mathfrak{Math}$; it simply substitutes a meaningful word for a definition expressed in the language $\mathfrak{Math}$. For example, "function," "integer," "quadratic form," "closed set" are designations for which no variable notations are fixed. On the other hand, there is the term "ordinal," for which variables from the beginning of the Greek alphabet are sometimes reserved (if it is a text from the foundations of mathematics), so that α, β, γ are always understood as arbitrary ordinals. At the same time, we often define a notation for a class, which can be considered a constant notation, i.e., a proper name. For example, the class of all ordinals is denoted by Ord, the class of all integers by $\mathbb{Z}$, etc. In natural language, classes of homogeneous objects are usually named by words in the plural, for example, "houses," "people," etc. This means that "a house" is an element of the class "houses." Compare: α is an element of the class Ord (an ordinal is an element of the class of ordinals).

However, most common nouns in natural language are notations for rather narrow classes of things. The same "house" is a very special class of objects, having a relatively small scope compared to the scope of all things described by the language.

At the same time, in mathematics (and programming), a variable usually ranges over a fairly general class (modulo the considered implementation of the language $\mathfrak{Math}$). This means that comparing variables only with common nouns from ordinary language is not entirely correct. And here pronouns come to our aid. After all, words such as "he," "she," "it," "this," "that," etc., are applicable in natural language to very broad classes of objects, as they distinguish them only by gender. Therefore, variables can be compared both with common nouns and with pronouns (but remember that any analogy is incomplete).

As we can see, both the concept of a variable itself and its correlation with concepts from natural language cause certain difficulties and are not unambiguous. This means that parallels between mathematical and natural languages should be drawn with caution.

Variables allow us to distinguish a special class of lexemes in the language $\mathfrak{Math}$ known as **binding constructs** or **variable-binding expressions**. In mathematics, these are typically expressions where an operator acts on a *dummy variable*, thereby binding it within a specific scope. However, for the sake of brevity and to emphasize their structural unity within our linguistic framework, we will refer to such lexemes simply as **bindings**. By a binding, we mean a lexeme in which one or more variables play the role of an iterator, over which some action is performed or a predicate is taken, with the result that this lexeme denotes an object that does not depend on the variables being iterated over. These iterated variables are called *bound* (or *dummy*) (by the given binding), and they can be replaced by other variables of the same type, but cannot be replaced by any other lexemes. A binding reduces the number of free parameters of the original expression by which it is defined. Let's provide some examples of lexemes that are bindings (separated by commas):

$$\int_0^1 e^{-x}dx, \quad \sum_{n=1}^{100} 2^n, \quad \{x \mid x^2 > 2\}, \quad \forall x\ (x = x), \quad \exists x\ (x > 0).$$

Here, the variables x and n are the bound variables over which some operation is iterated (in the case of the integral and the sum, summation; in the case of the set and the quantifiers, checking the truth of the corresponding predicate). In a specific mathematical semantics, each of these bindings yields a specific value that does not depend on x and n.

It is crucial here to distinguish our syntactic terminology from standard mathematical designations. For instance, in functional analysis, there is a concept called **convolution** (usually denoted as $f * g$), which is an operation on two functions that produces a third function. Although the definition of a convolution typically involves an integral (which is, syntactically, a binding structure), the designation "convolution" refers to the operation itself, not the syntactic mechanism of binding a variable. Similarly, in set theory, the designation **comprehension** (or the *Axiom of Comprehension*) is used to describe the formation of a set based on a logical formula, like the example $\{x \mid x^2 > 2\}$ above. While comprehension is indeed a classic example of a binding, we use the word **binding** as a general syntactic category to cover all such variable-binding operators, including integrals, limits, sums, and quantifiers.

Of course, we can supply bindings with parameters. For example, consider the definite integral $\int_a^b f(x)dx$. Here, the variable x is bound (it is the variable of integration), while a and b are *free variables* or *parameters*. The value of the lexeme depends on a and b, but not on x. Free variables are the parameters of a lexeme, and their role becomes especially clear when we revisit the concepts mentioned above:

- **Convolution of functions.** The convolution $(f * g)(t)$ is typically defined by the integral:

$$(f * g)(t) \overset{\text{def}}{\rightleftharpoons} \int_{-\infty}^{+\infty} f(\tau)g(t - \tau)d\tau.$$

Here, τ is the bound variable (the iterator) which is "consumed" by the integral. In contrast, t is a free variable (a parameter) within the binding structure. The resulting lexeme is a function of t, meaning its value changes as the parameter t varies.

- **Set comprehension with a parameter.** Consider the set of all real numbers strictly less than some number c:

$$M_c \overset{\text{def}}{\rightleftharpoons} \{x \mid (x \in \mathbb{R}) \wedge (x < c)\}.$$

The variable x is bound by the braces (the set-builder notation), serving only to describe the elements. The variable c is a free parameter; changing c changes the set itself (e.g., M_0 is the set of negative numbers, while M_{10} is a much larger set).

- **Quantifiers with parameters.** The formula $\exists x \ (x^2 = a)$ contains a bound variable x and a free parameter a. This formula asserts that a has a square root. The truth of this statement depends entirely on the value of the parameter a (in the domain of real numbers, it is true for $a \geqslant 0$ and false for $a < 0$).

Thus, free variables act as the "input interface" of a lexeme, linking it to the outer context, while bound variables ensure the internal operation of the binding mechanism.

Is this concept of binding purely a mathematical invention? Not at all. In fact, natural language actively uses a very similar mechanism, although it is less strictly formalized. When we use generalizing constructions, we are essentially creating a binding. A perfect example is the famous phrase that, according to legend, was inscribed over the entrance to Plato's Academy:

"***Anyone who*** *is ignorant of geometry must not enter here.*"

Syntactically, this is a clear binding structure via a quantifier. The phrase "Anyone who" essentially plays the role of the universal quantifier ($\forall$) combined with a bound variable. It does not refer to a specific person; rather, it creates a temporary placeholder for an arbitrary human being, let's call them x. If this arbitrary person x fits the description, then the prohibition applies to x.

In this specific phrase, the words "geometry" and "here" act as **constants** (specific values). They define the concrete instance of the rule. However, we can easily imagine the general logical *template* behind this phrase:

"***Anyone who*** *is ignorant of <u>Subject</u> must not enter <u>Location</u>.*"

In this template, "Subject" and "Location" become **parameters** (free variables). By assigning different values to these parameters, we change the scope of the rule without changing its internal binding logic. For instance, if we set Subject $\stackrel{\text{def}}{\rightleftharpoons}$ "the password" and Location $\stackrel{\text{def}}{\rightleftharpoons}$ "the server room", we obtain a standard security protocol. This is exactly analogous to a **formula with parameters**, such as the expression $\exists x(x^2 = a)$ we discussed earlier. The logical structure (the binding of the variable x) remains the same, but the specific assertion depends on the value assigned to the free parameter a.

Mathematics simply takes this natural linguistic capability to generalize and imposes strict rules on the *scope* of the variable (where it begins and ends) and its *type*, preventing the ambiguities often found in ordinary speech.

A1.7 Features of the Mathematical Language

The time has come to draw some interim conclusions from our investigations. In this section, we will note the main features of the language $\mathfrak{Math}$ that, in combination, make it, on the one hand, unique from the perspective of our knowledge of languages in general, and on the other hand, also emphasize a certain commonality with them:

1. Horizontal direction of the text.
2. The Matryoshka principle and the unambiguity of parsing.
3. Synonyms and reductions.
4. Finiteness of constructions.
5. Strict determinism of variables.
6. The indicative mood of narration.
7. The absence of usual conjugations and declensions.

The horizontal direction of the text can hardly be called a distinctive feature of the language $\mathfrak{Math}$, as it is also characteristic of natural languages. However, in both mathematics and programming languages, the general rule of reading lexemes from left to right is often embedded within a certain structural order of the text. On the one hand, within lexemes, very different directions are allowed (just recall the notation for a definite integral or a matrix), and the reading order depends on the semantics of the specific template. On the other hand, the overall logic of the text implies a more vertical orientation from top to bottom, which is also generally consistent with natural languages, but only partially. This is especially noticeable in the structure of a program listing, where a new line is determined not so much by the width of the text medium as by the rules for processing (reading) the given text. The same can be observed with respect to indentation at the beginning of a line: often, it is significant not only for the beauty and readability of nested program blocks but also carries a specific functional meaning (as, for example, in the Python language).

The Matryoshka principle is perhaps the main distinguishing feature of the language $\mathfrak{Math}$, as it is definitive for the concept of a lexeme in our language. This principle, on the one hand, allows for the potentially infinite generation of new lexemes. On the other hand, it makes it possible to parse each lexeme into its constituent parts (to build a parse tree) and understand how it was constructed. Moreover, as a rule, such a parse is unambiguous (even if alternative parsing options exist, they are in some sense equivalent).

This feature of the language $\mathfrak{Math}$ sharply distinguishes it from natural languages. Of course, some recursiveness or structurality can also be seen in natural languages: we all remember from childhood that a sentence can be broken down into a subject, a predicate, and secondary parts, which in turn can be represented by different parts of speech (noun, verb, adjective, adverb, etc.). These are then represented by specific words, and words by morphemes, and finally, by letters. However, we immediately see, first, the limitation of such a hierarchy, since we cannot increase the complexity of speech structure indefinitely, and second, the initially embedded typing of elements at each level of text "assembly" (for example, a verb or, less often, a compound predicate must always be used as the predicate). This is the very case of typing that we discussed above in connection with variables. However, in natural language, typing is more tied to the structure of speech than to semantics, whereas in mathematics, typing directly points to the semantics of the objects described by lexemes of a particular type.

In fairness, it is worth citing an exception that only proves the rule: in natural language, one can invent an arbitrary number of levels of nesting if we consider a chain of subordinate clauses:

This is the cock that crow'd in the morn,
That wak'd the priest all shaven and shorn,
That married the man all tatter'd and torn,
That kiss'd the maiden all forlorn,
That milk'd the cow with the crumpled horn,
That toss'd the dog that worried the cat,
That kill'd the rat that ate the malt
That lay in the house that Jack built.

And here we can see something like a power tower from arithmetic. It is noteworthy that although both constructions (both in this verse and in the case of a power tower) do not limit our creativity in any way, excessive fascination with them leads to the complete unreadability of the text.

Furthermore, most natural languages are not analytic, meaning that the lexemes used in constructing a natural language text can have a so-called base form and various variations formed with the help of dependent morphemes (endings, suffixes, prefixes, etc.), which change depending on the context. The language $\mathfrak{Math}$, if we do not include terms borrowed from natural languages, is, on the contrary, extremely analytic precisely because of the nature of its lexemes. Indeed, it is difficult to imagine mathematical lexemes whose parts would have a tendency to change depending on the surrounding graphemes. Usually, such changes are implemented by functions and conditional operators, which are themselves also extremely analytic.

In fact, the analytic nature of a language and the ability to build a parse tree using purely syntactic methods (for example, by analyzing parentheses in the text) are two sides of the same coin.

On a side note, despite all the complexity and non-analytic nature of natural languages, mathematics offers a number of methods for studying them using non-classical logics. For example, these include the substructural logics of Lambek, which formalize the syntactic structure of natural language sentences in a certain way (through non-commutative algebras). In the language $\mathfrak{Math}$, as in natural languages, one can find phenomena such as **synonyms** and **reductions**.

Reductions in mathematics are distributed no less generously than in natural languages, although, as a rule, they have a strictly regulated nature. This primarily concerns parentheses. For example, according to the rules for constructing arithmetic lexemes, we should write the expression $a + bc$ as $(a + (bc))$. However, such an abundance of parentheses causes difficulties in perceiving the text, and therefore there are standard operator precedences (both in mathematics and in programming) that allow for an unambiguous syntactic parsing of the given lexeme, after first restoring the reduced parentheses. In natural languages, reductions also take place, for example, the typical French contraction *du* for the combination *de le*, as well as the German *im* for *in dem*.

Unlike mathematical reductions, those in natural language do not have any general system of rules and are simply unique rules in themselves. However, in mathematics, there are also such specific reductions that are introduced more as re-designations "on the fly" for the convenience of locally performed calculations, with the intention of forgetting about the introduced re-designation upon completion of these calculations.

The term *synonym* requires separate clarification. Usually, synonyms are understood as two words that denote the same concept, or very close concepts. For example, "sofa" and "couch" are synonyms. However, when we discuss the language $\mathfrak{Math}$, we are not yet in a position to operate with the term "equality," let alone the term "close" or "similar," because for that we need to learn how to give definitions, as well as introduce certain axiomatics (at least for equality). The only equality we are able to present at the level of lexemes is the character-by-character identity of lexemes.

But in that case, only completely identical lexemes can be our synonyms, which means there are no synonyms as such. Nevertheless, if we recall reductions and the accompanying conventions on precedence, we can already construct a whole class of synonymous lexemes, considering as synonyms those lexemes that are obtained from each other by the reduction of parentheses.

Moreover, if we enlist the help of the language's semantics, we can give an indirect definition of synonyms as *lexemes that have the same semantics within a specific implementation of the language* $\mathfrak{Math}$. For example, if we are talking about the language of arithmetic, the lexemes 10 and 010 can be synonyms, since a leading zero in numbers is usually ignored, but in some cases, it can be useful for syntactic parsing. Or, for example, the letters E and M can be used as synonyms for mathematical expectation, and a fraction typically has two equivalent representations: a/b and $\dfrac{a}{b}$. However, synonyms of this kind are still rare in mathematics, and the "true" manifestation of synonyms is possible only after the introduction of equality into the language, when lexemes that are fundamentally different in appearance turn out to be equivalent due to some logical properties of the language under consideration.

The situation is slightly worse with the polysemy of lexemes, i.e., cases where the same lexeme can have different semantics depending on the context. It is enough to recall the letter ω, which can denote the first infinite ordinal, the period of a pendulum, or the Eisenstein integer $(-1+i\sqrt{3})/2$ (examples of different mathematical objects denoted by the same symbol); another example is the letter i, which often acts as a summation index and as the imaginary unit. Here we observe a complete analogy with natural languages, where one word can have many semantic shades depending on the context, and often simply denote completely different things (for example, homographs like "lead" (the metal) and "lead" (to guide), or the different meanings of "bank").

Note that in programming, such things are usually suppressed by the interpreter, since in this case the interpreter is a machine, whereas in natural languages and in mathematics, the interpreter is a human, who is prone to economizing on notations in favor of variations in context.

We will try to adhere to *context-free* lexemes in the future, so as not to confuse the reader even more.

Speaking of the **finiteness of constructions**, we mean the fact that all lexemes of the language $\mathfrak{Math}$ are finite in nature. In other words, each lexeme is made from a finite set of graphemes in a finite number of steps.

In natural language, of course, all constructions are also finite. And although they use such indefinite terms as "etc.," "and so on," this is a reference to a potentially infinite process described by the given finite text.[6]

Here we have a complete analogy with the mathematical language. Nevertheless, this finiteness of our language should, apparently, be considered a feature, since the mathematical objects themselves, which are described by this language, often have an actually infinite nature, and the diversity of these infinities is striking. And yet, it is precisely because of the finiteness of the language itself that in practice we operate only with a finite number of objects or their types, among which there are also various infinities. Such a paradox, or rather, an antinomy.

Another feature of the language $\mathfrak{Math}$ that we will note here is the **strict determinism of variables**. By this phrase, one should primarily understand the property of "isotropy" of variables, i.e., if a certain value (a specific one or a type) is assigned to a given variable, then in all its occurrences further down in the text, this variable has exactly the same value until it is explicitly redefined. In other words, if we have performed the assignment $a \overset{\text{def}}{\rightleftharpoons} 2$ somewhere, then all occurrences of a thereafter have the value 2, no matter how many times

[6] In mathematics, instead of etc., we use an ellipsis, but this is more an element of the metalanguage than of any formalism. In formal systems, infinite formulas or sequents are sometimes allowed, for example, the ω-rule in the language PA_ω, but all such infinities have a strictly regular, finitely describable construction.

this variable appears in the text until the next redefinition (note that the directionality of the text plays an important role here). We will assume this property of variables in the future when writing substitutions of the form $[a/x]$ (this is a homograph of a fraction, not a fraction!), which mean that in the given formula, all occurrences of the variable a (as substrings) must be replaced by x. Note that in natural languages, the most general analogues of variables, i.e., pronouns, do not obey this property. For example, in the text:

> "It was very hot from the scorching sun, for it was still high."

The referent of the word *it* changes right on the fly, yet the reader can unambiguously restore this meaning from the context. In mathematics and programming, such tricks are not permissible.

Nevertheless, the strict determinism of variables has a strictly regulated exception: the mechanism of **bindings** we discussed in Section A1.6. The key feature here is that a binding creates a *local scope* for its variable, temporarily overriding any global definitions. For example, in the lexeme $\forall x \in A\ (x = x^2)$, we forcibly restrict the values of the variable to the set A only within the parentheses. At the same time, we can then write another lexeme like $\exists x > 0\ (x^2 < 0)$, and it will be clear that the range of values for the variable x has now changed from the set A to the set of positive numbers. These ranges are relevant only for the formula inside the binding; once the binding ends, the variable x is free to be used again or reverts to its global meaning (if one exists). In programming, a similar limitation of a variable's scope is observed in loops and functions, where a variable has a strictly local context. Thus, a binding allows us to syntactically isolate a certain fragment of text and set its own permissible values for variables. However, this flexibility comes with a risk: **variable collision**. One can sometimes encounter nested bindings that use the same variable name. For example:

$$\sum_{k=1}^{\infty} k^2 \left(\sum_{k=-10}^{10} k a_k \right)^3 + k^3 \left(\sum_{k=-100}^{100} k^2 b_k \right)^5 .$$

In principle, it is clear that the "outer" k here has no relation to both "inner" k's, which are also not related to each other, i.e., we are dealing with bindings within a binding, in each of which the variable k takes different values, generally speaking. In this, one can see an analogue of the text phrase with the pronoun "it" that we cited above. It should be noted that lexemes of this kind are formally permissible (the inner variable scope shadows the outer one), but pragmatically disastrous for readability, and therefore are generally not used.

The fact is that they will work exactly until it becomes necessary to use the outer variable k together with an inner one in the same formula. Indeed, let's consider such a lexeme:

$$\sum_k \int_0^\infty e^{itk-t^2}\, dt.$$

Here, the bound variables k and t meet in the common arithmetic expression $itk - t^2$. What happens if we denote them identically? We get a lexeme like this:

$$\sum_x \int_0^\infty e^{ixx-x^2}\, dx.$$

And here we no longer understand which part of the formula should be integrated and what is a parameter of the sum. Therefore, the general approach to nested bindings is *that all bound variables should be different.* Moreover, in modern logic, it is generally accepted to divide variables into two classes: *free* and *bound.* Bound variables are used to form bindings, and free variables denote external parameters. This approach allows avoiding the described collisions with variable names. Finally, let's also note that in the language $\mathfrak{Math}$, all narration takes place in the indicative mood; that is, in mathematics, we describe things as they are. In the text of proofs, we may resort to using the subjunctive mood, for example, in a proof by contradiction, when we need to assume something and then refute it, but this does not manifest itself on a formal level.[7] Even less so is the imperative mood used in mathematics, except when assigning values to variables. We can say "let $x = 2$," and this looks like an imperative mood, but on a formal level, it is written as the premise of an implication.

The last thing worth noting as a feature of the mathematical language is the absence of changes in the names of objects (analogues of nouns) and operations (analogues of verbs) according to person, number, gender, case, tense, or aspect. Here it is appropriate to recall that not all of these attributes are mandatory for natural languages either; for example, English has no grammatical gender or cases, and fewer verbal aspects, but it has a large number of tenses. In mathematics, however, there are no such categories of change for objects and propositions. For example, the set intersection operation ($\cap$) cannot age over time and stop working as we defined it. Or the Pythagorean theorem cannot become obsolete in the sense that it was true yesterday but will cease to be

---

[7] More precisely, in formal logic, we use the rule of contraposition: $((\mathcal{A} \to \mathcal{B}) \wedge \neg\mathcal{B}) \to \neg\mathcal{A}$.

true tomorrow. There are, of course, changes in symbolism over time (such as the outdated notation for implication $\supset$), but this is more an evolution of the language than temporal forms of its units.

However, the absence of tenses and cases is no reason to breathe a sigh of relief, as mathematics has its own ways of changing names and operations. Most often, this is due to the fact that we economize on the number of symbols used so that the text does not look cumbersome, and therefore we define many symbols in the course of the exposition. For example, the sign $\cong$ is often used to denote an isomorphism, but which one specifically is determined in the text. The symbol ω can denote the first infinite ordinal, the Eisenstein integer, or be used as a notation for a differential form, etc. All such definitions of notations can depend both on the area of mathematics in which we are working and on the specific article or book. And common letters like a, b, c, x, y, z are redefined regularly in the course of a narration. The same can be said about such notations for operators as the prime for a derivative, the "hat," exponentiation, etc., the meaning of which varies from book to book, from one area of mathematics to another. And there are no uniform rules (like conjugation tables) for these changes.

Herein lies the absolute freedom of mathematics, and often this same thing serves as an obstacle for non-mathematicians to understand mathematical texts. On the other hand, any article or book on mathematics does not consist solely of formal text; usually, no less than half of its volume is composed of natural language lexemes, some of which are mathematical designations, and the vast majority are ordinary speech constructions of natural language, intended to clarify what is written in the formal part. This allows one to perceive purely formal material through informal images and concepts.

A1.8 Self-check Questions

Q1 What is the "Matryoshka principle"? [p. 7]

Q2 List at least three features of natural languages that the mathematical language ($\mathfrak{Math}$) possesses. [p. 5]

Q3 What is the main difference between the mathematical language and natural languages? [p. 5]

Q4 What are graphemes and rigid graphemes in the mathematical language? Provide examples. [p. 6]

Q5 Explain using the example of Cardano's formula (A1.1) how the "Matryoshka principle" is used to simplify complex expressions.
Q6 What is a parse tree and what role does it play in understanding the structure of mathematical formulas? [p. 10]
Q7 What is the difference between the syntax and semantics of the mathematical language? [p. 13]
Q8 What are predicates and terms in the mathematical language? Provide examples from the text. [p. 14]
Q9 What is a binding (lexeme-binding)? [p. 23]

Exercises

A1.1 Draw a parse tree for the formula of the roots of a quadratic equation:

$$x = \frac{-b + \sqrt{b^2 - 4ac}}{2a}$$

Indicate the hierarchy of templates used (fraction, addition, subtraction, square root, power, multiplication).

A1.2 In mathematics, parentheses are often omitted based on operator precedence. Restore all omitted parentheses in the following expressions to make the structure unambiguous:

1. $a \cdot b + c \cdot d$;
2. $x^2 + 2xy + y^2$;
3. $a/b/c$ (assume standard left-to-right associativity);
4. e^{x^2+y}.

A1.3 Draw (or describe) the parse tree for Euler's identity, identifying the hierarchy of templates used:

$$e^{i\pi} + 1 = 0$$

A1.4 Deconstruct the following nested expression into layers. What is the outermost "shell" (template), and what are the inner components?

$$\sqrt{1+\sqrt{1+\sqrt{1+x}}}$$

A1.5 In the expression below, identify which symbols act as *objects* (nouns/variables/constants) and which act as *actions/relations* (verbs/operators):

$$\frac{d}{dx}\left(x^3+3x\right)=3x^2+3$$

A1.6 Translate the following natural language sentences into the language of predicate logic using quantifiers ($\forall, \exists$), logical connectives ($\wedge, \vee, \rightarrow, \neg$), and appropriate predicates.

Example: "All Frenchmen speak French."

- Universe: All people.
- Predicates: $Fr(x)$ is "x is French", $Speak(x, \text{French})$ is "x speaks French".
- Formula: $\forall x(Fr(x) \rightarrow Speak(x))$.

Tasks:

1. "Some Frenchmen speak English."
2. "No Frenchman speaks Chinese." (Hint: Rephrase as "For any person, if they are French, they do not speak Chinese").
3. "There are physicists who love mathematics but do not love quantum mechanics."
4. "Anyone who studies mathematical logic must know set theory."
5. "There exists a physicist who proposed a theory that every mathematician understands."
6. "Every person has exactly one biological father." (Hint: "Exists exactly one" $\exists!x$ can be written as $\exists x(P(x) \wedge \forall y(P(y) \rightarrow y = x))$).
7. "Not all that glitters is gold."
8. "If a physicist does not know mathematics, they will understand neither General Relativity nor Quantum Mechanics."
9. "There is no theory that explains absolutely everything."

A1.7 Translate the following formal sentences into natural English (try to make it sound natural, not robotic):

1. $A \subseteq B \wedge B \subseteq A \rightarrow A = B$;
2. $\forall x \in \mathbb{R} : x^2 \geqslant 0$;

3. $n, m \in \mathbb{Z} \wedge n \neq 0 \to \exists q, r : (m = nq + r \wedge 0 \leqslant r < |n|)$.

A1.8 Translate the following statements into the language $\mathfrak{Math}$:

1. The square of the sum of a and b is equal to the sum of their squares plus twice their product.
2. The product of any number and zero is zero.
3. For every positive number ε, there exists a positive number δ.

A1.9 Determine which of the following strings are syntactically valid mathematical expressions and which are "gibberish". Explain why.

1. $5 + (3 \cdot 2)$;
2. $5 + \cdot(32)$;
3. $\sin(x) = \cos$;
4. $\int_a^b f(x)dx$.

A1.10 In the following expressions, identify the *bound* variable (the iterator/dummy variable) and the *free* variables (parameters):

1. $\sum_{k=1}^{n} k^3$;
2. $\lim_{x \to a}(x^2 - 3)$;
3. $\bigcup_{i \in I} A_i$.

A1.11 Consider the expression $\sum_{i=1}^{n}(i + \sum_{j=1}^{i} j)$. Why would it be confusing or incorrect to write it as $\sum_{i=1}^{n}(i + \sum_{i=1}^{i} i)$?

A1.12 Perform the syntactic substitution $[x/(a + b)]$ (replace x with the lexeme $a + b$) in the following expressions. Be careful with parentheses!

1. $3x$;
2. x^2;
3. $\sin x$.

A1.13 In the formula for the area of a circle $S = \pi r^2$:

1. Which symbols are constants?
2. Which symbols are variables?

A1.14 Restore parentheses in the expression 2^{3^4} according to standard mathematical convention (from top to bottom). Calculate the values of $(2^3)^2$ and $2^{(3^2)}$ to demonstrate the difference.

A1.15 Look at the expression $f^{-1}(x)$. Depending on context, this could mean two very different things syntactically. What are they? (Hint: function inverse vs arithmetic power).

A1.16 Classify the following operators as unary (1 argument) or binary (2 arguments):

1. $-$ in the expression -5;
2. $-$ in the expression $a - b$;
3. ! in the expression $n!$;
4. $\cup$ in the expression $A \cup B$.

Level A2
At the Threshold of Logic

But you will have to reconcile yourself to it, — Woland objected, a grin twisting his mouth, — you have hardly appeared on the roof when you have already uttered an absurdity, and I will tell you what it is, — it is in your intonations. You uttered your words as if you do not recognize shadows, or evil either. But would you be so good as to reflect on the question: what would your good do if evil did not exist, and what would the earth look like if shadows disappeared from it? After all, shadows are cast by things and people. There is the shadow of my sword. But there are also shadows of trees and of living creatures. Do you want to strip the whole globe by removing every tree and every living being from it, because of your fantasy of enjoying bare light?

— M. Bulgakov, "The Master and Margarita"

The preceding quote from the classic author shows us the importance of differentiating anything into at least two classes. In doing so, we do not focus on what these classes are called, whether these names have any semantic subtext, or are just gibberish. What is important is that if the totality of everything we could in principle describe with the language $\mathfrak{Math}$ (or any other language) could not be divided into categories, we would be fundamentally unable to distinguish between objects, events, and phenomena, or to see any relationships between them. Our entire universe would collapse into a single point.

In essence, **logic** as such is a *complex of rules and methods that allow us, in some regular way, to distinguish "light" from "darkness," to discover similarities and differences, to separate the essential from the non-essential, and to find connections and structures* in the body of knowledge to which we apply

N. I. Kazimirov, *Mathematics as a Foreign Language*, Mathematics in Mind,
https://doi.org/10.1007/978-3-032-30721-7_2

the language 𝔐𝔞𝔱𝔥 as a descriptive tool. In this section and the next, we will speak of logic in a general sense, understanding it as a certain inherent property of our minds that allows us to construct correct judgments.

The very first question a reader might ask is: why perform any division into classes if they already exist? Indeed, when we defined the language 𝔐𝔞𝔱𝔥, we learned to construct syntactically correct lexemes. Moreover, we presented a very specific algorithm for doing so, starting from certain predefined substitution templates and alphabetic symbols. Finally, for any given text, we can even recognize in finite time whether it is a well-formed lexeme or not. Let us then consider lexemes to be "true" texts, and all other combinations of symbols to be "false" texts.

This dogmatic approach conceals a number of problems. Firstly, we have thereby constructed nothing more than a *logic of lexemes*. That is, with this logic, we can identify correct lexemes, and that is where its capabilities end. Such a logic can be called *syntactic*.[1]

Secondly, if we want to apply the language 𝔐𝔞𝔱𝔥 to any external entities (be they mathematical abstractions or physical phenomena), we must be able to describe the properties of these entities in a positive way, i.e., describe them explicitly, rather than by negating or excluding other properties. Furthermore, the differences between lexeme-descriptions must correspond to actual differences in the described entities. If we do not learn to distinguish between well-formed lexemes, we will not learn to analyze the entities they describe—for us, the entire world will collapse into being and non-being, as we will be unable to describe specific elements of being with incorrect texts (it would simply be unclear how to relate one to the other).

Thirdly, knowledge of what is and what is not is not always given to us through our senses; therefore, we must have some rigorous scheme that would allow us, by a sequence of *correct* manipulations, to show whether a given lexeme denotes some entity from the subject domain or whether, while being syntactically correct, it is semantically incorrect, i.e., it does not describe actually existing objects.

[1] If desired, a syntactic logic can be developed by identifying new subclasses within the existing classes, whose definitions would be tied to their syntax. For example, one could leverage the fact that every well-formed lexeme can be represented as a parse tree, and trees can be classified in a wide variety of ways based on their height and branching characteristics. However, even such a *ramified* logic would be no more than a sham from the perspective of logic in general, for it would provide nothing beyond the syntactic analysis of a text. Such logics are convenient in highly specific cases but are not suitable as a general approach for revealing the very semantics referred to in the quote.

Roughly speaking, syntactically correct lexemes are correctly posed questions with the possible answers "Yes" or "No." The answer itself is determined by the semantics. But it is precisely in order to be able to compute the difference between "Yes" and "No" that the texts of the questions themselves must be grammatically (syntactically correctly) constructed.

Therefore, a correct, workable, and practically convenient description of any subject domain is reasonably produced within a fragment of our universal language $\mathfrak{Math}$ that has already been defined in some regular way, discarding all syntactically incorrect expressions.

Having defined such a fragment in each specific case, we gain the ability to write down in an algorithmically rigorous way a) descriptions of objects from the given subject domain and b) predicates describing the mutual relations of these objects. By and large, it is the formalized representation of predicates that is the main goal of formalizing a subject domain.

Why is this so? The fact is that any predicate describing a relationship between objects, in a sense, simplifies and generalizes the situation, translating domain-specific knowledge into a truth-value category. On the one hand, this is quite sufficient, since from a practical point of view we are always interested in the factology of the subject domain, namely, the answer to the question of whether a particular relationship between objects is a fact in it, as this allows us to predict the environment's reaction to our actions within it. On the other hand, the transition to predicates moves the investigation into a philosophical truth-value category of the most general character. Indeed, by appropriately tailoring the formalism to describe the objects and predicates of a subject domain, we reduce its study to some universal tools for all of humanity—logical inferences.

In turn, the application of an abstract logical apparatus implies operating with such concepts as Truth and Falsity, which in the context of a specific subject domain (semantics) are defined, respectively, as the satisfaction or non-satisfaction of certain predicates. More specifically, if some objects are related by a certain predicate (i.e., they jointly satisfy it with their properties), then such a predicate is considered true for these objects, and otherwise—false.

This specific definition of the Truth/Falsity pair for the purpose of formalizing a subject domain forms the basis for defining the logic of that subject domain and the corresponding formal system.

More precisely, by the **logic** of a certain subject domain, we will understand a *regularized (formal) scheme for determining the relative truth of formulas in the language of this subject domain.* The relativity of truth in this case means that the truth of formulas is established not in itself, but on the condition of the truth of other formulas.

Therefore, it is worth noting that logic is not the establishment of ultimate truth, but rather presents a rigorous, formalized mechanism for obtaining new knowledge (true predicates) from existing ones. The responsibility for choosing the initial true knowledge (facts) lies with the researcher (the human).

Thus, the task of formalizing a subject domain in the language $\mathfrak{Math}$ consists of the following stages (Table A2.1):

Table A2.1. Formalization Scheme

(1) construction of a fragment of the language $\mathfrak{Math}$ by describing initial graphemes and rules for their combination;
(2) definition, within this fragment, of lexemes corresponding to formulas and terms;
(3) definition of rules by which the truth of some formulas can be established by assuming the truth of others (rules of inference);
(4) specification of initial truths (axioms).

Note that while the construction of a language fragment (items 1 and 2) is usually tailored strictly to a specific subject domain, the definition of rules of inference (item 3) is of a more universal nature and belongs to the field of logic proper.

A special requirement for formalization is the **soundness of the logic** with respect to the described semantics (subject domain). Soundness means that from formulas corresponding to factual truths, i.e., verified facts, only (though not necessarily all) factual truths can be derived by the rules of logic, which can subsequently be verified by external means.

Note that the aforementioned logic of lexemes is, on the one hand, a special case of logic in the sense we have described, if by subject domain we mean precisely the lexemes of the language $\mathfrak{Math}$, and on the other hand, it is the basic space within which we will build all other logics of arbitrary subject domains. A similar situation is encountered in mathematics itself, when the syntactically very poor formal set theory is capable of describing all of mathematics.

A2.1 Concept

The question of the rigorous definition of concepts has concerned philosophers and mathematicians since antiquity. Plato in his dialogues often seeks the essence of concepts through the method of *maieutics* (Socratic method), and Aristotle in the "Organon" laid the foundations for the classification of concepts and methods of definition through genus and specific difference. The striving for precision in definitions is an eternal companion to the development of mathematical thought.

Strictly speaking, "concept" is an undefinable concept in the standard sense (yes, "concept" is also a concept!).

Every **concept** presupposes the existence of an **extension of the concept** and its **definition** (in logic books, the latter is often called **intension**, see, e.g., [14]). In addition, a verbal **designation** (a word or phrase) is attached to the concept to make it convenient to use in the future. Often, a **notation** is also assigned to the concept, i.e., a special symbol (or group of symbols) denoting this concept or its individual instances.

The definition of each specific concept can be either *direct*, when, starting from an already existing terminological apparatus, we introduce a new concept according to the formula "A is such-and-such", or *indirect*, when two or more concepts are introduced simultaneously through the context of their interrelations (or, in other words, through an enumeration of the properties of their interrelations, expressed by predicates).

But in any case, every concept is introduced in a predicative way, since its definition is written using a finite or infinite series of formulas in the chosen language. It should be especially noted that infinite definitions in the vast majority of cases are actually a finite scheme, the specific instances of which are strictly regulated.

An example of a direct definition of a concept: "a prime number is a number that has exactly two divisors".

An example of an indirect definition of a concept: the *Peano axioms of arithmetic*, which define the logical interrelation of the concepts "number", "addition", "multiplication", "equality", "successor". Here, several concepts are introduced at once by means of a finite list of axioms and an induction scheme.[2]

As soon as we have a definition of a concept, its extension immediately arises. In essence, the extension of a concept is everything that falls under its

[2] We are talking about first-order arithmetic, as it will be studied here later.

definition within the chosen semantics. In the case of the concept of a prime number, it is the set of all prime numbers. In the case of the concept of a neutron star, it is the set of all neutron stars in the entire universe, and so on. It is usually assumed that the extension of a concept is non-empty in any semantics, i.e., the concept is not contradictory. Although for the sake of completeness, one should also introduce a contradictory concept, whose extension is the empty set.

In contrast to a contradictory concept with an empty extension, there is the concept of the universe (the "null" concept), which includes all objects of the semantics currently under consideration (all living beings, all stars, all natural numbers, all geometric figures, etc.). It is important to understand that the concept of the universe itself is universal, although its content directly depends on the chosen semantics.

Note that any definition must be written down in some way in some language. Since we have already gone some way in formalization and have defined (albeit in the most general terms) the language we use, we have the opportunity to give a narrower and more precise definition of a concept for our specific needs.

Therefore, within this book, we will only define a mathematical concept. The definition that follows is of the nature of an indirect definition and resembles axioms or postulates, as it introduces several terms at once by describing their relationships.

A **Mathematical Concept** consists of three mandatory components: the extension of the concept, the definition of the concept, the designation for the concept, and one optional component: the notation for the concept.

The **extension of the concept** depends on the chosen semantics of the language and consists of entities belonging to that semantics, which are united by a common idea and are distinguished from other entities by a given list of properties, expressed by formulas.[3]

The **definition of the concept** is a formal syntactic construct, written using formulas of the chosen fragment of the language $\mathfrak{Math}$, which is independent(!) of the chosen semantics and expresses all those properties, the simultaneous possession of which distinguishes the entities belonging to the extension of this concept, and only them. Thus, the definition of a mathematical concept is always predicative.

[3] This means that every entity in the extension of the concept must possess all the given properties, and any entity outside the extension lacks at least one of the given properties.

The **designation for the concept** is a word or set of words from natural language, in combination with lexemes of the language $\mathfrak{Math}$, assigned to the given concept.

The **notation for the concept** is a lexeme of the language $\mathfrak{Math}$ that serves to denote specific instances of the concept's extension or the extension of the concept itself.

It should be noted that this definition, despite its modern form, has deep historical roots. The division of a concept into *extension* and *intension* (here — *definition*) is central even to Aristotelian logic. However, it was in the 20th century, in the works of proponents of *formalism*, such as David Hilbert, and especially in the works of the group of mathematicians under the pseudonym Nicolas Bourbaki, that the idea of separating a formal *structure* from its concrete instances was made the foundation of the entire architecture of modern mathematics. Our approach, where the *definition of a concept* is a purely syntactic construct, independent of semantics, and the *extension of a concept* is the collection of its models in one semantics or another, is a direct consequence of this structuralist point of view.

Let's illustrate how this definition works with examples. Let's start with a prime number. The definition of a mathematical concept requires us to provide, firstly, the idea of a prime number, and secondly, its formal definition and designation. It should be noted from the outset that we are now in the world of natural numbers, i.e., arithmetic, and are building the idea of the definition on a very specific semantics.

We know that some natural numbers can be factored, thereby representing a composite number through the product of simpler (smaller) ones. The idea is to be able to decompose any number into some elementary numbers that can no longer be decomposed into smaller factors. Thus, the general idea for prime numbers is the impossibility of decomposing them into smaller factors such that all the resulting factors are smaller than the original number.[4]

Note that the idea is somewhat redundant, since one can add one as a factor to any decomposition, and nothing will change. Therefore, we must immediately agree that we do not classify one as a prime number, since its use in factorization is meaningless. The same can be said about zero, since if it is in the decomposition, then the number being decomposed is also zero, and it cannot participate in the decomposition of other numbers.

[4] In higher algebra, this idea corresponds to the concept of an *irreducible element* of a ring, while *prime* elements are those which, being a divisor of a product, necessarily divide one of the factors. However, in a principal ideal domain (in particular, in $\mathbb{Z}$), these concepts coincide, so we neglect this distinction at an elementary level.

Consequently, a prime number can only be one that cannot be decomposed in a *nontrivial* way into smaller factors and is different from zero and one.

The formal definition of a prime number in the language of arithmetic is as follows: a prime number has exactly two divisors.

Why does the formal definition differ from the idea of a prime number given above? The reason is that the formal definition is given in a formalism that knows nothing about the idea of decomposability, nor what "less than" means. However, in this formalism (which we will study in detail later), it is quite easy to define the concept of divisibility, since the operation of multiplication is defined along with the concept of a natural number, i.e., it is immanently built into this formalism.

Next, it only remains to understand how to express our idea of a prime number through the concept of divisibility. We want it not to be decomposable into smaller factors, therefore, it should not have divisors smaller than itself, i.e., simply put, it should not have divisors other than itself. However, there is always such a divisor—it is one. But the point is that if we consider one as a divisor, then it is always paired with a divisor equal to the number itself, which means that by using one we do not achieve a decomposition into strictly smaller divisors. Therefore, we adjust the definition, allowing only trivial divisors: one and the number itself.

Note that we have thereby already excluded zero from the list of prime numbers, since it has more than two divisors.

However, that's not all. One itself falls under this definition, and we want to exclude it from the list of prime numbers. But one has a unique property—its list of divisors consists of exactly one number—one. Whereas all other numbers (even zero) have at least two divisors.

Then we take the final step and add the word "exactly" to the definition, and it follows that a prime number is a number that has exactly two divisors.

According to the definition, we must also give the new concept its own designation, but we introduced it right away when we formalized the definition, calling this concept a prime number.

Note that with the same idea, we can obtain different formal definitions, and therefore, different mathematical concepts. Thus, the idea of an elementary indecomposable building block in the integers will give a definition of a prime number as one that has exactly 4 divisors, since the presence of a sign in an integer must be taken into account. The same idea in Gaussian integers already gives a formal definition of a prime number as having exactly 8 divisors, and in the case of Eisenstein integers—exactly 12 divisors.

At the same time, it should be noted that all these definitions of concepts remain formal, which means they will work regardless of the semantics that can be described by this formalism. Thus, we see that, on the one hand, a given specific semantics can give rise to both the idea of a concept and its formal definition, and on the other hand, the formal definition of a concept itself can go beyond the original semantics, and even more so the idea of this concept can be widely applied to other cases.

Another example of a concept is the natural number, or more precisely, the system of natural numbers. At the heart of this concept lies the idea of some abstract quantities that can be used, on the one hand, to count and order things, and on the other hand, to perform ordinary arithmetic operations on them to calculate quantities.

In principle, this says it all, but one can come up with a more precise idea by turning to general algebra. Namely, we want the system of natural numbers to be a minimal ordered discrete commutative semiring with unity, in which induction can be used. Such an idea already practically turns into a formal definition, since if we write out all the concepts used axiomatically through their definitions, we will generally obtain Peano arithmetic. We will deal with these issues in detail in Chapter B2.1.

Above we said that one can define the concept of a universe as one that includes the entire subject domain without exception. In mathematics, the concept of a universe has a more specific meaning. By a mathematical universe, one should understand all the objects studied in a given subject domain, i.e., those entities on which some operations formalized in the given formal system are performed.

A formal definition of the mathematical universe can only be given in connection with the definition of variables. The idea of a **mathematical universe** is that it represents the scope of all variables of the formal fragment of the language $\mathfrak{Math}$ that we are currently considering. And the idea of a variable is that all variables collectively are anonymous designations for the elements of the universe.

The definition of the mathematical universe is the empty syntactic construct belonging to the language $\mathfrak{Math}$ (the empty string). Thus, the definition of the mathematical universe falls under the requirements for the definition of a concept.

Incidentally, we have introduced an indirect definition of variables, but variables have a general idea (denoting elements of the universe) and a formal definition, since the notations for variables in the language $\mathfrak{Math}$ are always strictly regulated. As a rule, these are letters or letters with accents and/or indices. Therefore, the concept of a variable can also be considered mathematical by virtue of concept definition.

It is customary to divide variables into bound and free variables with the same scope. Such a division is purely syntactic, but it also has all the features of concept definition, since the idea of bound variables is their purpose—they can only be used in binding-lexemes, while the idea of free variables is to denote the free parameters of a lexeme. Formally, bound variables are defined by their position in a lexeme after a special binding symbol (for example, a quantifier), while free variables cannot be in such a place in a lexeme.

If we use variables of types (for example, variables for sets and classes), then a certain part of the universe (sometimes it may coincide with the universe) and a certain part of the list of variables are assigned to each type of variable. By introducing variables of a certain type, we implicitly define, firstly, the extension of this type, secondly, we assign a common notation to its elements in the form of the variables of this type themselves, and thirdly, we give a strict formal description of the variables of this type (we present a list).

In this book, we will hardly touch upon typed variables (except for the division into logical and object types), so without loss of generality, we can assume that all variables in the formal systems under consideration are divided into only two classes: bound and free.

Like any top-level definition, our definition of a mathematical concept operates with words that have not been previously defined and are assumed to be understood from context or some general notions. Unfortunately, it is impossible to create a completely deterministic logic detached from all existence, because in trying to utter even a word, we inevitably find ourselves in a certain semantic context of natural language and are forced to appeal to it as the base of concepts and designations that is predefined for us, but our task is to make this predefined base of concepts as small and simple as possible.

It is worth noting that since we have introduced the definition of a mathematical concept, we would like to understand whether "mathematical concept" is a mathematical concept. That is, does definition of a concept possess the characteristics indicated in it? Or, in other words, does the concept "mathematical concept" belong to its own extension?

It turns out that it does not, because the definition of a mathematical concept itself is given in a natural language syntactic and semantic construct, and not within the apparatus of lexemes of the language $\mathfrak{Math}$ that must be used when defining any mathematical concept. If it were not for this limitation, the mathematical concept would satisfy all other criteria—it has a certain general idea (expressed by the definition), a definition, and a designation.

This immediately brings to mind such famous logical paradoxes as Russell's paradox and the barber paradox.[5]

But in fact, the definition of a mathematical concept simply belongs to the metalanguage, while its concrete instances, i.e., the mathematical concepts themselves, belong to that part of the language $\mathfrak{Math}$ that describes a specific area of mathematical knowledge.

A2.2 "Truth" and "Falsity"

As noted above, when formalizing the description of a subject domain in the language $\mathfrak{Math}$, we use a limited notion of truth, understanding it as the mutual correspondence of object properties to a predicate written in this language.[6]

It should be understood that the verification of correspondence between properties and a predicate is itself a certain process, which reduces the result of the verification to some initial truths given by a list of axioms. Thus, for us, truth is a relative concept.

Let there be some concept $\mathfrak{C}$, whose extension we denote by V, and its formal definition by $\mathcal{D}(a)$, where a is a free variable. The membership of a certain object v (which is either an object variable or a more complex term) in the extension of this concept can be written in the universal language $\mathfrak{Math}$ in two ways:

$$v \in V \quad \text{or} \quad \mathcal{D}[a/v].$$

The formula $v \in V$ effectively means that the object v belongs to the extension V, while the formula $\mathcal{D}[a/v]$, or more concisely, $\mathcal{D}(v)$, means that the object v satisfies the formal definition of the concept $\mathfrak{C}$.[7]

---

[5] The barber paradox: in a certain village, there is only one barber who shaves all those who do not shave themselves. The question is: does the barber shave himself? Obviously, if he shaves himself, then the barber does not shave him, i.e., himself. But if he does not shave himself, then the barber shaves him, i.e., himself. It turns out that the barber is not from this village, i.e., he falls out of the semantic universe to which he was assigned by the formulation of the problem.

[6] With this definition of the truth-category, we are expressing, in a simplified form, the classic *correspondence theory* of truth (i.e., with verification against some criteria), which dates back to Aristotle and Hegel.

[7] Recall that $[a/v]$ denotes the total substitution of the variable a with the term v in the formula $\mathcal{D}$.

Thus, the truth of the correspondence of the object υ to the concept $\mathfrak{C}$ is expressed by the truth of either of the formulas: $\upsilon \in V$ and $\mathcal{D}[a/\upsilon]$. If, however, the object υ does not actually correspond to the concept $\mathfrak{C}$, then both of these formulas are simultaneously false.

Note that, as already mentioned, we want to establish the truth of certain formulas in some regular way according to rules of inference, reducing it to the truth of axioms. In doing so, we require that the truths obtained by the rules of inference be sound with respect to the semantics, provided that the axioms correspond to that semantics. This means that the truth of formulas depends not only on the axioms but also on the chosen semantics.

For example, the property of a number being negative can be true in the universe of integers, while in the universe of natural numbers it is always false, and in the universe of complex numbers this property can be expressed in various ways, giving this concept different extensions. This is not surprising, since the listed semantics have different sets of axioms, and in one case the property of being negative can be derived from the axioms, while in others it cannot.

Nevertheless, some formal expressions are such that their truth does not depend on semantics. For example, the universe is defined by an empty formal definition. If we denote the extension of the universe by U, then the formula $\upsilon \in U$ will always be true regardless of what we understand by the universe, as long as this universe is not empty. The formula corresponding to the universe, which should be the empty string according to our definition, must also be a **tautology**. Therefore, this formula is denoted by the symbol $\top$ (from True) to distinguish it from empty strings used in other contexts (e.g., as the neutral element of the string concatenation operation).

The formula $\top$ has neither free nor bound variables, as it depends on nothing. The opposite situation is a contradictory concept, whose extension is by definition the empty set $\emptyset$, and the formula $\upsilon \in \emptyset$ is always false. The corresponding definition formula is denoted by the symbol $\perp$ and is called a **contradiction**.

The formula $\perp$ has neither free nor bound variables, as it depends on nothing.

As we can see, there is a close connection between expressions about sets and formal truth-expressions, and this is no accident. The fact is that the language of set theory is the semantic embodiment of the language of logic. We will show this as we define the symbols for logical connectives and set-theoretic operations and relations.

A2.3 Logic and Sets

In the context of studying logic in the classical truth-category, let us agree to use calligraphic letters $\mathcal{A}, \mathcal{B}, \mathcal{C}$, etc., to denote variables that range over statements and predicates, i.e., we assign them a *logical type*. Such variables are called *propositional*.[8] By a *propositional formula*, we will understand any lexeme that describes a transformation of objects of a logical type into an object of a logical type, i.e., it operates within the truth-category.

More precisely, among all the lexemes of the language $\mathfrak{Math}$, we want to single out a specific class of lexemes constructed according to the Matryoshka principle and corresponding to our ideas about the structure of a logical expression. To do this, we will define a basic limited set of symbols with one or two substitution places, after which we will declare that lexemes of a logical type are those and only those lexemes that are obtained by combining these basic symbols and propositional variables according to the Matryoshka principle. The basic logical symbols are:

- $\rightarrow$ **Implication**: a binary logical operator $\mathcal{A} \rightarrow \mathcal{B}$, meaning "$\mathcal{A}$ implies $\mathcal{B}$" or "if $\mathcal{A}$, then $\mathcal{B}$".[9]
- $\neg$ **Negation**: a unary logical operator $\neg\mathcal{A}$, meaning "not $\mathcal{A}$" or "the negation of $\mathcal{A}$".
- $\wedge$ **Conjunction**: a binary logical operator $\mathcal{A} \wedge \mathcal{B}$, meaning "$\mathcal{A}$ and $\mathcal{B}$".
- $\vee$ **Disjunction**: a binary logical operator $\mathcal{A} \vee \mathcal{B}$, meaning "$\mathcal{A}$ or $\mathcal{B}$".
- $\leftrightarrow$ **Equivalence**: a binary logical operator meaning "$\mathcal{A}$ is equivalent to $\mathcal{B}$" or "$\mathcal{A}$ if and only if $\mathcal{B}$".

The symbols $\rightarrow, \neg, \wedge, \vee, \leftrightarrow$ are called **logical connectives**. Thus, a propositional formula is a lexeme constructed from logical connectives and propositional variables (parentheses are implied as auxiliary separator symbols).

The notion of a propositional formula in the form presented can already be considered a mathematical concept in the sense of concept definition. Indeed, we know the extension of this concept—all lexemes constructed from logical connectives and propositional variables according to the Matryoshka principle. We have given a clear, algorithmic definition for these lexemes, so

[8] From the Latin *propositio* — a judgment, a statement.

[9] In some books, the equivalent notation $\mathcal{A} \supset \mathcal{B}$ is found, but we will not use it to avoid confusion with set-theoretic inclusion.

for any string of the language $\mathfrak{Math}$, we can recognize whether it is a propositional formula or not. Finally, we have a designation for it—"propositional formula".

As in the case of arbitrary lexemes built according to the Matryoshka rule, one can notice that propositional formulas can be constructed not only by definition but also, for example, by replacing propositional variables with propositional formulas or other propositional variables. An example of a propositional formula:

$$(\mathcal{A} \to (\mathcal{B} \to C)) \to ((\mathcal{A} \land \mathcal{B}) \to C),$$

In cases where an expression is so complex that the letters of the Latin alphabet are insufficient to denote its variables, various "add-ons" to the variable notations are used, for example, numerical indices ($\mathcal{A}_1$, $\mathcal{A}_{11}$, etc.).

Although we noted above that the notion of a propositional formula is already a mathematical one, it must be understood that for now, it is a purely syntactic notion, not filled with any semantics, despite the fact that we gave verbal explanations when defining the logical connectives. To make this notion truly valuable, we need to define the "rules of the game" with these formulas. This is usually done as follows: some propositional formulas are declared to be true beforehand, and to check the others for truth, certain rules called rules of inference are used. Thus, explicitly or implicitly, we must be able to give a truth-assessment to all propositional formulas.

Formally, the basic true propositional formulas (axioms) and rules of inference are defined in zero-order logic (propositional logic), where they are studied in their own right, i.e., as a separate algebraic structure. In this case, the mathematical concept of a "propositional formula" acquires a deeper definition, which allows giving the syntax a certain truth-meaning and linking formulas with the concepts of Truth and Falsity, which we discussed above. The mathematical theory dealing with the study of propositional logic (i.e., the logic of propositional formulas and logical connectives) is called **propositional calculus** (or **sentential logic**). We will consider it in more detail in Chapter B1.

Nevertheless, the truth-content of propositional formulas is only the tip of the iceberg. After all, the ability to operate with pure logic, with refined representations of thoughts and judgments, gives us only the most general rules for establishing truth relations between knowledge from an arbitrary subject domain. On the one hand, propositional logic is universal; on the other hand, it is too abstract.

Therefore, in order to infuse logic with a more concrete meaning, we must understand how the very statements and predicates that are the subject domain for logic and are denoted by propositional variables are formed. In other words, we must learn to write predicates in the language $\mathfrak{Math}$, so that by substituting them for propositional variables, we can obtain more complex logical judgments (formulas) about a specific subject domain.

A characteristic example of a predicate we had was the lexeme $\mathcal{A}(k, \text{"red"})$, which told us that the object k has some relation to the word "red" (for example, it has a red color). As we can see, by taking objects from the subject domain as input, the predicate outputs a truth value, which means it can be substituted for a propositional variable in a propositional formula.

Furthermore, with the appearance of object variables, another type of logical operation is added, namely quantifiers, which form binding-lexemes of a logical type and are needed to express generalizing judgments (the universal quantifier) as well as counterexamples to them (the existential quantifier). For example, $\forall k\ \mathcal{A}(k, \text{"red"})$ expresses that the predicate $\mathcal{A}(k, \text{"red"})$ is always true, i.e., any object k corresponds to the attribute "red".

Such a coupling of semantics with pure logic leads us to the need to consider a more complex syntactic apparatus: in addition to propositional formulas, we must be able to define predicates that express various properties of objects. A lexeme obtained as a result of substituting predicates into a propositional formula instead of propositional variables (again, the Matryoshka principle) or under quantifiers is called a **formula** (of predicate logic).

To simplify the terminological apparatus, predicates are also classified as formulas, so a predicate is a special case of a formula and is called an **atomic formula**. Some predicates can be *expressed* by more complex formulas through other predicates, logical connectives, and quantifiers (for example, the equality of sets can be expressed through the membership relation).

If we look at the parse tree of a formula, it is a parse tree of a propositional formula, extended by the parse trees corresponding to quantifiers and predicates. In other words, the parse tree of a formula associated with some non-logical subject domain, as it were, grows its branches into that subject domain.

However, when describing a subject domain, the matter is not limited to just predicates and formulas. As already noted, there is also such a concept as a **term**. This is a lexeme of an object type, representing an algorithm for constructing some elements of the subject domain from others by combining (in accordance with the Matryoshka principle) operator symbols and variables. Operator symbols are roughly the same as logical connectives, only they transform an object type into an object type. These symbols are much more diverse than logical ones, as they are directly related to the subject domain.

The mathematical theory that deals with the study in a general form (without specifying the subject domain) of the properties of formulas with predicates, terms, and quantifiers is called **predicate calculus** (or **predicate logic**). We will consider it in more detail in Chapter B1.

In order not to immediately drown in the diversity of mathematical formalisms, we will limit ourselves here to considering one subject domain, namely, sets. Moreover, sets are, in fact, a universal subject domain suitable for describing any mathematical and logical constructs. In this particular subject domain, there is a very specific list of operator symbols, which we will list below.

As in the case of formulas, there are also atomic terms among the terms. These are terms that do not contain operator symbols in their notation. For now, as atomic terms, we will use object variables $A, B, C, \ldots, Z, a, b, c, \ldots, z$ (possibly with indices), as well as special constant symbols assigned to specific objects. As usual, terms (object lexemes) are assembled according to the Matryoshka principle, so that only sets and object lexemes of the same type can be substituted for variables (Fig. A2.1).

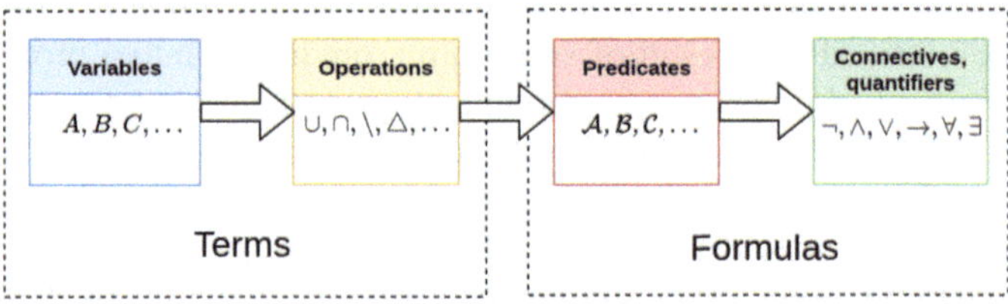

Fig. A2.1. Levels of a formula's parse tree

In summary, note that every formula has four levels in its parse tree: the level of propositional formulas and quantifiers, the level of predicates, the level of terms, and the level of atomic terms, i.e., object variables.

The most commonly used set-theoretic operations and predicates are given in Table A2.2.

Regarding the last predicate, note that in set theory, "everything is a set," so the elements of sets are also sets. From the perspective of ordinary ideas, this seems absurd, but to simplify our narrative, we will assume that elements of an arbitrary nature are encoded in some way using sets, so as not to multiply entities.

Finally, let us not forget the constant symbols:

Table A2.2. Operations and Predicates on Sets

Symbol	Description
$\cup$	The *union of sets* A and B is a binary operation, denoted by $A \cup B$, and represents the set containing all elements of set A as well as all elements of set B, and no others.
$\cap$	The *intersection of sets* A and B is a binary operation, denoted by $A \cap B$, and represents the set containing all elements of set A that are simultaneously elements of set B, and only them.
$\setminus$	The *difference of sets* A and B is a binary operation, denoted by $A \setminus B$, and represents the set containing all elements of set A that are simultaneously NOT elements of set B, and only them.
$\vartriangle$	The *symmetric difference of sets* A and B is a binary operation, denoted by $A \vartriangle B$, and represents the set containing all elements of sets A and B, except for those that are simultaneously elements of both sets, and only them.
$\subseteq$	The *inclusion of sets* A and B is a binary predicate, denoted by $A \subseteq B$, and means that all elements of set A are simultaneously elements of set B.
$=$	The *equality of sets* A and B is a binary predicate, denoted by $A = B$, and means that sets A and B consist of the same elements, i.e., $A \subseteq B$ and $B \subseteq A$ simultaneously.
$\in$	The *membership* of an object A in a set B is a binary predicate, denoted by $A \in B$, and means that A is an element of set B.

$\mathfrak{U}$ — the entire universe of sets[10], defined by the property-formula: $A \in \mathfrak{U}$;

$\emptyset$ — the empty set, which is defined by the property-formula: $\neg(A \in \emptyset)$.

Here, in both cases, A is an arbitrary set, other than the universe itself. Recalling our example (A1.2) with neutron stars and crocodiles, we note that we have already used the notation $K \cap M$, which denoted crocodiles in Loch Ness or neutron stars, depending on the chosen semantics.

[10] Looking ahead, we note that the universe of all sets cannot be a set due to well-known paradoxes, but the specified operations and predicates can also be extended to it, so for now, we will consider it as a special kind of set as well

A2.4 Logic and Semantics via Concept Extension

Having introduced all the necessary notations and concepts, let us now see how logical connectives are related to set-theoretic operations and predicates. Let us again have some concept $\mathfrak{C}_1$, which is given by a definition in the language $\mathfrak{Math}$ using the formula $\mathcal{D}_1(a)$. Further, let V_1 be the extension of this concept. Similarly, let there be a concept $\mathfrak{C}_2$ with definition $\mathcal{D}_2(a)$ and extension V_2. Here, V_1 and V_2 are sets.

Consider the formula $\mathcal{A}_1(a) \overset{\text{def}}{\leftrightarrow} \mathcal{D}_1(a) \wedge \mathcal{D}_2(a)$. Here, we introduce the symbol $\overset{\text{def}}{\leftrightarrow}$ to denote **equivalence by definition**, distinguishing it from $\overset{\text{def}}{\rightleftharpoons}$, which is used for the equality of objects. The formula $\mathcal{A}_1(a)$ tells us that the object a must simultaneously possess property $\mathcal{D}_1$ and property $\mathcal{D}_2$. Let $\mathcal{A}_1$ be the definition of a new concept $\mathfrak{C}$. What then is the extension of this concept? It is easy to see that the extension V of the concept $\mathfrak{C}$ contains those and only those objects that are simultaneously in the extension V_1 (since they satisfy property $\mathcal{D}_1$) and in the extension V_2 (since they satisfy property $\mathcal{D}_2$). But then, by the definition of the intersection of sets and the equality of sets, we get that $V = V_1 \cap V_2$.

Thus we see a clear correspondence between the logical connective $\wedge$ and the symbol $\cap$, since the former generates the definition of the new concept $\mathfrak{C}$, and the latter—its extension. In other words, logical and set-theoretic notations are two sides of the same coin.

Exactly the same connection can be traced between the logical connective $\vee$ and the symbol $\cup$, as well as between $\neg$ and $\setminus$. In the latter case, however, we have to involve the notation for the universe, since if $\mathcal{A}_2(a) \overset{\text{def}}{\leftrightarrow} \neg\mathcal{D}_1(a)$, then for the corresponding extension V, the equality $V = \mathfrak{U} \setminus V_1$ will hold. However, often, if the universe does not change during the investigation, its notation can be omitted and the *complement* of V_1 can be written simply as $\setminus V_1$ or, even simpler, $\overline{V_1}$.

The matter is a bit more complicated with implication. Consider the definition $\mathcal{A}_3(a) \overset{\text{def}}{\leftrightarrow} \mathcal{D}_1(a) \rightarrow \mathcal{D}_2(a)$. It expresses the following property of an object a: if the object a belongs to the concept $\mathfrak{C}_1$, then it also belongs to the concept $\mathfrak{C}_2$. Thus, the new concept $\mathfrak{C}$ must necessarily include all those objects that, being elements of V_1, are also elements of V_2, i.e., the extension of the new concept $\mathfrak{C}$ contains at least the intersection $V_1 \cap V_2$.

However, the formula $\mathcal{A}_3(a)$ tells us nothing about those objects that do NOT satisfy $\mathcal{D}_1(a)$, i.e., that belong to the extension $\mathfrak{U} \setminus V_1$. And here we encounter a key point that defines mathematical (material) implication: if an object a does NOT satisfy the premise of the implication (i.e., the "if" part is not fulfilled for it), then the entire implication as a whole is considered true for it. We will talk about this property of implication a little later, but for now, we conclude that in this case, the extension V of the new concept $\mathfrak{C}$ with the defining formula $\mathcal{A}_3(a)$ must also include the complement of V_1, i.e., $\mathfrak{U} \setminus V_1$.

And since we have considered all cases (when a satisfies $\mathcal{D}_1$ and when it does not satisfy $\mathcal{D}_1$), we have completely calculated the extension of the concept $\mathfrak{C}$, which consists of such a that belong either to the extension $V_1 \cap V_2$ or to the extension $\mathfrak{U} \setminus V_1$, i.e., in this case

$$V = (\mathfrak{U} \setminus V_1) \cup (V_1 \cap V_2).$$

In other words, the definition $\mathcal{A}_3(a)$ can be rewritten, using the previously obtained correspondences between logical connectives and set-theoretic operators, as follows:

$$\neg\mathcal{D}_1(a) \vee (\mathcal{D}_1(a) \wedge \mathcal{D}_2(a)).$$

Later we will see that this formula can be simplified to $\neg\mathcal{D}_1(a) \vee \mathcal{D}_2(a)$, since $V_2 = (V_1 \cap V_2) \cup (V_2 \setminus V_1)$, but the second part of this union is already contained in $\mathfrak{U} \setminus V_1$, so that $(\mathfrak{U} \setminus V_1) \cup V_2 = (\mathfrak{U} \setminus V_1) \cup (V_1 \cap V_2) = V$.

Thus, the logical connective $\rightarrow$ corresponds to a more complex set-theoretic operator composed of the union of V_2 and the complement of V_1.

However, in the case that the definition $\mathcal{A}_3(a)$ turns out to be a tautology, i.e., it corresponds to the empty concept with an extension equal to the entire universe, then the formula $\mathcal{A}_3(a)$ simply tells us that every instance of the first concept is an instance of the second concept, i.e., $V_1 \subseteq V_2$. In this special case, we obtain a correspondence between the implication $\rightarrow$ and the inclusion $\subseteq$.

A2.5 Logical Reasoning

Here we gradually transition to the part of the formalization process (see p. 40) that expresses logic itself, i.e., the rules for obtaining new truths from already known ones. Below, we will discuss how judgments are constructed in mathematics and why they differ from those commonly accepted in natural language.

A2.5.1 Understanding Logical Connectives

Above, we provided several definitions of logical connectives and set-theoretic operations in which we used conversational parts of speech, such as AND, OR, NOT. It must be understood that in mathematics, we fix a specific meaning and properties for them, although in natural language, AND and NOT can change their behavior depending on the context.

For example, the conjunction AND ($\wedge$) in mathematics denotes a commutative logical operation, i.e., to say $\mathcal{A} \wedge \mathcal{B}$ is the same as saying $\mathcal{B} \wedge \mathcal{A}$, and such a substitution is valid regardless of context. In natural language, however, the conjunction AND is not always commutative, as AND can often imply a logical dependence between its components. For example, the phrase "the sun came out, AND it became warm," despite its formal appearance as a conjunction, has an implicit implicative connotation (the sun came out, as a consequence of which it became warm). You would agree that if you swap the members of this conjunction, the phrase is perceived differently: "it became warm, and the sun came out." Here, properties of ordinary language that are absent in mathematics become apparent—the reflection of processes extended in time, such that we unwittingly imply that first the sun came out, and then it became warm, although it is not explicitly stated that the second is a consequence of the first (post hoc non est propter hoc), but we implicitly believe in such an implication.

In programming, conjunction is also not entirely commutative. A program is executed by a computer over time, and therefore executing the code in the order $\mathcal{A} \wedge \mathcal{B}$ is not equivalent to executing it in the reverse order $\mathcal{B} \wedge \mathcal{A}$. This feature is often used to save on computations and avoid certain errors. For example, checking the conjunctive condition `len(a)>0 and a[0]==1` in the case of an empty list a will stop at the first condition and return **False** without checking the second. However, if we swap the members of the conjunction (`a[0]==1 and len(a)>0`), then in the case of `a[0]==1`, the second condition will still be checked (which would be an extra waste of time), and in the case where the list a is empty, we will get an error, as accessing the element `a[0]` becomes impossible.

Thus, the behavior of the logical operator **and** in program code may differ when the operands are rearranged due to the direction of reading and executing the program text.

Such deterministic, non-commutative logic is especially important in robotics. The control system of a robotic manipulator or an autonomous vehicle is, in essence, a highly complex logical formula executed in real time. The condition "check for an obstacle AND take a step forward" is safe, while the reverse—"take a step forward AND check for an obstacle"—is catastrophic. In this world, where the cost of an error is physical destruction, the language $\mathfrak{Math}$ in its not-quite-mathematical, *algorithmic dialect* serves as the only means of guaranteeing predictability and reliability.

The particle NOT (non-/un-) in natural language also does not behave exactly as we assume in mathematics and formal logic. Very often, the negative particle can indicate a clearly negative meaning. For example, "unhappy" is the opposite of "happy," but this does not mean that all objects can be divided into happy and unhappy. In this case, the particle NOT plays the role of an arithmetic negation rather than a logical exclusion.

The law of double negation, which assumes that $\neg\neg\mathcal{A}$ is the same as $\mathcal{A}$, usually does not work in natural language. For instance, saying someone is "not unhappy" does not at all mean they are "happy"; it is a much weaker statement, often implying mere contentment rather than joy. The structure not (un-happy) serves as an understatement, not a full return to the original positive. In other words, the operator NOT (non-/un-) in natural language is not *involutive*.

In order to unambiguously fix the meaning we put into logical operators, we need to use truth tables, which prescribe the calculation of the values of the expressions $\mathcal{A} \wedge \mathcal{B}$, $\mathcal{A} \vee \mathcal{B}$, $\mathcal{A} \to \mathcal{B}$, $\neg\mathcal{A}$ under the condition that the values of both expressions $\mathcal{A}$ and $\mathcal{B}$ have already been calculated. Usually, these tables resemble the multiplication tables from a school notebook, but with only two numbers: 0 and 1, which are the algebraic notations for the concepts of Falsity and Truth, respectively (see Table A2.3).

Table A2.3. Truth Table

$\mathcal{A}$	$\mathcal{B}$	$\neg\mathcal{A}$	$\mathcal{A} \wedge \mathcal{B}$	$\mathcal{A} \vee \mathcal{B}$	$\mathcal{A} \to \mathcal{B}$	$\mathcal{A} \leftrightarrow \mathcal{B}$
0	0	1	0	0	1	1
0	1	1	0	1	1	0
1	0	0	0	1	0	0
1	1	0	1	1	1	1

If any two logical expressions of the same variables behave identically according to the truth table, we identify them, which is written with the symbol $\equiv$. For example, $\mathcal{A} \wedge \mathcal{B} \equiv \mathcal{B} \wedge \mathcal{A}$ (the property of commutativity of conjunction).

By composing new propositional formulas according to the Matryoshka principle, we can obtain more and more new logical expressions, some of which will be equivalent. For example,

$$\neg\neg\mathcal{A} \equiv \mathcal{A}, \qquad \mathcal{A} \to \mathcal{B} \equiv \neg\mathcal{A} \vee \mathcal{B},$$
$$\neg(\mathcal{A} \wedge \mathcal{B}) \equiv \neg\mathcal{A} \vee \neg\mathcal{B}, \qquad \neg(\mathcal{A} \vee \mathcal{B}) \equiv \neg\mathcal{A} \wedge \neg\mathcal{B}.$$

Thus, we have a purely algebraic apparatus for manipulating logical expressions, and at this point, the *algebra of logic* is born. Indeed, we have variables that can only take logical values (which we denote by 0 and 1), which means we can understand these variables as arbitrary truth-bearing statements like "snow is black" or "Socrates is a man." In addition, we have logical connectives ($\wedge$, $\vee$, $\neg$, $\to$, $\leftrightarrow$), with which we can construct quite complex logical expressions and manipulate them in much the same way as we do in school arithmetic. The only difference is that the variables conceal statements instead of numbers, and the operations are logical operations instead of addition and multiplication.

It should be understood that all our variables are independent. In the expression $\mathcal{A} \wedge \mathcal{B}$, the variables $\mathcal{A}$ and $\mathcal{B}$ are in no way related to each other until we substitute something more complex for them. This means that all logical identities that we can obtain using a truth table do not depend on the nature of the statements that we could substitute for these variables. For example, the law of double negation $\neg\neg\mathcal{A} \equiv \mathcal{A}$ will be true regardless of what we mean by $\mathcal{A}$ (the property of being "happy" or the verb "go").

Herein lies perhaps the most significant difference between mathematical judgments and any others (both in natural language and in the languages of natural sciences like physics or biology): *the form of a judgment prevails over its content*, because judgments in the mathematical language are constructed according to strict (algebraic) rules of manipulation with logical forms (or, more precisely, *formulas*).

Of course, the choice of the direction of these manipulations is no longer a category of pure form and is largely determined by the content. For example, if we want to prove Cantor's theorem, we select the sequence of purely formal manipulations in such a way as to obtain Cantor's theorem, and not something completely different. In other words, the technology of mathematical judgments (proofs) is purely formal in nature, while their final result is understood by mathematicians less formally and sometimes contains a very deep meaning.

In physics, for example, this is not quite the case, since in all manipulations with formulas in physics, the subject of these manipulations (a theoretical physicist) is always guided by physical meaning when choosing each next step of formal manipulations. Thus, for example, if a mathematician, solving a quadratic equation, obtains complex roots, he will consider them in the course of further manipulations as possible solutions, because logic requires considering all formally permissible branching options of the reasoning procedure. A physicist in such a situation will immediately discard those roots of the equation that have no physical meaning and will limit further reasoning only to the roots that are permissible from his point of view. Thus, it turns out that a physicist, despite the constant use of mathematical language in his search for truth, employs some additional constraints, discarding cases that are incorrect from his point of view. A mathematician, however, stubbornly checks all possible options, even the most seemingly absurd ones, simply because they are formally permissible.

It should be noted that although Aristotle is credited with laying the foundations of logic, the systematic study of connectives between statements (propositional logic) began later, with a significant contribution from the Stoics (for example, Chrysippus in the 3rd century BC). However, the modern symbolic form and algebraic approach to logic were developed in the 19th century by George Boole ("An Investigation of the Laws of Thought," 1854) and Augustus De Morgan, whose works laid the foundation for mathematical logic.

George Boole

A2.5.2 The Principle of Reduction

The next technique in mathematical reasoning is that we prefer to somehow denote the most frequently used procedures and results and then refer to them to save space and time, without any fear that they might "spoil" from frequent use or cease to be relevant.

This approach is well described by the famous joke about a kettle. A physicist and a mathematician are given the same task: to heat water. In the first version of the problem, the conditions are: there is a stove, an empty(!) kettle, and a tap with water. Both solve this version in the same way: pour water into the kettle, put it on the stove, and heat the water. In the second version, the conditions are different: the water is already in the kettle! The physicist solves the problem thus: put the kettle on the stove and heat it. The mathematician solves it differently: pour the water out of the kettle, and thus reduce the problem to the one already solved.

More formally, this approach is expressed by mathematicians (and programmers) as follows: if you need to regularly perform some procedure on data, the easiest way is to write a separate function (or at least introduce a special notation for an operator) and then insert into the current text not the transformed data, but a function of the source data. The result is concise and clear. If you try to repeat this in ordinary language, you get something like this: let's introduce a function Perf(x) that takes a verb in the infinitive form as input and returns that verb in the perfect tense. Then instead of the phrase "John arrived yesterday," we could write the phrase "John Perf(to arrive) yesterday."

For a programmer, this *functional* way of presenting text is natural, while for a layperson, it looks abnormal.

And although sometimes such things may seem ridiculous to outsiders, they predictably work (or don't work) with equal force and universality.

A2.5.3 Quantifiers

If we want to construct logical expressions not only in the form of pure logical formulas with propositional variables or predicates substituted for them, then we need quantifiers.

A little earlier we used predicates related to sets, for example, $x \in A$ or $A = B$. These two predicates can be considered atomic formulas in formal set theory, and all other properties can be described by more complex formulas. However, equality can be expressed through membership by a more complex formula, but with the use of quantifiers.

At this point, **predicate logic**, or *predicate calculus*, or first-order logic, is born.

Quantifiers are logical operators that have the property of binding, i.e., reducing the number of free variables in a logical lexeme (formula). The main quantifiers are: $\forall$—the universal quantifier, and $\exists$—the existential quantifier.[11] Their semantics are as follows: $\forall x$ means "for all x," and it is followed by some formula that depends on the variable x; $\exists x$ means "there exists an x such that," and it is followed by some formula that depends on the variable x. The variable

[11] Although the symbols for quantifiers appeared later, the conceptual breakthrough in formalizing statements of generality and existence belongs to Gottlob Frege. In his work [23], he first introduced a clear system for writing judgments that included analogs of quantifiers, which allowed for the analysis of the structure of mathematical statements with unprecedented precision.

x thereby becomes *bound*, but besides x, there may be other free variables in the subsequent formula.

The following formula is the definition of the predicate of set equality $A = B$:

$$\forall x\ (x \in A) \leftrightarrow (x \in B),$$

where the variable x is bound, and the variables A, B are free. The formula tells us that two sets are equal if they consist of the same elements.

Using quantifiers and the membership predicate, we can express other predicates of set theory. For example, the inclusion of sets $A \subseteq B$ is by definition expressed by the formula $\forall x\ (x \in A) \to (x \in B)$, which states that every element of A is also an element of B (in full accordance with the definition of the relation $\subseteq$).

Subsequently, from these two definitions, we will be able to derive another definition of the equality of sets A and B: $(A \subseteq B) \wedge (B \subseteq A)$.

In mathematical logic, so-called *universal closure* is often used, where a true formula with parameters (e.g., an axiom) is replaced by the same formula but with a universal quantifier over all variables, i.e., a logical transition is made from the formula $\mathcal{A}(a, b, c)$ to the formula $\forall a \forall b \forall c\ \mathcal{A}(a, b, c)$.

Let's give another humorous, but perfectly rigorous, example of mathematical reasoning. Three mathematicians walk into a bar, sit down at a table, and the waiter who approaches them asks:

— Will everyone have beer?

— I don't know, — replies the first mathematician, after thinking.

— I don't know, — replies the second mathematician.

— Yes! — replies the third.

Why did they all answer this way? The first one didn't know what the others wanted, but he himself wanted a beer, so he answered "I don't know." If he hadn't wanted one, he would have immediately answered "no," as it would have been clear that not everyone wanted beer (since "not everyone wants" means "at least one does not want"). The second, after the first's answer, understood that the first one would have a beer, but he didn't know about the third, and he himself wanted one too. So he also answered "I don't know." Finally, the third mathematician understood from their answers that they wanted beer (otherwise one or both of them would have answered "no"), and he himself wanted one too, so he could safely conclude that everyone would have beer. Therefore, he answered "Yes!" to the waiter's question.

Note that the waiter's question was understood in a purely mathematical way: is it true that $\forall x\ \ Wants\_beer(x)$, where the predicate $Wants\_beer$

expresses the property of an object x "wanting beer." Therefore, when the third mathematician calculated the truth of this expression, he answered Yes.

And if one of them had not wanted beer, the statement that is the negation of the original would have been true:

$$\neg(\forall x\ Wants\_beer(x)),$$

which is mathematically written using the existential quantifier as:

$$\exists x\ \neg Wants\_beer(x).$$

The universal and existential quantifiers are dual to each other through the logical negation operator: [12]

$$\neg(\forall x\ \mathcal{A}) \equiv \exists x\ \neg\mathcal{A}(x) \text{ and } \neg(\exists x\ \mathcal{A}) \equiv \forall x\ \neg\mathcal{A}(x).$$

Therefore, to prove that a statement of the form "all..." is not true, it is sufficient to provide a counterexample, i.e., to provide at least one example of the opposite. In the case of the beer, if one of the mathematicians had not wanted it, he would have been such a counterexample to the statement about everyone, and thus his single "no" would have been enough to give the waiter the answer "no." And if, for example, the first of them had answered "no," the others would have remained silent, as the answer to the waiter's question would have already been found.

What is essential in the given example is that all those answering the question about beer were mathematicians and, consequently, reasoned in the same mathematical way. If the question had been addressed to non-mathematicians, then surely each of them would have answered something like "Yes, I think so," and this would have been perceived by the waiter as "I will have one," i.e., he would have answered not the question that was asked about everyone, but the implied question addressed to each of them individually, "will you have a beer?".

[12] The quantifiers $\forall$ and $\exists$ can be understood, respectively, as a conjunction and a disjunction over an unlimited number of values of a variable. For example, $\forall x\ \varphi(x)$ is $\varphi(x_1) \wedge \varphi(x_2) \wedge \ldots$, where x_i is an enumeration of all possible values of the variable x, if there is a finite number of them.

A2.5.4 On Implication

Implication ($\mathcal{A} \to \mathcal{B}$) in mathematics is the so-called **material implication**. Its definition, like that of all other formulas of propositional logic, does not take into account the relationship between the cedents (i.e., the premise and the conclusion). And as we have already seen in the truth table, the implication $\mathcal{A} \to \mathcal{B}$ is false in only one case: when $\mathcal{A}$ is true and $\mathcal{B}$ is false. In all other three cases, the implication is true.

For example, the following statements are considered true: "If there is life on the Sun, then two times two is four," "If the Nill is a lake, then New York is a large city," "If the Sun is a cube, then the Earth is a triangle," "If two times two is five, then New York is a small city," etc. In ordinary reasoning, all these statements would hardly be considered meaningful, and even less so as true.[13] Obviously, material implication does not align well with the ordinary understanding of a conditional connection, but it works excellently where the calculus of forms needs to be performed in a way that is independent of semantics, i.e., in mathematics.

In fact, we often use such statements in ordinary life to emphasize the absurdity of a premise. For example: "if he is a mathematician, then I am the Pope," which literally means that a certain person absolutely cannot be a mathematician. Note that this construction uses precisely material implication, since the premise and the consequence are in no way related, and only by the rules of pure logic through contraposition (reasoning by contradiction) can we obtain the falsity of the premise (since I am not the Pope, he is not a mathematician), for:

$$\mathcal{A} \to \mathcal{B} \equiv \neg\mathcal{B} \to \neg\mathcal{A}.$$

In everyday language, we use at least two different implications: 1) material and 2) relevance. The second implies that there is some meaningful connection between the premise (antecedent) $\mathcal{A}$ and the consequence (succedent/consequent) $\mathcal{B}$. *Relevance implication* does not permit such rules of inference that are defining for material implication, namely: from a false premise, any statement follows; a true statement follows from any premise.

An example of a relevance implication: if the notation of a number a ends in 0, then it is divisible by 5 (we are talking about integers). Here we know nothing about the number a, but if the premise holds for it, then the conclusion also holds. And if the premise does not hold, the truth of the judgment itself is

[13] Such judgments are also called *alogisms*.

not affected in any way. In this case, both the premise and the conclusion have a common object variable, which in the premise is represented by the variable a, and in the conclusion by the pronoun "it." Mathematically, it would be more correct to write this judgment as: if the notation of a number a ends in 0, then the number a is divisible by 5.

Moreover, we know that other numbers are also divisible by 5, and this means that one must never confuse the premise and the conclusion! After all, it is not always true that if a number is divisible by 5, its notation ends in 0. This example clearly demonstrates that one must not confuse implication ($\rightarrow$) with equivalence ($\leftrightarrow$)!

In mathematics, we try to clearly determine different implications in different ways. For example, in predicate logic, in addition to purely logical implication, we use deductive inference. And if logical implication derives absolute logical truths, then deductive inference is more needed to obtain relevant conclusions (theorems) from non-logical truths (axioms). Of course, one cannot fail to mention here the deduction theorem, which states that $\mathcal{A} \rightarrow \mathcal{B}$ is true if and only if $\mathcal{B}$ is derivable from $\mathcal{A}$, i.e., that formally these two implications are equivalent (in classical predicate logic). But in essence, this means that a formula $\mathcal{B}$ that is true in pure logic is considered deductively true as well, and vice versa. Most mathematical theorems, however, are formulated more like relevance implications, since their direct proof takes place in one chosen context of axioms of a given theory.

For comparison, here are two more examples of theorems from Analysis:

(1) if a function is continuous on a closed interval, then it is integrable on it; (2) if the set of natural numbers is countable, then every continuous function on a closed interval is integrable on it.

In the first case, we see a clear connection between the premise and the consequence, since in both cases we are talking about the same function. In the second case, we simply connected two true statements that have nothing in common. The first example presents us with a relevance implication, the second—a material one. And if we look closely at the deductive proof of the second statement, it turns out that the premise about the set of natural numbers is used in it fictitiously, i.e., it can be eliminated without any loss to the conclusion.

So the deduction theorem itself can be regarded merely as an additional formal technique that helps to shorten a deductive proof based on purely formal features.

There are other ways in mathematics to study different types of implications and other logical connectives (e.g., another kind of negation). Here we are referring to so-called non-classical logics, i.e., propositional logics in which the logical operations are initially structured in a more complex way. For example, intuitionistic logic implies a prohibition of the law of the excluded middle, so that in it, negation ceases to be involutive (it is not always true that $\neg\neg\mathcal{A} \equiv \mathcal{A}$, although it remains true that $\mathcal{A} \to \neg\neg\mathcal{A}$). In relevance logic, the implication $\mathcal{A} \to \mathcal{B}$ is required to have at least one common object variable in the formulas $\mathcal{A}$ and $\mathcal{B}$ (in our example (1), this is the function whose integrability we are asserting).

In addition, new connectives and new quantifiers can be added to logic, resulting in non-classical logics that can describe various algebraic structures. These include, for example, modal, substructural, non-monotonic, and other logics.

In other words, mathematics is fully aware that classical formal logic has certain features that are not always understandable or applicable in a cultural or scientific context, and mathematics analyzes these features within such disciplines as mathematical logic and general algebra.

A2.6 Constructing Inferences

Above, we discussed how to correctly construct logical lexemes (formulas) within a given language $\mathfrak{Math}$, how they can be linked to mathematical concepts through the logical apparatus of set theory, and how to understand them in a stricter sense than in everyday life. However, in order to investigate a subject domain and obtain non-trivial results, one needs skills not only in writing and understanding mathematical text but also in knowing certain techniques for synthesizing new judgments that are as reliable as the basic (axiomatic) premises. The most general of these techniques lie in the field of logic, and below we will consider some of them.

Here, we will operate not only with mathematical concepts to simplify the perception of logical judgments, but all the examples given can, if desired, be transferred into a purely set-theoretic language by appropriately defining the extensions of the concepts.

The typical structure of a judgment is:

$$\textbf{Premises} \vdash \textbf{Conclusion} \tag{A2.1}$$

Here we do not use the implication symbol ($\rightarrow$), as it is a purely logical operator expressing part of a statement. The inference symbol ($\vdash$), called the "turnstile," shows the logical connection between separate statements, concealing a certain sequence of reasoning.[14]

The premises are usually written separated by commas, although formally they should be connected by the conjunction symbol ($\wedge$), as it is assumed that either all premises must be fulfilled to proceed to the conclusion, or at least one of them must be false (the material implication mentioned earlier is at work again!). "At least one" in this case means precisely that the falsity of the Premises is the falsity of the conjunction of the premises or, equivalently, the truth of the disjunction of their negations:

$$\neg(\text{Premise1} \wedge \text{Premise2} \wedge \text{Premise3}\dots) \equiv \neg\text{Premise1} \vee \neg\text{Premise2} \vee \neg\text{Premise3}\dots$$

A2.6.1 Syllogisms and Abduction

Aristotle

The first example of a non-trivial judgment is the most common one in logic textbooks and dates back to the times of Socrates and Aristotle.[15] Here we are talking about a two-premise syllogism, or more accurately, a composition of two implications.

Let us be given two premises: 1) All birds are animals, 2) All sparrows are birds. These premises are themselves implications. Indeed, the first tells us that every bird is an animal. More formally, this is written as follows: let $\mathfrak{B}$ be the concept "bird" and assume its definition is given by the formula $\mathcal{B}(x)$; let also $\mathfrak{A}$ be the concept "animal" with the defining formula $\mathcal{A}(x)$, and, furthermore, $\mathfrak{S}$ be the concept "sparrow" with the defining formula $\mathcal{S}(x)$. We are not currently focusing on what these definitions look like or whether they

[14] However, as we know from the deduction theorem in classical predicate calculus, pure mathematical formalism allows these two symbols to be considered equivalent.

[15] The theory of syllogisms was first developed in detail by Aristotle in his "Prior Analytics" and for centuries remained the core of logical education. Modern predicate logic includes syllogistics as a special case.

can be formalized in the language $\mathfrak{Math}$ at all, as this is irrelevant for general logic.

In these notations, the first premise is written as the implication $\mathcal{B}(x) \rightarrow \mathcal{A}(x)$, which can be universally closed, i.e., have a universal quantifier attached: $\forall x\ (\mathcal{B}(x) \rightarrow \mathcal{A}(x))$, which reads exactly as we need—all birds are animals. Similarly, the second premise will have the form: $\forall x\ (\mathcal{S}(x) \rightarrow \mathcal{B}(x))$.

Finally, if we denote the corresponding extensions of the concepts by A, B, S, we get two inclusions of extensions: $B \subseteq A$ (birds are animals) and $S \subseteq B$ (sparrows are birds). A visual representation of the inclusion of these concept extensions is shown in Fig. A2.2 using a so-called **Euler-Venn diagram**, where the extensions of concepts (and any sets in general) are depicted as circles.

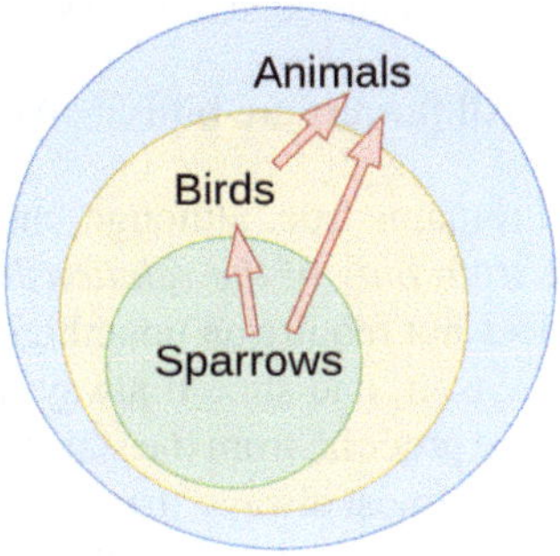

Fig. A2.2. A simple syllogism

The obvious conclusion from these premises is the inclusion $S \subseteq A$, which corresponds to the formula $\forall x\ (\mathcal{S}(x) \rightarrow \mathcal{A}(x))$ and the verbal formulation "All sparrows are animals." Thus, the structure of this syllogism is written in its entirety as:

All birds are animals, All sparrows are birds $\vdash$ All sparrows are animals,

or

$$\forall x\ (\mathcal{B}(x) \rightarrow \mathcal{A}(x)),\ \forall x\ (\mathcal{S}(x) \rightarrow \mathcal{B}(x)) \vdash \forall x\ (\mathcal{S}(x) \rightarrow \mathcal{A}(x)),$$

or

$$B \subseteq A,\ S \subseteq B \vdash S \subseteq A.$$

The last two records of the inference are written purely in the language of mathematical logic, or more precisely, in the language of predicate calculus. The only difference between these inferences is the choice of a specific language, namely: in the first case, the language of pure predicate calculus is used, as it uses no special notations other than logical ones; and in the second case, the language of set theory is used.

The common feature of these records is that the two premises share a common cedent, which is the formula $\mathcal{B}(x)$ expressing the property of being a bird, as well as the set B of all birds. This cedent is an **interpolant**,[16] as they say in logic, for the implication $\mathcal{S}(x) \longrightarrow \mathcal{A}(x)$. In essence, we see here an implicit manifestation of relevance implication.

The inference drawn above is valid regardless of whether the premises are correct. We can modify it, for example, as follows:

All birds are animals, All flowers are birds $\vdash$ All flowers are animals.

This judgment itself still remains true, although one of its premises is false. The judgment shows the truth *only of the relationship* between the premises and the conclusion, but does not require an unambiguous truth-assessment for its components. For this reason, one should always separate the work of the "turnstile," i.e., the inference process, from the statements involved in it. Recall that material implication is "almost always" true in the sense that there is only one of the four possible value combinations of the antecedent and consequent for which the implication is false (see Table A2.3). Thus, from a false premise, any statement can be derived, and a true statement can be derived from any premise, no matter how absurd it may be. This realizes the principle of the stability of truth with respect to derivability.

Consider an example of a similar, but incorrect, judgment: if *some galaxies in the Virgo Cluster are elliptical* and *some spiral galaxies are in the Virgo Cluster*, then *some spiral galaxies are elliptical*. Here, both premises are true, but the conclusion is false by definition. The set of spiral galaxies that are in the Virgo Cluster might be completely disjoint from the set of elliptical galaxies that are also in the cluster.

This conclusion is merely a supposition, not a rigorous logical argument. Fig. A2.3 shows a diagram where we can see that the set of "Spiral Galaxies" and the set of "Elliptical Galaxies" can both overlap with the "Galaxies in the

[16] A reference to the well-known Craig's interpolation theorem from proof theory.

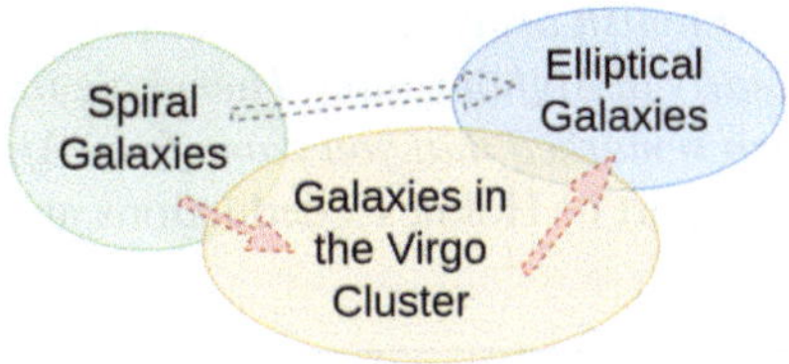

Fig. A2.3. An invalid syllogism

Virgo Cluster" set without overlapping each other. The objects that fall into these intersections are those characterized by the word "some."

In this example, it is precisely the link between the premises and the conclusion that is broken, as the two premises are not joined by a distributed middle term (they have no valid interpolant).

We previously noted that reasoning in the reverse direction—from conclusion to premises—is invalid. However, very often it is partly valid. For example, we know the following conclusion: if a number ends in 0, then it is divisible by 5. Based on this, we cannot prove for certain, but we **can suppose**, that if a number is divisible by 5, it probably might end in 0. As we know, this is true in about half of the cases. If such an *inversion of implication* were always absolutely impossible, then deduction would be the simplest case of inference, where a falsehood implies any proposition. This is completely useless for building theories.

The method of *reasoning backwards*, towards a known premise, is called **abduction**. It is this method, as a rule, that Sherlock Holmes used in his deductions. That is why his conclusions are always probabilistic in nature and are accompanied by words like "probably," "most likely," etc. The art of Sherlock Holmes lies in choosing the most probable premise from all possible ones in a given situation, based on experience.[17]

For example, quoting from "A Study in Scarlet" (by Conan Doyle):

"Here is a gentleman of the medical type, but with the air of a military man. Clearly an army doctor, then. He has just come from the tropics, for his face is dark, and that is not the natural tint of his skin, for his wrists are fair. He has undergone hardship and sickness, as his haggard face says clearly. His left arm has been injured. He holds it in a stiff and unnatural manner. Where

[17] The term 'abduction' as a special type of logical inference (inference to the best explanation) was introduced and thoroughly investigated by the American philosopher and logician Charles Sanders Peirce in the late 19th and early 20th centuries.

in the tropics could an English army doctor have seen much hardship and got his arm wounded? Clearly in Afghanistan." The whole train of thought did not occupy a second. I then remarked that you came from Afghanistan.

Let's consider only a part of Holmes's deductions and compare them with an arithmetic example:

Watson is a haggard, tanned army doctor. Those who fought in Afghanistan are haggard, tanned army doctors.	The number 30 is divisible by 5. A number ending in 0 is divisible by 5.
Conclusion: Watson came from Afghanistan.	Conclusion: 30 ends in 0.

As we can see, we are given two premises. One provides a connection between properties (those who fought are army doctors, etc., and numbers ending in 0 are divisible by 5), and the other gives a property of a specific object (Watson and the number 30). This property is common to both premises, but they cannot be joined into a syllogism because the property is always at the end of the premise. But Holmes knows that practically all haggard, tanned army doctors are those who fought in Afghanistan (although this is not 100% true), and based on this, he presumes(!) that Watson is the same, since he possesses the same property.

In the example of the number 30, this also worked. However, if we were to substitute 25 for 30, the entire chain of reasoning would collapse!

Therefore, abductive inferences cannot be considered mathematical, but they can lead to a correct deductive inference, resulting in either a theorem (*All haggard army doctors fought in Afghanistan*) or the discovery of a counterexample (in our case, the number 25, which refutes the assumption that all numbers divisible by 5 end in 0).

Note that abductive reasoning is widely used in machine learning when, based on an incomplete set of properties and possessing an array of data with similar properties, we classify the object currently being tested (Watson or the number 25) with a certain degree of probability. But in fairness, it should be noted that adding probabilities to the abduction scheme actually brings us back to classical logic, since the judgments now relate not to the original statements, but to statements about their probability. Moreover, one can even consider a non-classical version of logic with probabilistic implications. But that is a whole other story.

A2.6.2 Modi

From the inference $(\mathcal{S} \to \mathcal{B})$, $(\mathcal{B} \to \mathcal{A}) \vdash (\mathcal{S} \to \mathcal{A})$ discussed above, a simpler method of constructing inferences can be extracted, which is called **Modus Ponens** and is one of the rules for eliminating implication:

$$\mathcal{B},\ (\mathcal{B} \to \mathcal{A}) \vdash \mathcal{A}.$$

It can be justified as follows. If we substitute the tautology $\top$ for the statement $\mathcal{S}$ in the previous rule, we get $(\top \to \mathcal{B})$, $(\mathcal{B} \to \mathcal{A}) \vdash (\top \to \mathcal{A})$. But an implication of the form $(\top \to \mathcal{A})$ is equivalent to the statement $\mathcal{A}$ itself, due to the property of implication that only truth can follow from truth. Thus, we can replace the formula $\top \to \mathcal{B}$ with $\mathcal{B}$, and the formula $\top \to \mathcal{A}$ with $\mathcal{A}$. The result is the Modus Ponens rule.

This rule can also be verified using the truth table A2.3, by considering all possible logical values of the variables $\mathcal{A}$ and $\mathcal{B}$ for which the premises of the rule are true, and confirming that the conclusion is also true in every such case.

And of course, this rule is easily verified using sets, by replacing implication with inclusion. Indeed, assuming that B is the extension of the "empty" concept, i.e., the universe, and also that $B \subseteq A$, it is easy to see that A must also coincide with the universe, i.e., it is the extension of the "empty" concept (defined by the formula $\top$), since it contains the universe and cannot be larger than the universe by definition.

Modus Ponens is a key rule of inference when defining the properties of the turnstile ($\vdash$). A surprising fact is that this rule is sufficient to derive all logical truths (tautologies), including more complex rules of inference.

We mentioned above that the turnstile ($\vdash$) means that we have moved from premises to conclusions, possibly hiding a chain of reasoning. But rules like Modus Ponens represent a single-step inference, used in logic as an elementary step of derivation. To distinguish a multi-step inference from an elementary one, it is customary to use a different notation when writing down the basic rules of inference, namely, a vertical one:

$$\frac{\mathcal{B},\ \mathcal{B} \to \mathcal{A}}{\mathcal{A}},$$

where the horizontal bar plays the same role as $\vdash$, while simultaneously indicating the elementary nature of the logical transition. Therefore, we will always write the basic rules of inference in this way.

Another well-known rule of inference is Modus Tollens:

$$\frac{\mathcal{A} \to \mathcal{B},\ \neg\mathcal{B}}{\neg\mathcal{A}},$$

which means that if the implication $\mathcal{A} \to \mathcal{B}$ is true and $\mathcal{B}$ is false, then $\mathcal{A}$ must also be false. In essence, this rule illustrates the method of reduction to absurdity.

Closely related to Modus Tollens is the rule of proof "by contradiction" (*contraposition*):

$$\frac{\neg\mathcal{A} \to \bot}{\mathcal{A}},$$

which states that if, by assuming the falsity of $\mathcal{A}$, we arrive at a contradiction (derive falsity), then $\mathcal{A}$ must be true.

There are many rules in logic that allow one to move from truths to truths. They are all properties of derivability, i.e., the relation between formulas denoted by $\vdash$.

Modern mathematical logic studies various types of derivability, as well as different types of inference rules and dependencies between them, with the aim of identifying a minimal sufficient set of inference rules (a kind of basis for derivability) in different specific instances of this relation.

Of course, all these complex matters relate to non-classical logics, but here we are strictly following the classical path, so in our case, derivability will have a rather simple form: the basic rule of inference is only Modus Ponens (not counting the unconditional rule of substitution, which will be discussed later), and all other rules are derived from it.

The choice of Modus Ponens, and not some other rule, as the basis is a matter of tradition, i.e., it is essentially arbitrary.

A2.6.3 Deduction and Induction

The construction of rigorous inferences from given or previously obtained true statements and predicates is called **deduction** and is the primary method of reasoning in obtaining mathematical theorems. Sometimes, to construct the required inference, one needs to go through hundreds of combinations of previously proven premises. But often, auxiliary illustrations or the experience of a researcher immersed in the topic are sufficient to find the right chain of proof.

Deduction is always a transition from the general to the particular. For example, the axioms of geometry describe in a general way the relationships between points, lines, and planes, while the theorems of geometry are the application of these axioms in specific cases for different classes of figures (triangles, trapezoids, etc.).

In contrast to deduction, there is the method of **induction**, which involves a transition from the particular to the general. More precisely, an inductive judgment is constructed as follows: we observe the same result in many homogeneous cases and expect the result to be the same in the next case of the same kind.

Let's get specific. Consider an arithmetic example of a statement:

($\mathcal{A}$) If a is an even number, then it is divisible by 4.

This statement is false, as not all even numbers are divisible by 4. Conversely,

($\mathcal{B}$) If a number is divisible by 4, then it is even.

This statement is true.

Statement ($\mathcal{B}$) is *deductive* because it is based on a rigorous inference from the properties of natural numbers. It is **derived** from axioms or previously proven theorems.

Statement ($\mathcal{A}$) can only be a supposition. If we were to take the numbers $4, 16, 32, 64, 128, 256, 512, 1024$, it would turn out that they are all not only even but also divisible by 4. A suspicion may arise that this is always the case, and we form a hypothesis: if a number is even, then it is divisible by 4. Such an inference is called inductive.

But then we discover that, for example, 6 is not divisible by 4, although it is even (i.e., as mathematicians say, we have found a **counterexample**). The hypothesis turns out to be false.

Another example of inductive reasoning: no matter how many odd numbers we have taken, their square has always been of the form $4k + 1$. We then form the hypothesis that the square of any odd number is of the form $4k + 1$. We can try to find a counterexample if we doubt the correctness of our inductive conclusion. But we will not succeed, because this time induction has led us to the correct answer.

And to deductively prove that this is indeed the case, we will need another method, namely: **mathematical induction**, which is a cornerstone of arithmetic, if not of all mathematics.

Mathematical induction is a judgment of the following kind. Let us have some property $\mathcal{A}(n)$ of natural numbers. Suppose that $\mathcal{A}[n/0]$ is true and that whenever $\mathcal{A}[n/x]$ is true, $\mathcal{A}[n/x+1]$ is also true. Then $\mathcal{A}(n)$ is totally true (i.e., for all instances of the variable n in the natural numbers). Formally, the axiom of induction looks like this:

$$\mathcal{A}[x/0] \wedge (\forall x\ (\mathcal{A}[n/x] \rightarrow \mathcal{A}[n/x+1])) \vdash \forall x\ \mathcal{A}[n/x].$$

Recall that the slash in this case means replacing the original variable n with the corresponding term that follows the slash.

Thus, mathematical induction is not induction in the general logical sense, but a deductive method that, depending on the chosen theory, is either an axiom or is derived as a theorem.

Several other principles are associated with mathematical induction, which we will discuss in the context of arithmetic:

- the principle of infinite descent;
- the principle of well-foundedness;
- the pigeonhole principle.

In general, it can be said that mathematics employs all logical methods, but only considers deductive ones as reliable ways of obtaining results. Methods such as induction (not mathematical) and abduction are used by mathematicians to search for refutations or as an attempt to find a general rule that must then be proven deductively.

A2.7 On Relations and Functions

Previously, we have consistently dealt with properties of objects of the form $\mathcal{A}(x)$, which can be characterized as *unary relations* on the universe. Unary relations correspond to the extensions of concepts, i.e., subsets of the universe.

However, we can also consider more complex properties of objects, namely, their joint properties, or *relations*. Thus, if a property $\mathcal{R}(x, y)$ asserts something simultaneously about objects x and y, we say that the pair $\langle x, y \rangle$ is in the relation $\mathcal{R}$.

For example, if we consider people as our subject domain, by a relation $\mathcal{R}(x, y)$ we can mean, for instance, the relation of friendship (x and y are friends) or a kinship relation (father-son, brother-sister, ancestor-descendant, etc.).

In this case, the extension of such a concept as a relation becomes more complex. It is not a subset of the universe $\mathfrak{U}$, but a subset of some power of the universe, $\mathfrak{U}^n$. By the n-th power of the universe $\mathfrak{U}$, we mean the collection of all possible ordered tuples of elements of $\mathfrak{U}$ of length n (sequential selections with replacement). These tuples are usually written as a vector: $\langle x_1, \ldots, x_n \rangle$, and often the shorthand $\vec{x}$ is used if n is known from the context.

Thus, the extension of an n-ary relation is a subset $\mathcal{R} \subseteq \mathfrak{U}^n$, but more often this extension itself is called an n-**ary relation** (for $n = 2$, binary; for $n = 3$, ternary; and for $n = 1$, unary) on $\mathfrak{U}$. The number n is then called the *arity* (or *valency*) of the relation $\mathcal{R}$.

Thus, the set of all pairs of people $\langle x, y \rangle$ in which x is the brother of y is a binary relation; the relation $B(x, y, z)$, meaning that point y lies between points x and z, is a ternary relation; and the relation $Par(a, b, c, d)$, meaning that the line passing through points a, b is parallel to the line passing through points c, d, is a quaternary (four-place) relation.

If $\mathfrak{U}^n$ is considered as a new universe, n-ary relations will correspond to certain extensions of concepts, so that any mathematical relation can be defined as a mathematical concept with its own definition (a formula with n arguments), extension (the relation itself), designation, and, optionally, notation.

Moreover, one can go even further and extend the universe to the sum of all possible powers $\mathfrak{U}^n$ for all natural numbers $n \geqslant 1$, i.e., include in our universe all ordered tuples of any finite length, consisting of elements of the original universe $\mathfrak{U}$. Such a universe is denoted by $\mathfrak{U}^{<\infty}$. All possible unary, binary, ternary, etc., relations are subsets of this universe.[18]

This technique of converting arbitrary relations into sets is widely used in mathematics and completely translates logic into set theory, since every logical formula with n arguments defines some n-ary relation (although the converse, generally speaking, is not true). In a sense, the set-theoretic approach even extends the natural-logical one.

Let us specify some types of binary relations:

1. $\mathcal{R}(x, y)$ is *reflexive* if $\mathcal{R}(x, x)$ is true for any $x \in \mathfrak{U}$;
2. $\mathcal{R}(x, y)$ is *symmetric* if $\mathcal{R}(x, y)$ always implies $\mathcal{R}(y, x)$ for any $x, y \in \mathfrak{U}$;

[18] We will leave aside the question of why the reverse statement is not true.

3. $\mathcal{R}(x, y)$ is *transitive* if $\mathcal{R}(x, y) \wedge \mathcal{R}(y, z)$ always implies $\mathcal{R}(x, z)$ for any $x, y, z \in \mathfrak{U}$;
4. $\mathcal{R}(x, y)$ is an *equivalence relation* if it is reflexive, symmetric, and transitive;
5. $\mathcal{R}(x, y)$ is a *preorder* if it is reflexive and transitive.

As an illustration, consider the "father-son" relation, i.e., $\mathcal{R}(x, y)$ is true if and only if x is the father of y. It is obvious that such a relation is not reflexive; moreover, it is *irreflexive*, i.e., the formula $\mathcal{R}(x, x)$ is false for all $x \in \mathfrak{U}$. It is also not symmetric, because if x is the father of y, the reverse cannot be true (at least while observing the principle of causality). And it is certainly not transitive, because if x is the father of y, and y is the father of z, then x cannot be the father of z. Consequently, the relation $\mathcal{R}$ is neither a preorder nor an equivalence relation.

However, based on it, one can define two new relations: 1) $\mathcal{P}(x, y)$ is true if and only if either $\mathcal{R}(x, y)$ or there exist people $x_1, \ldots, x_n$ such that $\mathcal{R}(x, x_1) \wedge \mathcal{R}(x_1, x_2) \wedge \cdots \wedge \mathcal{R}(x_{n-1}, x_n) \wedge \mathcal{R}(x_n, y)$, i.e., there is a chain of people connected by the father-son relation in the specified order; 2) $\mathcal{Q}(x, y) \overset{\text{def}}{\leftrightarrow} \exists z\ (\mathcal{R}(z, x) \wedge \mathcal{R}(z, y))$.

The first relation $\mathcal{P}(x, y)$ tells us that x is an ancestor of y, i.e., it is the "ancestor-descendant" relation. This relation is still irreflexive, but it is transitive, since if x is an ancestor of y, and y is an ancestor of z, then x is an ancestor of z. An irreflexive, transitive relation is called a **strict (partial) order**.

Finite partially ordered sets are often visualized using **Hasse diagrams**. In such diagrams, elements are represented as vertices, and if $x < y$, then x is drawn lower than y. An edge connects x and y only if $x < y$ and there is no z such that $x < z < y$ (i.e., edges representing implied transitivity are omitted).

For instance, consider the "ancestor" relation $\mathcal{P}$ we defined earlier. If we establish an order where "lower" means descendant and "higher" means ancestor, we get a classic family tree (see Fig. A2.4). Here, the Son (x) is related to the Grandfather (z) by the relation $\mathcal{P}(z, x)$, i.e., z is an ancestor of x. However, there is no direct line drawn between them. The connection is implied through the Father (y), since $\mathcal{P}(z, y)$ and $\mathcal{P}(y, x)$. It is worth noting that the *edges* of the Hasse diagram correspond exactly to the original relation $\mathcal{R}$ ("Father"), while the *paths* in the graph correspond to the transitive relation $\mathcal{P}$ ("Ancestor"). This visual omission of transitive lines is the key feature that makes Hasse diagrams clean and readable compared to standard graph representations of relations.

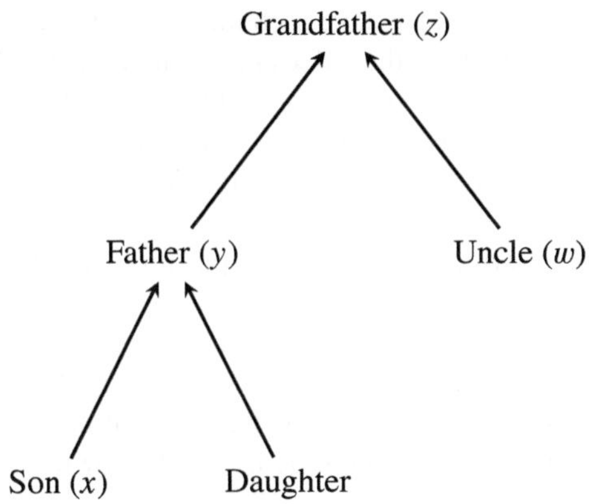

Fig. A2.4. Hasse diagram for the "Descendant $\leqslant$ Ancestor" relation

If desired, we can complete the relation $\mathcal{P}$ to a reflexive relation by adding to the set $\mathcal{R}$ the *diagonal* of the universe $\mathfrak{U}^2$, namely, the set of all pairs $\langle x, x \rangle$. As a result, we obtain a non-strict ancestor-descendant relation, which allows every x to be its own ancestor. As strange as this may sound when applied to people, such "closures" of mathematical constructs often turn out to be very convenient from the point of view of simplifying the language and theorems.

Here one might recall a legal scenario. The supplier-client relation is by default assumed to be irreflexive, i.e., no one can be their own supplier (of services, goods). However, we can well imagine a case where such a relation could be reflexive. For example, a hairdresser can cut their own hair (and not just a hairdresser), in which case they are both the supplier (of the service) and the client simultaneously within the same relation (the haircutting service). But in that case, it must be recognized that supplier-client relations can be at least partially reflexive. Then the legal aspect comes into play, involving a) payment for services and b) levying taxes on this type of activity. And since there is no restriction on applying tax legislation to reflexive financial relationships, strictly speaking, anyone who cuts their own hair must document this service and pay taxes on it. Of course, this doesn't work in practice, but the aftertaste of this kind of purely logical reasoning remains.

Next, the relation $Q(x, y)$ tells us that x and y have a common father. Such a relation is, firstly, reflexive (x and x do indeed have a common father),[19]

[19] With the caveat that everyone in the considered universe has one.

secondly, symmetric (if x and y have a common father, then y and x do), and thirdly, transitive (if w is the father of x and y, and w' is the father of y and z, then w and w' are the same person, who is the father of x and z). Thus, Q is the *sibling relation*, which is an example of an equivalence relation.

In fact, given a preorder $\mathcal{R}(x, y)$, one can always define an equivalence relation by the formula $\mathcal{R}(x, y) \wedge \mathcal{R}(y, x)$.[20]

An equivalence relation allows one to partition the entire universe into disjoint *equivalence classes* such that in each class are those and only those objects that are pairwise related by this relation. In the previous example, the relation Q partitions all people into classes by direct blood kinship. In each such class are brothers and sisters who have a common father (note that if "father" is replaced by "parent," the scheme breaks down—the relation ceases to be transitive).

To define the following types of relations, we need to bring in one of the fundamental relations—**equality**. This is a binary equivalence relation on the elements of the universe $\mathfrak{U}$, which has the property that each equivalence class contains only such objects that are indistinguishable from each other by the means of the language we have. In section B1.2.5, we will define equality in a more rigorous way, based on the language and logic of predicates. Equality is usually denoted by the symbol = and instead of $=(x, y)$, it is written as $x = y$.

Let us give a few more definitions:

6. $\mathcal{R}(x, y)$ is an *antisymmetric* relation if $\mathcal{R}(x, y) \wedge \mathcal{R}(y, x)$ implies $x = y$;
7. $\mathcal{R}(x, y)$ is a (non-strict) *partial order* (or simply an *order*) if it is an antisymmetric preorder;
8. $\mathcal{R}(x, y)$ is a (non-strict) *linear order* if it is a partial order that has the property of *connexity*: for any $x, y \in \mathfrak{U}$, either $\mathcal{R}(x, y)$ or $\mathcal{R}(y, x)$ holds.

It is interesting to note that if we have a preorder relation, then, as we have already noted, it is not difficult to construct an equivalence relation from it. After which, by considering the equivalence classes as individuals in a new subject domain, it is easy to induce the original preorder relation on them, which will then turn out to be a partial order.[21] Thus, quotienting (moving from the original universe to the universe of equivalence classes) improves the properties of the original relation by *coarsening* the original universe.

And a couple more definitions:

---

[20] Verify that this is so.

[21] This simple exercise is left for the reader to do independently.

9. a relation $\mathcal{R}(x, y)$ is *total* if for every x there exists a corresponding y such that $\mathcal{R}(x, y)$ holds;
10. a relation $\mathcal{R}(x, y)$ is *functional* (or simply a *function*) if it is total and from the condition $\mathcal{R}(x, y) \wedge \mathcal{R}(x, z)$ it always follows that $y = z$, for all $x, y, z \in \mathfrak{U}$.[22]

For example, a reflexive relation is total. And if it is restricted to the diagonal (leaving only pairs $\langle x, x\rangle$), it represents a function that is usually denoted by id.

Above, we only considered relations on the entire universe. However, it is more common in mathematics to consider narrower cases.

First, we can restrict a relation to be on a single set A (where $A \subseteq \mathfrak{U}$). In this case, both variables x and y in the lexeme $\mathcal{R}(x, y)$ are assumed to range only over the elements of A.

Second, we can generalize this further to define relations *between* two different sets, A and B. For such a relation, we assume that the first variable, x, ranges over the elements of set A, while the second variable, y, ranges over the elements of set B.

The same generalization applies to arbitrary n-ary relations, when we assume that the relation $\mathcal{R}(x_1, \ldots, x_n)$ connects variables that have their own independent domains of variation: $x_k \in A_k$. Such a relation itself is a subset of the set of all possible n-tuples $\langle x_1, \ldots, x_n\rangle$, which is denoted by $A_1 \times \cdots \times A_n$ and is called the Cartesian product of these sets.

It is easy to see, however, that any narrower n-ary relation defined on sets (between sets) is, nevertheless, a relation on the entire universe, although it is not defined everywhere.

The same considerations apply to functions as a special case of relations. Namely, we can say that $\mathcal{F}$ is a functional relation on $A \times B$, where A and B are some sets in our universe. In this case, the totality of $\mathcal{F}$ is limited only to the set A, and in this case, one writes $\mathcal{F} : A \to B$ as a notation for the statement "$\mathcal{F}$ is a functional relation that is total on A."

Moreover, one can take a more complex set as A: $A_1 \times \cdots \times A_n$, and obtain a function $\mathcal{Y} : A_1 \times \cdots \times A_n \to B$. Such a function is called n-ary, its arguments are n-tuples of the form $\langle x_1, \ldots, x_n\rangle$, where $x_k \in A_k$. Note that if you look at such a function as a relation, then, strictly speaking, it is a binary relation

[22] Unless explicitly stated otherwise, the designation **function** in this book refers to a *total* function, defined on every element of its domain. This is the standard convention in Set Theory. However, readers with a background in Computer Science or Computability Theory should be aware that in those fields, partial functions are the norm.

between the sets $A_1 \times \cdots \times A_n$ and B, but it can also be considered as an $(n+1)$-ary relation on the set $A_1 \times \cdots \times A_n \times B$.

In summary, we note that relations in the most general form are practically the same as the predicates we spoke of earlier. The only difference is that a predicate is a lexeme of a logical type that participates in the construction of formulas of the language, i.e., it is a property of objects written formally, while a relation is the domain of truth of a predicate or, in other words, the extension of the concept-property that is defined by the predicate. Therefore, in books, one can often find an identification between the set-theoretic entity "relation" and the formal-logical "predicate." One can also say about predicates that they, as a rule, have their own names and notations, i.e., each of them is a mathematical concept, while not all relations can be named (if only for cardinality reasons), and therefore not all can be a mathematical concept, but they are all elements of the extension of the general concept "relation."

Our further goal is a rigorous, but as simple as possible, exposition of propositional logic at level B1 and then, on its basis, predicate logic. We will adhere to the scheme A2.1. The exposition of the formal theory itself will be done in a sufficiently rigorous language, which can be considered part of the language $\mathfrak{Math}$ and which already contains the definitions of logical connectives and set-theoretic notations we have given above as mathematical concepts.

A2.8 Self-check Questions

Q1 What is the main function of logic as applied to the language $\mathfrak{Math}$? [p. 38]

Q2 Why is syntactic logic (the logic of lexemes) insufficient for a full analysis of a subject domain? [p. 38]

Q3 List the main stages of formalizing a subject domain in the language $\mathfrak{Math}$. [p. 40]

Q4 What is a "mathematical concept"? Name its mandatory and optional components. [p. 42]

Q5 Explain using the example of a prime number how the idea of a concept can differ from its formal definition in different semantics. [p. 48]

Q6 What is the connection between the truth of formulas, axioms, and the chosen semantics? Explain the connection between logical connectives

($\wedge, \vee, \neg, \rightarrow$) and set-theoretic operations ($\cap, \cup, \setminus$) through the extension of a concept. [p. 49]

Q7 What are the universal ($\forall$) and existential ($\exists$) quantifiers? How are they related through the logical negation operator? [p. 60]

Q8 Explain the difference between the material implication used in mathematics and the relevance implication characteristic of everyday language. [p. 63]

Exercises

A2.1 Classify the following expressions into three categories: True proposition, False proposition, or Predicate (truth depends on the value of variables).

- $2 + 2 = 5$
- $x + 1 = 5$
- All prime numbers are odd.
- $5 > 3$

A2.2 Write the negation of the following statements (without just adding "It is not true that..."):

- All cats are black.
- Some students like mathematics.
- If it rains, the ground is wet.

A2.3 Let $A = \{1, 2, 3\}$ and $B = \{2, 3, 4\}$. **Determine the truth value** of the following formulas:

- $\forall x \in A \quad (x < 4)$
- $\exists x \in B \quad (x \text{ is even})$
- $\forall x \in A \cap B \quad (x \geqslant 2)$

A2.4 Determine whether the "if... then..." construction is a *material implication* or a *relevance implication*. Use the criterion from Section A2.5.4: does the consequent logically or semantically follow from the antecedent, usually sharing a common variable?

- If a number n is divisible by 6, then n is divisible by 3.
- If the Eiffel Tower is in Paris, then snow is white.
- If a person is a grandmother, then she is female.

A2.5 Check if the following logical inference is valid using a Venn diagram or logical reasoning:

All squares are rectangles.
All rectangles are quadrilaterals.
Therefore, all squares are quadrilaterals.

A2.6 (Assume non-empty sets) Given two premises, select **all** conclusions that logically follow.

P1: All Faxi are Krolo.
P2: No Winda is Krolo.

a) At least one Faxi is not Winda.
b) At least one Krolo is a Faxi.
c) At least one Krolo is not Winda.
d) At least one Winda is not Faxi.
e) No Faxi is Winda.

A2.7 (Assume non-empty sets) Given two premises, select **all** conclusions that logically follow.

P1: All Donas are Sudr.
P2: At least one Donas is not Kalsi.

a) At least one Sudr is not Kalsi.
b) At least one Kalsi is not Sudr.
c) At least one Sudr is Donas.
d) No Sudr is Kalsi.

A2.8 (Assume non-empty sets) Given two premises, select **all** conclusions that logically follow.

P1: No Teka is Arati.
P2: All Teka are Fato.

a) All Arati are Fato.
b) At least one Fato is not Arati.

c) At least one Teka is Fato.

d) At least one Arati is not Teka.

A2.9 (Assume non-empty sets) Given two premises, select **all** conclusions that logically follow.

P1: All Bron are Hanto.
P2: At least one Dax is Bron.

a) All Hanto are Bron.

b) At least one Dax is Hanto.

c) At least one Hanto is Bron.

d) No Dax is Hanto.

A2.10 Using the connection between logical connectives and set operations (e.g., $A \cap (B \cup C)$ corresponds to $P \wedge (Q \vee R)$), simplify the following set-theoretic expressions.

1. $A \cup (A \cap B)$;
2. $(A \cup B) \cap (A \cup \overline{B})$;
3. $A \setminus (A \setminus B)$;
4. $(A \cap B) \cup (A \cap \overline{B}) \cup (\overline{A} \cap B) \cup (\overline{A} \cap \overline{B})$.

A2.11 Draw Venn diagrams to verify whether the following equalities hold:

1. $A \setminus (B \setminus C) = (A \setminus B) \cup C$;
2. $A \cap (B \vartriangle C) = (A \cap B) \vartriangle (A \cap C)$.

A2.12 Let R be a binary relation on a set M. Determine whether R is reflexive, symmetric, antisymmetric, or transitive in the following cases:

1. M is the set of all people, xRy means "x is a brother of y".
2. $M = \mathbb{R}$, xRy means "$|x - y| \leqslant 1$"
3. M is the set of words in a dictionary, xRy means "word x comes after word y".
4. $M = \mathbb{N}$, xRy means "x divides y".

A2.13 Let $M = \{a, b, c\}$. Construct (list the pairs of) a relation R on M that is:

1. Reflexive but not transitive;

2. Symmetric but not reflexive;
3. A strict partial order (irreflexive and transitive).

A2.14 Draw the Hasse diagram (a graph where x is connected to y if $x < y$ and there is no z such that $x < z < y$) for the relation of divisibility on the set $D_{12} = \{1, 2, 3, 4, 6, 12\}$.

A2.15 Let $M = \mathbb{Z}$ (integers). Check if the following relations are equivalence relations. If yes, describe the equivalence classes.

1. $x \sim y \overset{\text{def}}{\leftrightarrow} x + y$ is even;
2. $x \sim y \overset{\text{def}}{\leftrightarrow} x^2 = y^2$;
3. $x \sim y \overset{\text{def}}{\leftrightarrow} x \leqslant y$.

A2.16 Based on the definition in Section A2.7 (a relation is a function if it is total and functional), determine which of the following relations on $A = \{1, 2, 3\}$ are functions $f : A \to A$:

1. $R_1 = \{(1, 2), (2, 3), (3, 1)\}$;
2. $R_2 = \{(1, 2), (2, 3)\}$;
3. $R_3 = \{(1, 2), (1, 3), (2, 1), (3, 1)\}$;
4. $R_4 = \{(1, 1), (2, 1), (3, 1)\}$.

A2.17 On an island, Knights always tell the truth, and Knaves always lie. You meet two inhabitants, A and B.

1. A says: "At least one of us is a Knave." What are A and B?
2. A says: "If B is a Knight, then I am a Knave." What are A and B?

(Hint: Formalize the statements using propositional logic variables).

A2.18 Determine whether the following deductions are valid.

1. Premise 1: If it rains, the street is wet.
   Premise 2: The street is wet.
   Conclusion: Therefore, it rained.
2. Premise 1: All men are mortal.
   Premise 2: Socrates is a man.
   Conclusion: Therefore, Socrates is mortal.

3. Premise 1: If x is divisible by 4, then x is even.
   Premise 2: x is not even.
   Conclusion: Therefore, x is not divisible by 4.

A2.19 Classify the following arguments as *Deductive* (derived from axioms/general rules) or *Inductive* (generalized from observations):

1. "The sun has risen in the east every morning for recorded history. Therefore, the sun will rise in the east tomorrow."
2. "All triangles have angles summing to 180°. Figure ABC is a triangle. Therefore, its angles sum to 180°."
3. "Every number ending in 0 is divisible by 5. Number 30 ends in 0. Therefore 30 is divisible by 5".

A2.20 In mathematics, $\mathcal{A} \wedge \mathcal{B}$ is equivalent to $\mathcal{B} \wedge \mathcal{A}$. Give an example of a sentence in natural language or a command in programming where swapping the parts of an "AND" construction changes the meaning or causes an error.

A2.21 Let the universe be $\mathbb{N} = \{0, 1, 2, \ldots\}$. List the elements (or describe the set) corresponding to the extension of the following concepts/formulas:

1. $\mathcal{A}(x) : x < 5$;
2. $\mathcal{B}(x) : \exists y \in \mathbb{N}(x = 2y)$;
3. $C(x) : (x < 3) \wedge (x > 5)$;
4. $\mathcal{D}(x) : x \cdot 0 = 0$.

A2.22 Using the base relation $Parent(x, y)$ ("x is a parent of y"), write formal definitions (formulas) for:

1. $Grandparent(x, y)$ (x is a parent of a parent of y);
2. $GreatGrandparent(x, y)$;
3. $CommonParent(x, y)$ (x and y share a parent).

Level B1
Formal Logic

God, grant me the serenity to accept the things I cannot change, courage to change the things I can, and wisdom to know the difference.

— The Serenity Prayer by Reinhold Niebuhr

B1.1 Propositional Logic

Gottlob Frege

The formalization of logic itself—transforming it into an object of mathematical study—became a key objective at the turn of the 20th century. The works of Gottlob Frege, Bertrand Russell, and Alfred North Whitehead (especially their foundational work *Principia Mathematica*, 1910–1913) and David Hilbert were aimed at creating rigorous axiomatic systems for logic, which would then allow for the rebuilding of all of mathematics on this foundation.

At Level B1, we begin discussing logic in a strictly formal manner, as is done in textbooks. The purpose of this approach is to compare our previous semi-formal investigations with the modern conception of mathematical logic.

N. I. Kazimirov, *Mathematics as a Foreign Language*, Mathematics in Mind,
https://doi.org/10.1007/978-3-032-30721-7_3

B1.1.1 The Language of Propositional Logic

Propositional logic (or calculus) is characterized by the absence of a distinction between terms and formulas, as its domain of discourse is the truth-value category, i.e., the universe of propositions. By propositions, we understand any expressions (independent of language) that are either true or false. Within propositional logic, it is assumed that propositions have no parameters; that is, each proposition has a single, specific logical value. Thus, we study a uniform universe whose elements are denoted by propositional variables $\mathcal{A}, \mathcal{B}, \ldots$, possibly with numerical indices. The word "propositional" will only acquire its full meaning when we consider predicate logic—the logic in which the domain of discourse is no longer part of the truth-value category.

Furthermore, we use the logical connectives $\neg, \wedge, \vee, \rightarrow$ as operator symbols to construct the formulas of propositional logic (propositional formulas, or simply formulas). Finally, we use the letters φ, ψ (possibly with numerical indices) to denote arbitrary formulas. The symbols φ, ψ are called *non-terminal* because they do not participate in the construction of any specific formula.[1] They are necessary to write down the rules for formula construction as a single schema. If we did not use non-terminal symbols, we would have to write an infinite number of rules for each specific set of formulas and variables. This technique is frequently used in mathematical logic, and indeed in mathematics as a whole. The rules for constructing formulas are as follows:

1. Every variable is a formula.
2. If φ and ψ are formulas, then so are $\neg(\varphi)$, $(\varphi) \wedge (\psi)$, $(\varphi) \vee (\psi)$, and $(\varphi) \rightarrow (\psi)$.
3. Any formula is constructed according to rules 1 and 2.

The third point is sometimes replaced in logic textbooks by the requirement that the set of formulas is the *minimal* set of lexemes closed under rules 1 and 2.

[1] We borrow the designation "non-terminal" from the theory of formal grammars, where it describes symbols used as placeholders in the rules for generating a language's strings. In our case, these symbols, such as φ and ψ, belong to the *metalanguage* we use to describe our *object language* $\mathbb{L}$; they do not appear in any final, concrete formula itself. The process of generating a specific formula is 'non-terminal' (i.e., unfinished) as long as these placeholders remain. This connects to our broader point: since our domain of discourse consists of propositions, we view formulas as 'propositional terms'—algorithms for building propositions. In this framework, a non-terminal symbol is thus one that does not participate in the final, written form of these terms and formulas.

Thus, we have defined the initial graphemes (variables and logical connectives) and the rules for combining them. As a result, we have obtained an infinite (countable) set of well-formed lexemes, which we will call formulas. We denote the entire set of formulas by $\mathbb{L}$ and call it the language of propositional logic. Simply put, the language of propositional logic consists of all formulas that can be constructed by combining propositional variables using logical connectives. Note that the second rule for formula generation requires that all operands of logical connectives be enclosed in parentheses, which results in unwieldy "sausages" such as:

$$(((\mathcal{A}) \to (\mathcal{B})) \wedge ((\mathcal{A}) \vee (\neg(\mathcal{B})))) \to ((\mathcal{A}) \wedge (((\mathcal{B}) \wedge (\mathcal{C})) \vee (\neg(\mathcal{D})))).$$

To simplify the perception of formulas, rules for parenthesis reduction are adopted by assigning priorities to the operations. The highest priority is given to existing parentheses, followed by $\neg$, then $\wedge$, then $\vee$, and finally $\to$. Consecutive operations with the same priority are grouped from left to right. This allows the previous expression to be simplified to:

$$(\mathcal{A} \to \mathcal{B}) \wedge (\mathcal{A} \vee \neg\mathcal{B}) \to \mathcal{A} \wedge (\mathcal{B} \wedge \mathcal{C} \vee \neg\mathcal{D}).$$

Often, the operators $\wedge$ and $\vee$ are considered to have equal precedence, in which case they are evaluated from left to right in the absence of parentheses. However, we will treat them as analogous to the arithmetic operations $\cdot$ and $+$, respectively, which will determine their priority. The convention on priorities allows for the unique restoration of all reduced parentheses, thus making it possible to present a unique, correct parse tree for any formula. For example, the formula

$$\mathcal{A} \vee \mathcal{B} \wedge \mathcal{C} \to \mathcal{D} \to \mathcal{A} \vee \mathcal{B} \vee \neg\mathcal{C} \wedge \mathcal{D}$$

is restored to the well-formed formula

$$(((\mathcal{A}) \vee ((\mathcal{B}) \wedge (\mathcal{C}))) \to (\mathcal{D})) \to (((\mathcal{A}) \vee (\mathcal{B})) \vee ((\neg(\mathcal{C})) \wedge (\mathcal{D}))).$$

B1.1.2 Axioms and Rules of Inference

Now that we have learned to generate the lexemes corresponding to the well-formed formulas of propositional logic (the language $\mathbb{L}$), we must decide how this language should correspond to the truth-value semantics that we assume in truth table A2.3. The issue is that although we can determine the value of any formula using a truth table, acting purely algebraically as if applying a "multiplication table," this external knowledge is unknown to the language $\mathbb{L}$ we have created. Its alphabet includes neither the elements of this table, nor the concept of a "table," nor the rules for calculating the values of formulas using it.

Therefore, over this language (i.e., on the set $\mathbb{L}$), we will define a kind of demiurge of truth: a system of rules consisting of two simple components: axioms and a derivability relation between formulas. Axioms arise in a natural way for mathematics: in them, we simply fix the basic properties of the logical operators, which we can extract from the truth table or from the considerations we deemed defining for logical connectives in Section A2.3. For example, we want truth to follow from anything; this is one of the defining rules for material implication. We can then say the following: if $\mathcal{A}$ is true, then the implication $(\mathcal{B} \to \mathcal{A})$ is true, regardless of whether $\mathcal{B}$ is true or false. In our language $\mathbb{L}$, we write this requirement as the formula

$$\mathcal{A} \to (\mathcal{B} \to \mathcal{A}) \tag{B1.1}$$

This axiom formalizes the principle that a true statement ($\mathcal{A}$) follows from any premise ($\mathcal{B}$). Another fundamental axiom, often called Frege's axiom, allows us to distribute implication:

$$(\mathcal{A} \to (\mathcal{B} \to C)) \to ((\mathcal{A} \to \mathcal{B}) \to (\mathcal{A} \to C)). \tag{B1.2}$$

It is easy to verify with a truth table that both formulas (B1.1) and (B1.2) are **tautologies**.[2] This confirms that our choice of axioms is sound with respect to the classical truth-value semantics.

A symmetric axiom for implication states that anything follows from a false premise (Ex falso quodlibet), and it is written as:

$$\neg\mathcal{A} \to (\mathcal{A} \to \mathcal{B}). \tag{B1.3}$$

In a similar way, we describe axioms for conjunction, for example:

$$(\mathcal{A} \wedge \mathcal{B}) \to \mathcal{A}; \quad (\mathcal{A} \wedge \mathcal{B}) \to \mathcal{B},$$

[2] In more scientific terms: the formulas have a valuation of 1 for any valuation of the variables. A valuation is an assignment of the values 0 or 1 to the variables.

which essentially mean that if a conjunction is true, then its components are true. The next axiom allows for the disjunctive combination of alternative premises:

$$(\mathcal{A} \to C) \to ((\mathcal{B} \to C) \to (\mathcal{A} \vee \mathcal{B} \to C)), \tag{B1.4}$$

which literally states that if C follows from $\mathcal{A}$, then if C also follows from $\mathcal{B}$, then it surely follows from $\mathcal{A} \vee \mathcal{B}$ as well.

The rule for negation is the **law of the excluded middle** (Tertium non datur):

$$\mathcal{A} \vee \neg\mathcal{A}, \tag{B1.5}$$

which states that for any proposition, there are only two alternatives: either it is true, or its negation is true.

We will not write out all the axioms of propositional logic here, referring the reader to Section D2.1 of Part II for them. For a general understanding, it is sufficient to know that there are 11 such axioms (in the classical version of logic), all of them are tautologies in the sense indicated, and their set is necessary and sufficient to derive all tautologies of the language $\mathbb{L}$.

It remains to be understood how our demiurge of truth can produce new truths from existing ones (e.g., axioms). It turns out that it has exactly two methods for this: the rule of substitution and Modus Ponens.

The rule of substitution has the form:

$$\frac{\varphi}{\varphi[\mathcal{A}/\psi]}, \tag{B1.6}$$

where φ is an arbitrary tautology, and ψ is an arbitrary formula of the language $\mathbb{L}$. This rule works as follows. First, it is a schema, because with a single lexeme we define an infinite series of inference rules, each obtained from this schema by replacing the non-terminal symbols φ and ψ with specific formulas from the language $\mathbb{L}$. Second, suppose we have already obtained some true formula φ. Then the formula obtained from φ by replacing the variable $\mathcal{A}$ everywhere with the formula ψ (regardless of whether it is true or false) will also be true. If the formula φ does not contain the variable $\mathcal{A}$, then such a substitution is fictitious—the result of the substitution remains the formula φ. It should be emphasized that the choice of the variable $\mathcal{A}$ is irrelevant here; any other propositional variable can be used instead, so we can consider that we have as many substitution rule schemas as there are variables. Here is an example. Let $\varphi \overset{\text{def}}{\leftrightarrow} \mathcal{A} \vee \neg\mathcal{A}$ and $\psi \overset{\text{def}}{\leftrightarrow} \mathcal{A} \to \mathcal{B}$. Then, by the rule of substitution, we obtain that the formula $(\mathcal{A} \to \mathcal{B}) \vee \neg(\mathcal{A} \to \mathcal{B})$ is true, because φ is.

The rule of substitution is easily verified by a truth table. From an algebraic point of view, it simply states that if we substitute any functions into the rguments of a function that is identically equal to 1, the value will not cease to be equal to 1. This may seem obvious, but even such things must be programmed into the "firmware" of our demiurge of truth, so that it knows exactly how to *compute* it.

From the truth table, we can also easily understand the limits of applicability of the substitution rule. If a formula φ is not a tautology, then substituting an arbitrary formula into it can lead to a situation where the formulas φ and $\varphi[\mathcal{A}/\psi]$ yield different results for the same variable assignments. For example, given the formula $\mathcal{A} \vee \mathcal{B}$, which is true for $\mathcal{A} = 1$ and $\mathcal{B} = 0$, and substituting the formula $\mathcal{A} \wedge \neg\mathcal{A}$ for the variable $\mathcal{A}$, we get the formula $(\mathcal{A} \wedge \neg\mathcal{A}) \vee \mathcal{B}$, which yields false for the same variable assignments. Thus, a rule of inference has led us from a true formula to a false one, and such rules are worthless. Therefore, *the application of the rule of substitution is always restricted to tautologies*. In this regard, propositional logic is often formulated without the rule of substitution at all. Instead, axioms are treated not as individual formulas but as axiom schemas, so that an axiom can be turned into a more convenient formula for a proof, while being guaranteed to be true.

Finally, the rule of inference already known to us is Modus Ponens (or MP):

$$\frac{\varphi, \quad \varphi \to \psi}{\psi}, \tag{B1.7}$$

which is also a schema, where any formulas of the language $\mathbb{L}$ can be substituted for the non-terminal symbols φ and ψ.

This rule is also easily verified by a truth table: assuming that the formulas φ and $\varphi \to \psi$ are true, it is easy to see that in this case ψ must also be true, regardless of the values of the variables appearing in these formulas. Here we see the universality of this rule (unlike substitution)—it does not matter whether the input formulas of the rule are true for certain variable assignments; the output formula will necessarily be true whenever the premises are true.

By combining these two rules repeatedly, our demiurge of truth can endlessly generate tautologies, starting from the axioms of propositional logic.

Moreover, we can now fully define the derivability relation $\vdash$ (the "turnstile") on the set of formulas $\mathbb{L}$. We say that a formula φ is **derivable** from a set of formulas Γ, written as $\Gamma \vdash \varphi$, if, starting from the formulas in the set Γ and the axioms, the formula φ can be obtained in a finite number of steps by applying only the Modus Ponens rule to preceding formulas and the substitution rule to axioms. The formulas from Γ are in this case called *hypotheses* or *premises of the derivation*. Note that in this definition, we have resorted to a set-theoretic concept by specifying that Γ is a subset of the language $\mathbb{L}$, i.e., a certain set of well-formed formulas. However, due to the finiteness of derivations, in each specific case, instead of $\Gamma \vdash \varphi$, we can write the expression $\psi_1, \ldots, \psi_n \vdash \varphi$, which does not require set-theoretic connotations.

If the set of hypotheses Γ is empty, the derivation $\Gamma \vdash \varphi$ represents a derivation purely from the axiomatics of propositional logic. A formula derived only from axioms is called a **theorem**. If Γ is empty, $\Gamma \vdash \varphi$ is written more concisely as: $\vdash \varphi$.

As an example, let's show a derivation of the theorem $\mathcal{A} \rightarrow \mathcal{A}$. We write the following sequence of formulas, where each is either an axiom or is obtained from the preceding formulas by one of the two rules of inference:

1. $(\mathcal{A} \rightarrow ((\mathcal{A} \rightarrow \mathcal{A}) \rightarrow \mathcal{A})) \rightarrow ((\mathcal{A} \rightarrow (\mathcal{A} \rightarrow \mathcal{A})) \rightarrow (\mathcal{A} \rightarrow \mathcal{A}))$ [substitution in axiom (B1.2): $\mathcal{B}/(\mathcal{A} \rightarrow \mathcal{A}), C/\mathcal{A}$];
2. $\mathcal{A} \rightarrow ((\mathcal{A} \rightarrow \mathcal{A}) \rightarrow \mathcal{A})$ [axiom (B1.1) with substitution $\mathcal{B}/(\mathcal{A} \rightarrow \mathcal{A})$];
3. $(\mathcal{A} \rightarrow (\mathcal{A} \rightarrow \mathcal{A})) \rightarrow (\mathcal{A} \rightarrow \mathcal{A})$ [from 1. and 2. by Modus Ponens];
4. $\mathcal{A} \rightarrow (\mathcal{A} \rightarrow \mathcal{A})$ [axiom (B1.1) with substitution $\mathcal{B}/\mathcal{A}$];
5. from 3. and 4. by Modus Ponens we get $\mathcal{A} \rightarrow \mathcal{A}$. Done!

The existence of such a derivation is written concisely as: $\vdash \mathcal{A} \rightarrow \mathcal{A}$. Any derivation $\Gamma \vdash \varphi$ can be depicted as a tree where the root is the derived formula φ, the leaves are axioms and/or hypotheses from Γ, the intermediate nodes are formulas derived along the way, and the edges correspond to the rules of inference (one edge is used for the substitution rule, two for Modus Ponens). The derivation tree for the theorem $\mathcal{A} \rightarrow \mathcal{A}$ is shown in Fig. B1.1.

B1.1.3 Basic Facts about Propositional Logic

Soundness and Completeness of Derivations

The first thing to note is that any theorem of propositional logic is a tautology, i.e., when its variables are replaced by the values 0 or 1 according to truth table A2.3, the result is always 1. This property of the logic is called **soundness** (it means that "the turnstile doesn't lie to us").

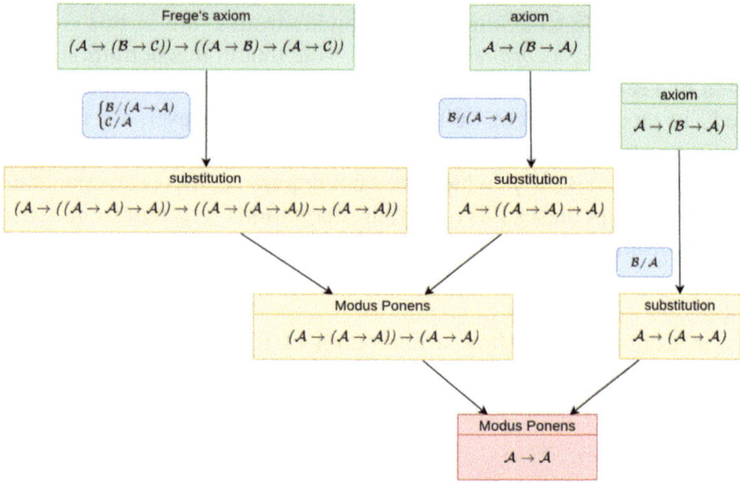

Fig. B1.1. Example of a derivation tree

The converse is also true: if a formula of propositional logic is a tautology (identically true as a function of 0 and 1), then it is a theorem of propositional logic. This property of the logic is called **completeness**, but it should be specified that this is a property of the derivability relation $\vdash$, to avoid confusion with other types of completeness (see below).

On Metatheory

The second important fact we can observe is that all definitions of concepts, as well as the processes of formula generation and derivation construction, are given not within the logic $\mathbb{L}$, but outside it, i.e., in some metatheory, with the help of which all these concepts are formalized as mathematical concepts. This metatheory must be rich enough to operate with things like sets (finite and some countable), functions (on finite sets), and recursive definitions. Various formal theories that allow all this to be programmed within them can be considered as such a metatheory. These include some weakened versions of set theory and Peano arithmetic. In other words, when we create a formalism, even a very primitive one, we do not create it from scratch, but within the framework of some "ordinary" mathematics, which itself can be formalized, and in relatively weak forms (weak in terms of what can be proven). Accordingly, the properties of propositional logic, such as soundness, completeness, and many others (see below), are proved within *this* metatheory; they are not theorems of propositional logic itself. Such theorems are often called metatheorems (relative to the given formalism). Therefore, when studying formal theories, a mathematician works simultaneously on three levels of abstraction: the formal level (within the formalism being studied), the meta-level (within the metatheory, but outside the formalism), and the general level (at which they explain their work, using natural language and various allusions and allegories).

The Deduction Theorem

Theorem B1.1 (Deduction Theorem). *Let $\varphi_1, \ldots, \varphi_n, \psi$ be formulas from $\mathbb{L}$. Then*

$$\varphi_1, \ldots, \varphi_n \vdash \psi \quad \Leftrightarrow \quad \vdash (\varphi_1 \wedge \cdots \wedge \varphi_n) \rightarrow \psi.$$

In other words, a derivation of the formula ψ exists if and only if the corresponding implication is a tautology—naturally, within the axiomatic system of propositional logic. Note that here we used the symbol $\Leftrightarrow$ instead of $\leftrightarrow$, since the stated assertion is a metatheorem, not a formula of the language $\mathbb{L}$.

Functional Completeness

The truth table, to which we often appeal, allows us to compute the value of any formula of propositional logic for given variable assignments. In particular, for any formula, one can check whether it is a tautology (identically true) or not. The problem of checking a formula for tautology is a famous problem in computational complexity theory. To be precise, it is co-NP-complete. This means that its complementary problem—checking if a formula is *not* a tautology (i.e., if it is falsifiable)—belongs to the class NP. For our purposes, the essential takeaway is that no efficient (i.e., polynomial-time) algorithm is known for solving it. For very large formulas, this makes a brute-force check an extremely lengthy computational process. At the same time, obtaining tautologies by applying rules of inference is a relatively simple and fast process. But there is a nuance: if you are *given* an arbitrary formula, the task of checking it for tautology can be equally laborious, whether by functional means (the truth table) or by searching for a derivation. Therefore, while considering logic as a calculus, i.e., a system of inference rules and axioms, we do not dismiss its functional interpretation. Moreover, there is a very close connection between these two approaches, which is called **functional completeness**. We present the theorem on the functional completeness of propositional logic in Part II (see Section D2.1). In a simplified form, it can be stated as follows: every formula corresponds to a truth table and, conversely, every truth table corresponds to some formula.

Functional completeness allows us to introduce an equivalence relation on the set $\mathbb{L}$. We say that formulas φ and ψ are **equivalent**, denoted as $\varphi \equiv \psi$, if they have the same truth tables. For example,

$$\mathcal{A} \to \mathcal{B} \equiv \neg\mathcal{B} \to \neg\mathcal{A}; \quad \text{but (!)} \quad \mathcal{A} \to \mathcal{B} \not\equiv \neg\mathcal{A} \to \neg\mathcal{B}.$$

In Part II, we provide a whole series of useful formula equivalences; here we give only De Morgan's laws:

$$\neg(\mathcal{A} \vee \mathcal{B}) \equiv \neg\mathcal{A} \wedge \neg\mathcal{B}; \quad \neg(\mathcal{A} \wedge \mathcal{B}) \equiv \neg\mathcal{A} \vee \neg\mathcal{B},$$

the law of double negation:

$$\neg\neg\mathcal{A} \equiv \mathcal{A},$$

the expression of equivalence via implication:

$$\mathcal{A} \leftrightarrow \mathcal{B} \equiv (\mathcal{A} \to \mathcal{B}) \wedge (\mathcal{B} \to \mathcal{A}).$$

It is also useful to mention the distributive laws for $\wedge$ and $\vee$:

$$\mathcal{A}\wedge(\mathcal{B}\vee\mathcal{C})\equiv(\mathcal{A}\wedge\mathcal{B})\vee(\mathcal{A}\wedge\mathcal{C}), \quad \mathcal{A}\vee(\mathcal{B}\wedge\mathcal{C})\equiv(\mathcal{A}\vee\mathcal{B})\wedge(\mathcal{A}\vee\mathcal{C}).$$

Using the Deduction Theorem and the truth table for equivalence, one can obtain the following metatheorem.

Theorem B1.2.

$$\varphi \equiv \psi \quad \Leftrightarrow \quad \vdash \varphi \leftrightarrow \psi \quad \Leftrightarrow \quad \varphi \dashv\vdash \psi,$$

where the last part means that the formulas are mutually derivable from each other.

A Real-Life Example

Suppose we want to design a lighting system that can be controlled from two locations (e.g., at the bottom and top of a staircase). We can start by constructing the corresponding logical formula.

Let $\mathcal{S}_1$ and $\mathcal{S}_2$ be variables representing the state of the first and second switches, respectively. The switches have two states: 0 (e.g., contact down) and 1 (contact up). Let φ define the state of the light bulb (1 for on, 0 for off). We define the truth table for φ based on the physical requirement: if both switches connect to the same wire (both "up" or both "down"), the circuit is closed. If they connect to different wires, the circuit is open. (See table on the right).

$\mathcal{S}_1$	$\mathcal{S}_2$	φ
0	0	1
0	1	0
1	0	0
1	1	1

To find the formula φ, we recall the truth table for $\leftrightarrow$ from Table A2.3, from which it is easy to see that $\varphi \equiv (\mathcal{S}_1 \leftrightarrow \mathcal{S}_2)$. From here, by expanding $\leftrightarrow$ into Disjunctive Normal Form (checking lines where $\varphi = 1$), we find that:

$$\varphi \equiv (\neg\mathcal{S}_1 \wedge \neg\mathcal{S}_2) \vee (\mathcal{S}_1 \wedge \mathcal{S}_2).$$

This logical structure translates directly into a circuit:

- $(\neg\mathcal{S}_1 \wedge \neg\mathcal{S}_2)$ corresponds to current flowing through the bottom wire (both switches in position 0).
- $(\mathcal{S}_1 \wedge \mathcal{S}_2)$ corresponds to current flowing through the top wire (both switches in position 1).
- $\vee$ corresponds to the fact that these two paths are parallel (either one works).

The circuit diagram is shown on the right. We use two SPDT switches (Single Pole Double Throw).

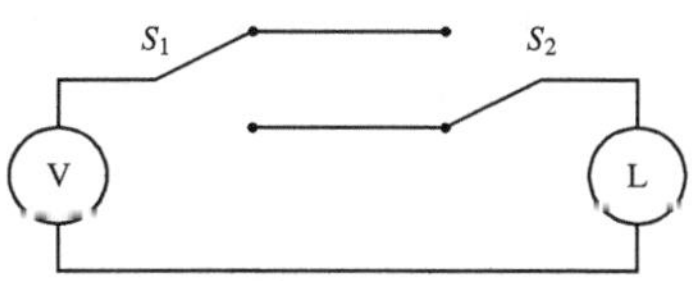

As an exercise, the reader is invited to think about a circuit for controlling a light from three locations.

Logical Constants

In the previous chapter, we used the constant symbols $\top$ (truth) and $\bot$ (falsity). Here, following the classical presentation of logic, we omitted them to simplify the language. It is now time to bring them back. We define

$$\top \overset{\text{def}}{\leftrightarrow} \mathcal{A} \vee \neg\mathcal{A}; \quad \bot \overset{\text{def}}{\leftrightarrow} \mathcal{A} \wedge \neg\mathcal{A}.$$

It is easy to check that the first formula is a tautology, and the second is a contradiction (identically false). It does not actually matter which propositional variable is used here; the value of the formulas will remain the same. For the constants, one can derive, for example, the following equivalences:

$$\top \wedge \mathcal{A} \equiv \mathcal{A}; \quad \top \vee \mathcal{A} \equiv \top; \quad \bot \wedge \mathcal{A} \equiv \bot; \quad \bot \vee \mathcal{A} \equiv \mathcal{A}.$$

Furthermore, it can be noted that a formula is a tautology if and only if it is equivalent to $\top$.

B1.2 Predicate Logic

In Section A2.3, we have already dealt with predicates and terms, but our narrative was quite informal. Now the time has come to build a solid foundation for all the ideas that were presented earlier.

In Section A2.7, we emphasized that there is a subtle but important difference between a syntactic object (a predicate symbol or a formula) and its semantic interpretation (a relation). In strict logic, these concepts are kept separate. However, in mathematical practice, when the semantics are fixed and clear from the context, a symbol is often identified with its meaning.

Henceforth in this book, we will adhere to the following approach: when discussing the general properties of formal languages, we will distinguish between a *predicate symbol* (syntax) and a *relation* (semantics). However, in the context of specific theories (arithmetic, set theory), where the interpretation of symbols is obvious, we will, following the common mathematical tradition, use the word "predicate" for convenience to denote both the formula itself and the property or relation it expresses. This will make the exposition more natural and less cumbersome.

Thus, we can consider each specific **predicate** as a full-fledged *mathematical concept*. Its designation is a predicate symbol (a grapheme) of a logical type; its definition is either a direct definition by a formula or a list of formulas containing the given predicate symbol (an indirect definition via axioms). The extension of this concept in any specific semantics is the corresponding relation. Each predicate usually has its own name.

A predicate symbol whose argument places are filled with permissible terms of the language under consideration is called an *atomic formula.*

Having a dedicated designation for predicates allows us to distinguish atomic formulas from arbitrary formulas (which also define relations).

B1.2.1 Terms

Suppose we have a certain homogeneous domain of discourse, the objects of which are generally denoted by variables $a, b, c, \ldots, x, y, z$. Due to this homogeneity, all these variables have the same sort, so they can be freely substituted for one another without sort agreement. If, as now, only one sort (or kind) of objects is assumed in the domain of discourse, such a domain is called *one-sorted* (or unsorted). It is customary to divide variables into two classes: *free* and *bound.* Letters from the beginning of the alphabet are usually considered free variables, and those from the end, bound variables. We will also use numerical indices for variables to make the text more comfortable to read.[3]

Besides variables, we can use constant symbols, which are assigned to specific objects (e.g., 0 and 1 in arithmetic, or the numbers π and e in Analysis). We denote the set of all constant symbols by $\mathfrak{C}$.

Next, we need **function symbols**, which denote some basic operations on the universe of objects. In Section A2.3, we introduced notations for set-theoretic operations: $\cap$, $\cup$, $\setminus$, etc. In arithmetic, these would be the operations of addition and multiplication. Function symbols play the same role in the construction of terms as logical connectives do in the construction of formulas. We will use the notations $F, F_1, F_2 \ldots$ for function symbols. It should be noted that for each function symbol, its arity, i.e., the number of argument places, must also be specified. This is usually done by writing the required number of free variables in parentheses after the function symbol, for example, $F(a, b, c)$. A programmer defines a function with arguments in a similar way.[4] We denote the set of all function symbols by $\mathfrak{F}$. It is usually considered that $\mathfrak{C} \subseteq \mathfrak{F}$, since constants can be seen as 0-ary (nullary) functions. We will

[3] A more rigorous approach would prescribe a recursive rule for forming new variable names, for example, by adding a prime.

[4] If the domain of discourse contained two or more different sorts of objects, writing the free variables after the function symbol would also indicate the sort of each argument place, as we would also have divided all variables by sorts.

use the non-terminal symbols $T_1, T_2, \ldots$ as general notations for terms (analogous to the notations φ and ψ for arbitrary formulas).

The rules for constructing **terms** are as follows:

1. every free variable is a term;
2. every constant is a term;
3. if F is a function symbol of arity n and $T_1, \ldots, T_n$ are arbitrary terms, then the expression $F(T_1, \ldots, T_n)$ is a term;
4. any term is constructed according to rules 1–3.

Thus, in the realm of term construction, we observe the same recursive process: the initial (smallest) terms are free variables and constants, and all other terms are obtained by combining already constructed terms with function symbols. For example, $(a + b) \cdot c$ is an arithmetic *term* (and not a formula!). It should be noted, however, that in this example, instead of the symbols $F_1(a, b)$ and $F_2(a, b)$, we use the more familiar $a + b$ and $a \cdot b$. Such liberal notations are not used in the general case when we study an arbitrary formalism, but they are constantly encountered in every specific formalism, as we will see later. For now, let us just note that the operation $F(a, b)$ can have a rather arbitrary notation in each specific case ($a + b$, $a \in b$, a^b, a/b, etc.). A term is called *closed* if it contains no free variables. For example, constants, as well as compositions of function symbols and constants, contain no free variables and are effectively algorithms for constructing some specific object in the domain of discourse (e.g., $((1 + 1) \cdot 0) + 1$). An open term contains free variables and effectively describes an algorithm for constructing a function that transforms one or more objects into a single object of the domain of discourse (e.g., $(a + b) \cdot c$).

B1.2.2 Formulas

We will call the symbols $P, Q, P_1, Q_1, P_2, Q_2 \ldots$ predicate symbols. Their semantic difference from function symbols is that they map objects to the truth-value category, i.e., they participate not in term construction but in formula construction. For predicate symbols, it is also required to specify the arity by listing the required number of free variables, for example, $P(a, b, c)$. We denote the set of all predicate symbols by $\mathfrak{P}$. The collection of sets $\mathfrak{P}$, $\mathfrak{F}$, $\mathfrak{C}$ is called the **signature** of the formal language of predicate logic and is usually denoted by the letter Σ.

We will use the non-terminal symbols φ and ψ (possibly with numerical indices and an indication of free variables in parentheses) to denote arbitrary formulas. The rules for constructing **formulas** in predicate logic are as follows:

1. if P is a predicate symbol of arity n and $T_1, \ldots, T_n$ are arbitrary terms, then the expression $P(T_1, \ldots, T_n)$ is a formula (called an atomic formula);
2. if φ and ψ are formulas, then so are $\neg(\varphi)$, $(\varphi) \wedge (\psi)$, $(\varphi) \vee (\psi)$, $(\varphi) \rightarrow (\psi)$, and also, if the text of formula φ does not contain the variable x, then

$$\forall x\ (\varphi[a/x]), \quad \exists x\ (\varphi[a/x])$$

are formulas, where $[a/x]$ denotes the substitution of all occurrences of the free variable a in the formula φ with the bound variable x.[5]
3. any formula is constructed according to rules 1–2.

Here we have practically repeated the definition of formulas from propositional logic, only instead of propositional variables, we have used predicate symbols, and we have added quantifiers to the logical operations. The quantifiers necessitated the division of variables into free and bound. But such a step is a very modest price to pay to avoid explaining what variable collisions under quantifiers are (although we have already briefly discussed this on p. 30 in our discussion of binding lexemes). The construction of terms and formulas fully complies with the principles of lexeme construction (in particular, the Matryoshka principle) that we discussed in Section A1.2. So all terms and formulas are well-formed lexemes of the language $\mathfrak{Math}$ and have parse trees. Furthermore, every formula has four abstract levels of construction: object variables and constants, function symbols, predicate symbols, and logical connectives and quantifiers (Fig. B1.2).

Examples of predicate logic formulas:

$$P(a, b, 0),$$
$$P(a + b, b \cdot c, 1) \wedge Q(a, 0, 1),$$
$$\exists x \forall y\ P(x, y, a) \rightarrow \forall y \exists x\ P(x, y, a),$$

[5] Strictly speaking, one should have used non-terminal symbols A and X to denote an *arbitrary* free and bound variable, respectively, so as not to give the impression that only the specific letter a can be replaced by the specific letter x.

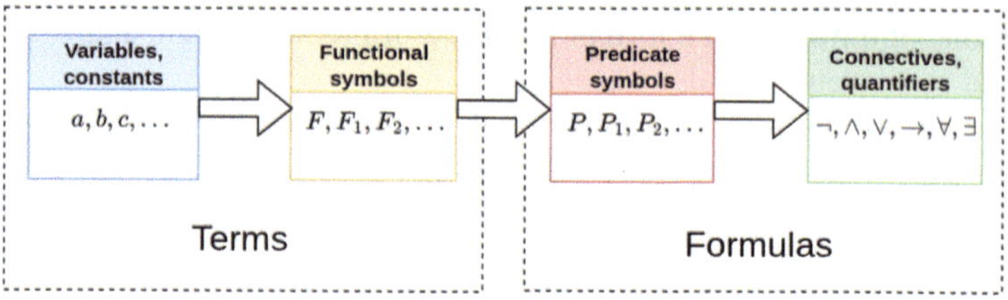

Fig. B1.2. Levels of a formula's parse tree

where P and Q are some ternary predicate symbols. Here, as in propositional logic, we use operator priorities to reduce parentheses, but in the case of a formula under a quantifier, we recommend doing so with caution, only in the most obvious cases. The formula within the parentheses under a quantifier is the *scope* of that quantifier, and when the parentheses are removed, it can sometimes be difficult to determine where this scope ends. Note that if we remove the use of quantifiers from the definition of formulas, all formulas can be described even more simply. They are everything that can be obtained from propositional logic formulas by replacing the propositional variables with atomic formulas. Such formulas are called *quantifier-free*; their text contains no bound variables. Moreover, if we assume that our language has only one predicate symbol—the empty one—and no function symbols, i.e., every atomic formula has the form (a), where a is a free variable, then the quantifier-free formulas are exactly the formulas of propositional logic. A formula is called **closed** if its text contains no free variables. The *universal closure* of a formula is a closed formula in which all free variables are bound by universal quantifiers ($\forall$). For example, the formula $\forall x \forall y\ P(x, y)$ is the universal closure of the formula $P(a, b)$, where P is a binary predicate symbol.

Further on, we will need the following notations: $\mathbb{F}$ for the set of all formulas, and $\mathbb{T}$ for the set of all terms. It is clear that the content of these sets varies depending on which specific function symbols, constants, and predicate symbols we use. The set of all formulas $\mathbb{F}$ is called a **first-order formal language.**[6]

[6] There are so-called higher-order languages (logics), which we do not consider here. For example, in second-order logic, our non-terminal symbols φ, ψ are encapsulated as variables over all first-order properties (i.e., they become terminal), which allows for the use of second-order formulas containing quantifiers over first-order properties and functions. And so on.

B1.2.3 Axioms and Inference Rules of Predicate Logic

Following the general logic of constructing logics, we will now define the truth basis, i.e., the axioms of predicate logic, as well as the derivability relation on the set of formulas $\mathbb{F}$. The axioms of predicate logic (*logical axioms*) are the following formulas:

PC1 the universal closures of all formulas obtained by substituting formulas from $\mathbb{F}$ into all tautologies of propositional logic;

PC2 axioms for quantifiers:

$$(\forall x\ \varphi[a/x]) \rightarrow \varphi[a/T]; \quad \varphi[a/T] \rightarrow \exists x\ \varphi[a/x],$$

where φ is any formula from $\mathbb{F}$ not containing the bound variable x, and T is any term from $\mathbb{T}$. Here, as before, we assume that any free variable can be used instead of a, and any bound variable not present in φ can be used instead of x.

In the first point, we have in fact specified an infinite list of axioms, which is determined by the sets $\mathbb{L}$ (more precisely, its subset of tautologies) and $\mathbb{F}$. It could have been expressed more formally if we had used non-terminal variables ranging over all formulas of propositional logic, all propositional variables, and all formulas from $\mathbb{F}$. The requirement of universal closure informs us that we want to see only closed formulas as axioms, as they correspond to the notion of a proposition. However, axioms are often written as formulas with free variables, with the understanding that they should be closed by universal quantifiers. The quantifier axioms state roughly the following: if a formula φ is true for any value of a, then any term can be substituted for a; if some term is a witness for the formula φ, then there exists a value of a for which the formula φ is true. Here we no longer require the formulas to be closed, to leave room for maneuver: to allow for the addition of an existential quantifier or the removal of a universal quantifier. Inference rules:

$$(MP) \quad \frac{\varphi, \quad \varphi \rightarrow \psi}{\psi};$$

$$(R1) \quad \frac{\varphi \rightarrow \psi}{\varphi \rightarrow \forall x\ \psi[a/x]};$$

$$(R2) \quad \frac{\psi \rightarrow \varphi}{(\exists x\ \psi[a/x]) \rightarrow \varphi},$$

where for rules $(R1)$ and $(R2)$, the variable a does not occur in the formula φ. $(R1)$ and $(R2)$ are called **Bernays' rules**. As usual, we note that a here implies any free variable, and x implies any bound variable. In the inference rules, we also do not require the formulas to be closed, since quantifiers can only be applied to a formula with a free variable. We say that a set of hypotheses $\Gamma \subseteq \mathbb{F}$ **derives** the formula φ, denoted $\Gamma \vdash \varphi$, if, starting from the formulas in Γ and the axioms, the formula φ can be obtained in a finite number of steps using the inference rules. In the case of an empty set Γ, i.e., when the formula φ is derived only from the axioms of predicate logic, we simply write $\vdash \varphi$.

Theorem B1.3 (Properties of Derivability).

1° *(monotonicity) if* $\Delta \subseteq \Gamma$ *and* $\Delta \vdash \varphi$, *then* $\Gamma \vdash \varphi$;

2° *(compactness) if* $\Gamma \vdash \varphi$, *then there exists a finite subset* $\Delta \subseteq \Gamma$ *such that* $\Delta \vdash \varphi$;

3° *(transitivity) if* $\Gamma \vdash \varphi$ *for all* $\varphi \in \Delta$ *and* $\Delta \vdash \psi$, *then* $\Gamma \vdash \psi$.

Theorem B1.4 (Deduction Theorem).

$$\Gamma \cup \{\varphi\} \vdash \psi \quad \Leftrightarrow \quad \Gamma \vdash (\varphi \rightarrow \psi).$$

Technical condition: in the direction from left to right, when applying generalization rules in the derivation of ψ *from* $\Gamma \cup \{\varphi\}$, *one must not generalize over variables that occur free in* φ.

Note: instead of $\Gamma \cup \{\varphi\}$, one most often writes Γ, φ or $\Gamma + \varphi$. A proof of this theorem can be found, for example, in [16].

The Deduction Theorem, which connects derivability and implication, is an important metalogical result. It was independently proven by Jacques Herbrand (1930) and Alfred Tarski (in works from the 1920s-30s). This theorem is a convenient way to replace a derivation with an implication, and vice versa. It often allows for a significant shortening of the formal proof of a theorem. Now is the time to give precise definitions for such mathematical concepts as theory, axiom, and theorem. A **theory** is a set of closed formulas that is closed under the derivability relation. The formulas of a theory are called **theorems**. For a theory, one can identify a generating (in the sense of derivability) subset of formulas, which is called the **axiomatics** of that theory, and its elements are called **axioms**. It is desirable that the axioms be *independent*, i.e., that none of them can be derived from the others. In practice (most often due to tradition), it turns out that the axioms of a theory can be partially dependent. Nevertheless, an effort is made to choose the axiomatics to be as compact as possible.[7]

If only the axioms of predicate logic are taken as the axiomatics of a theory, then the corresponding theory (the set of all derived closed formulas) consists of the *theorems of predicate logic* and is called a *pure predicate theory*. It is clear that such a theory depends on the language $\mathbb{F}$, i.e., on the set of function and predicate symbols. Therefore, unlike propositional logic, there is a huge variety of specific pure predicate theories.

Note that since the axioms of predicate logic are always involved in any derivation by default, when speaking of the *axiomatics of a theory*, one usually means only the *non-logical axioms*, i.e., formulas that are not theorems of the pure predicate theory.

Further study of theories requires the introduction of the concept of a model, which we will consider a little later. For now, let's look at a couple of simple examples of a language and a theory.

B1.2.4 Examples of Theories

Theory of Strict Partial Orders

Consider a language $\mathbb{F}$ with only one binary predicate symbol $<$ and no function symbols or constants. Instead of $<(a, b)$, we will write $a < b$. The atomic formulas in such a language have the form $(a < b)$. We define the theory of strict partial orders with two non-logical axioms:

$$\neg(a < a); \quad (a < b) \wedge (b < c) \to (a < c),$$

the first of which is called the axiom of irreflexivity, and the second, the axiom of transitivity.

By tradition, we do not use universal closure on the axioms, but we imply it. In this theory, one can prove, for example, the following theorem:

$$(a < b) \to \neg(b < a).$$

[7] Axioms are theorems by virtue of these definitions. This is convenient, as every axiom can prove itself by virtue of the tautology $\varphi \to \varphi$.

Indeed, let's assume the opposite, i.e., that the negation of this implication, $\neg((a < b) \to \neg(b < a))$, is true. Then we use propositional logic, in which the equivalence $\neg(\mathcal{A} \to \neg\mathcal{B}) \equiv \mathcal{A} \wedge \mathcal{B}$ holds. Then from our hypothesis follows the formula $(a < b) \wedge (b < a)$, from which, by the axiom of transitivity, we get $a < a$. This, together with the axiom $\neg(a < a)$, derives $\perp$. Consequently, one of the hypotheses is not true. But our hypotheses were only the two axioms and the assumption $\neg((a < b) \to \neg(b < a))$, i.e., we have the derivation:

$$\varphi_1, \varphi_2, \neg\psi \vdash \perp,$$

where φ_i are the axioms and $\psi \overset{\text{def}}{\leftrightarrow} (a < b) \to \neg(b < a)$. By the Deduction Theorem, we get that $\varphi_1, \varphi_2 \vdash \neg\psi \to \perp$, from which, with the help of propositional logic, we get that $\varphi_1, \varphi_2 \vdash \psi$, since $\neg\mathcal{A} \to \perp \equiv \mathcal{A}$. In essence, we have just carried out a formal proof by contradiction within the given axiomatic system.

Pure Theory of Equality

Consider a language $\mathbb{F}$ with only one binary predicate symbol $=$ and no function symbols or constants. Instead of $= (a, b)$, we will write $a = b$. The atomic formulas in such a language have the form $(a = b)$. The pure theory of equality is defined by three non-logical axioms:

$$a = a; \quad (a = b) \to (b = a); \quad (a = b) \wedge (b = c) \to (a = c), \qquad \text{(B1.8)}$$

which we write without universal closure and which express the properties of reflexivity, symmetry, and transitivity of equality. When using the equality predicate, the inequality predicate $a \neq b$, defined as $\neg(a = b)$, is usually used along with it. The theory of equality holds a central place in predicate logic, as equality is part of the vast majority of formal theories. Whenever we include the equality predicate in a language, we automatically imply the presence of the equality axioms in the theory, just as we automatically imply the presence of logical axioms. Moreover, in addition to the axioms of pure equality, we also add congruence axioms, which tell us that substituting an equal term for an equal term does not change the value of either a term or a predicate. More on this shortly.

Semigroup theory

Consider a language $\mathbb{F}$ with one binary predicate symbol = and one function symbol $\circ$. Instead of $= (a, b)$ and $\circ(a, b)$, we will write $a = b$ and $a \circ b$. The theory of semigroups is defined by the following list of non-logical axioms: the axioms of equality (B1.8), and also:

$$(a \circ b) \circ c = a \circ (b \circ c); \quad (T_1 = T_2) \wedge (T_3 = T_4) \rightarrow (T_1 \circ T_3) = (T_2 \circ T_4),$$

where $T_i \in \mathbb{T}$, i.e., they are arbitrary terms of the language of the theory of semigroups (i.e., all possible combinations of variables and the operation $\circ$). The first axiom is the semigroup axiom proper and tells us about the associativity of the operation $\circ$. The second is an example of a *congruence axiom*; it says that under the operation $\circ$, one can freely substitute equal terms for each other, and the value of the operation will not change (with respect to equality). Without the congruence axiom, many of the identical transformations we are accustomed to would simply be impossible. From the axiom of associativity for three variables and the congruence axiom, one can derive the general law of associativity for an arbitrary n variables, which in words is expressed as: the order of placing parentheses in the expression $x_1 \circ \cdots \circ x_n$ does not matter (and thus the parentheses can be omitted altogether). Formally, however, this would have to be written as a set of identities with all correct placements of parentheses for each specific n (since there are no natural numbers in this theory). The pure theory of semigroups is a good example for us of a formalism that includes all the necessary components: predicate logic, a function symbol, and the theory of equality with a congruence axiom. The formal theory of semigroups should not be confused with the algebraic theory of semigroups, which is essentially concerned with studying the class of models of semigroups and all possible morphisms between them.

B1.2.5 Congruence of Equality

The **congruence of equality** means the property of a logic not to distinguish between equal objects. That is, if we have once established that two objects are equal in the sense of the equality predicate, then they can be interchangeable in any terms and formulas. In other words, to paraphrase a famous phrase, if something is a duck, then it looks like a duck, swims like a duck, and quacks like a duck. Thus, the equality predicate is endowed with the property of distinguishing only those objects that the given formalism is capable of distinguishing. In essence, equality is the lowest level of distinction between objects that a formal language can provide.[8]

Formally, the **congruence axioms** in their general form are formulated as follows. Let $T, T_1, T_2 \in \mathbb{T}$ be arbitrary terms and $\varphi \in \mathbb{F}$ be an arbitrary formula, then

$$\begin{aligned} T_1 = T_2 &\to T[a/T_1] = T[a/T_2]; \\ T_1 = T_2 &\to (\varphi[a/T_1] \leftrightarrow \varphi[a/T_2]), \end{aligned} \tag{B1.9}$$

where any free variable of the language can be considered instead of a. It is clear that the given axioms are actually an infinite series of formulas, each obtained by replacing the non-terminal symbols T, T_1, T_2, φ with specific formulas and terms, and also the variable a is replaced by any other free variables.

From this, in particular, follows the congruence axiom of the theory of semigroups, in which the function symbol with variables, $a \circ b$, played the role of the term T.

Most often, the axioms (B1.9) are given in a basic form, where all function symbols of the language's signature are considered as the term T, and the predicate symbols of the language's signature are considered as the formula φ. With this assumption, the general formulas can be derived by induction on the complexity of formulas. However, such an approach forces us to resort to so-called external induction from the metatheory, which is a rather strong requirement for the metatheory. So we will stick to the schema (B1.9) in cases where we reason at the most abstract level, and in specific cases like the theory of semigroups, arithmetic, or set theory, we will formulate these axioms in a simplified form, relying only on atomic formulas and terms.

Important note: when defining a formal theory that uses the equality predicate (=), the congruence axioms (B1.9) (in full or abbreviated form), as well as the equality axioms (B1.8), are always implied along with the axioms of predicate logic. Therefore, they may not always be found in the list of a theory's axioms.

---

[8] In natural languages, we can observe a phenomenon: some northern peoples have several dozen names for shades of color that more southern peoples denote with the single word "white". That is, the language itself presupposes a sufficient number of different constants that are identified in another language. In physics, the language allows for even greater detail, providing a continuous spectrum.

B1.2.6 Model of a Language and a Theory

Here we move to the synergy of formal logic and set theory. More precisely, we assume set theory as our metatheory, modulo which many results about formal theories become accessible, such as, for example, the Löwenheim-Skolem theorems or the Łoś-Vaught test. In doing so, we do not specify the requirements for the metatheory, although it is most often assumed to be ZFC. We only need to clearly understand that the metatheory can be formalized just like all other theories, and that this formalization itself does not require any strong assumptions like the axiom of choice or even induction; it can be carried out, for example, by means of a sufficiently weak set theory or arithmetic.

In other words, by taking a sufficiently weak theory as our foundation, we gain the ability to construct formal languages and formal theories of any complexity. This does not mean that we can interpret any theory within this weak base theory; it is merely a technical tool with which we can describe more complex theories. Then, by choosing more complex theories as metatheories, we can obtain various results about various formal theories. Thus, a process unfolds that is something like scrambling up the walls of a well, where, starting from the first easy step, we complicate the structure of formalism, logic, and theories step by step, gradually strengthening them, while the solid ground under our feet gets farther and farther away.

The next concept we will consider here is the model of a theory. So, let us have a language $\mathbb{F}$, constructed according to the rules of predicate logic, where $F_1, F_2, \ldots$ are function symbols, $C_1, C_2, \ldots$ are constant symbols, and $P_1, P_2, \ldots$ are predicate symbols. Together, all these symbols form a set Σ, called the *signature* of the language $\mathbb{F}$. Let us now suppose that we have a non-empty set $\mathfrak{m}$ (we use gothic letters to denote meta-theoretic objects). We can consider this set $\mathfrak{m}$ as a basic universe within which, using the language $\mathbb{F}$, we can introduce various concepts: the formulas of the language $\mathbb{F}$ will be their definitions, and the subsets of $\mathfrak{m}$ will be their extensions. Next, we can consider the set $\mathfrak{m}^n$ of all n-tuples of elements of the set $\mathfrak{m}$, which is called the Cartesian power of the set $\mathfrak{m}$. This will allow us to define relations and functions on $\mathfrak{m}$. We will use the concepts of relation and function that were introduced in Section A2.7.

Let us assign to the symbols of our signature Σ specific "values" in the set $\mathfrak{m}$: to each constant symbol C_k, we assign a specific element $\mathfrak{c}_k$ from $\mathfrak{m}$; to each function symbol F_k, a specific function $\mathfrak{f}_k$; to each predicate symbol P_k, a specific relation $\mathfrak{p}_k$. We will denote this assignment by the symbol $\dashrightarrow$, i.e.,

$C_k \dashrightarrow \mathfrak{c}_k$, $F_k \dashrightarrow \mathfrak{f}_k$, $P_k \dashrightarrow \mathfrak{p}_k$. Such an assignment defines an **interpretation** of the signature in the set $\mathfrak{m}$ (for details, see Section D2.2).

Furthermore, to determine the truth of formulas with free variables, we will need a **valuation** (or *assignment*)—a function that assigns an element from $\mathfrak{m}$ to each free variable.

Given an interpretation of symbols and a valuation of variables, we can recursively define the value of any term and the truth value of any formula of the language $\mathbb{F}$. The structure $\mathcal{M} \overset{\text{def}}{\rightleftharpoons} \langle \mathfrak{m}, \mathfrak{C}, \mathfrak{F}, \mathfrak{P} \rangle$, where $\mathfrak{C}$ is the set of constants $\mathfrak{c}_k$, $\mathfrak{F}$ is the set of functions $\mathfrak{f}_k$, and $\mathfrak{P}$ is the set of relations $\mathfrak{p}_k$, and the correspondence between the symbols of Σ and the elements of these sets is fixed by the interpretation ($\dashrightarrow$) is called a **model** of the language $\mathbb{F}$.

Since axioms and theorems are always considered as **closed formulas** (without free variables), their truth in a model does not depend on the specific choice of valuation for the free variables. A formula is considered *true in a model* if it is true under *any* valuation of the variables. If a certain theory $\mathfrak{T}$ is defined in the language $\mathbb{F}$, and all of its axioms (and theorems) turn out to be true in a given model under a given interpretation of the signature's symbols and any valuation of the free variables, then the structure $\mathcal{M}$ is called a *model of the theory* $\mathfrak{T}$. This fact is written as: $\mathcal{M} \models \mathfrak{T}$.

B1.2.7 Examples of Interpretation

Theory of Strict Linear Orders

To avoid repetition, let's consider a new example of a theory that extends two previously considered examples at once: the theory of partial orders and the pure theory of equality. Let our language $\mathbb{F}$ have two predicate symbols: $=$ and $<$. Accordingly, the atomic formulas will be $(a = b)$ and $(a < b)$. A model for such a language would be any non-empty set of objects with two binary relations. For example, the elements of the set could be pencils in a pencil case, and the binary relations could be: 1) two objects are in relation $\mathfrak{p}_1$ if they have the same color, 2) two objects are in relation $\mathfrak{p}_2$ if they have the same length. Then we can make the assignments:

$$= \dashrightarrow \mathfrak{p}_1; \quad < \dashrightarrow \mathfrak{p}_2.$$

Note that so far we have no requirements for the predicates, as we have no axioms defining the properties of the symbols = and <. Therefore, the only thing that limits us in the interpretation is the arity of the relations. Now let's add axioms, i.e., consider the theory of strict linear orders in the language $\mathbb{F}$ (as usual, without universal closure by quantifiers):

$$\begin{array}{ll} a_1 : a = a & a_6 : \neg(a < a) \\ a_2 : (a = b) \to (b = a) & a_7 : (a < b) \wedge (b < c) \to (a < c) \\ a_3 : (a = b) \wedge (b = c) \to (a = c) & a_8 : (a < b) \vee (a = b) \vee (b < a) \\ a_4 : (a = b) \to (c < a \leftrightarrow c < b) & \\ a_5 : (a = b) \to (a < c \leftrightarrow b < c) & \end{array}$$

In the first three lines are the axioms of equality (reflexivity, symmetry, transitivity) and the axioms of linear order (irreflexivity, transitivity, connexity). Axioms a_4 and a_5 are the embodiment of the congruence axioms (B1.9), since in our theory, any term is a variable (due to the absence of function symbols). Usually, as already noted, the axioms of equality and congruence are implied automatically, so the proper non-logical axioms of the theory of linear order are axioms a_6, a_7, a_8.

Now we can check if our assignments $= \dashrightarrow \mathfrak{p}_1$ and $< \dashrightarrow \mathfrak{p}_2$ work. It turns out they do not, in both cases. The relation $\mathfrak{p}_1$, based on the identity of pencil colors, might at first glance seem to model equality, as it satisfies the equality axioms a_1, a_2, a_3. But together with the interpretation $< \dashrightarrow \mathfrak{p}_2$, a problem arises: the congruence axiom a_4 requires that if two pencils (a and b) have the same color ($\mathfrak{p}_1(a, b)$), then for any third pencil c, if it has the same length as the first ($\mathfrak{p}_2(c, a)$), it must also have the same length as the second ($\mathfrak{p}_2(c, b)$). This is, of course, not true in general, as the color of a pencil places no restriction on its length.

The interpretation of the predicate $(a < b)$ by the relation $\mathfrak{p}_2(a, b)$ is also incorrect, as this relation is reflexive, in contradiction to axiom a_6. Therefore, let's consider another relation on the set of pencils, $\mathfrak{p}_3(a, b)$, which will mean that pencil a is shorter than pencil b. In this case, the assignment $< \dashrightarrow \mathfrak{p}_3$ will be correct, as comparison by length satisfies axioms a_6, a_7, a_8. However, axiom a_5 is still not true, as the color of a pencil is not related to its length.

The question then arises as to how to correctly model the equality relation in the pencil box. To avoid conflict with the axioms of order and congruence, we can assign the symbol = to the relation $\mathfrak{p}_2$, which we previously tried to associate with the symbol <. In other words, we will identify pencils that have the same length. Then all the axioms will be satisfied, despite the fact that pencils of equal length may differ in their other attributes, such as color.

Thus, a model of the theory described by axioms $a_1 - a_8$ in the language with predicates $=, <$ is the set $\mathfrak{m}$ of pencils with predicates $\mathfrak{p}_2$ and $\mathfrak{p}_3$ under the correspondence $= \dashrightarrow \mathfrak{p}_2$, $< \dashrightarrow \mathfrak{p}_3$.

A question arises: can the symbol = be modeled as true identity of objects in the model? In our case, this is not possible, because if we find two different pencils a and b of the same length, then, on the one hand, $a \neq b$, and on the other hand, neither $a < b$ nor $b < a$ holds for them. This directly contradicts the connexity axiom a_8, which requires that for any pair of distinct elements, one must be less than the other.

Therefore, models in which = can be interpreted as identity within the model itself are distinguished into a separate class and are called **normal models**.

Thus, we see that **a)** the process of adapting a model of a language to the axioms of a theory is not always trivial, and **b)** equality does not always have to be interpreted as identity; what matters is that it partitions the carrier of the model (the set $\mathfrak{m}$) into equivalence classes such that within these classes, all properties of the objects are the same from the perspective of the interpreted language.

Group theory

In group theory, we use a language with one constant 1, one binary function symbol $\circ$, one unary function symbol $()^{-1}$, and one binary predicate =. We will write out only the non-logical axioms of group theory (also excluding the axioms of equality and congruence):

$$g_1 : a \circ (b \circ c) = (a \circ b) \circ c \text{ [semigroup]};$$
$$g_2 : a \circ 1 = a; \quad 1 \circ a = a \text{ [monoid]};$$
$$g_3 : a \circ a^{-1} = 1; \quad a^{-1} \circ a = 1 \text{ [group]}.$$

It is not difficult to see that a model of this language and theory would be, for example, the set of integers $\mathbb{Z}$ with the operations +, – (not subtraction, but the additive inverse), and the constant 0 (which interprets the symbol 1 of the language $\mathbb{F}$), where equality is interpreted as identity (i.e., the model is normal). If we want to use multiplication of integers as a model for the operation $\circ$, we should consider the multiplication of positive integers (the universe

$\mathfrak{m}$) modulo a prime $\mathfrak{p}$ ($a \circ b$ is the remainder of dividing ab by $\mathfrak{p}$). In this case, the symbol 1 will be interpreted as the usual unit, and the multiplicative inverse a^{-1} will be calculated as the remainder of dividing $a^{\mathfrak{p}-2}$ by $\mathfrak{p}$ (*Fermat's Little Theorem*). But in this case, equality will be interpreted as equivalence modulo $\mathfrak{p}$, so as not to "break" the axioms of equality.[9]

There are also "more physical" models of group theory, for example, the addition of time intervals on a clock is performed modulo 12 and represents a group under addition of remainders. The rotations of regular polyhedra (crystals) are an example of a group where the operation $\circ$ is interpreted as the composition (successive application) of rotations. Group theory has a huge number of fundamentally different interpretations in various fields of science and technology.

Divisible Torsion-Free Abelian Groups

Let's consider the theory of groups with additional axioms (besides $g_1 - g_3$ above), but in a slightly modified language. Namely, the binary group operation will now be denoted as $+$, the operation of taking the inverse element as $-$, and the neutral element as 0. This is the traditional notation for abelian (commutative) groups.

$$g_1 : a + (b + c) = (a + b) + c \text{ [semigroup]};$$
$$g_2 : a + 0 = a; \quad 0 + a = a \text{ [monoid]};$$
$$g_3 : a + (-a) = 0; \quad (-a) + a = 0 \text{ [group]};$$
$$g_4 : a + b = b + a \text{ [abelian/commutative]};$$
$$g_5 : \forall n > 0\ \forall x\ \exists y\ x = \underbrace{y + \cdots + y}_{n} \text{ [divisibility]};$$
$$g_6 : \forall n > 0\ \forall x \neq 0 : \underbrace{x + \cdots + x}_{n} \neq 0 \text{ [torsion-free]}.$$

The theory of divisible abelian groups, defined by axioms $g_1 - g_5$, is usually denoted by DAG (Divisible Abelian Groups), and DAG without torsion is denoted by DTFA (Divisible Torsion-Free Abelian groups). Note that axioms

[9] Simple calculations show that, for example, $6 = 6 \circ 1 = 6 \cdot 1 \pmod 5 = 1$, whereas in a normal model it should be $6 \neq 1$.

g_5 and g_6 are axiom schemas with parameter n, which is why the quantifier over n is highlighted in red; it belongs to the metatheory.

Axiom g_5 tells us that any element x can be represented as an n-fold sum of an element y, and in this case, we write that $y = x/n$. Furthermore, the sum of a single element m times is also customarily written as mx (for $m = 0$, we set $mx = 0$), and for a negative m, mx is understood as $-((-m)x)$. So, together with the divisibility axiom, we get the ability to multiply any element of the group by an arbitrary rational number m/n. In the torsion-free case (axiom g_6), this means that DTFA describes what is called in algebra a $\mathbb{Q}$-module, or a vector space over the field $\mathbb{Q}$.

Axiom g_6 states that no non-zero element of the group, when added to itself a positive number of times, is equal to 0. This excludes, for example, cyclic groups, in particular, the groups of addition of integers modulo some number, and makes such groups necessarily infinite, since under this condition all nx are pairwise distinct (for $x \neq 0$).

The standard interpretation of DTFA is the set of rational numbers $\mathbb{Q}$, as well as the set of real and complex numbers (with the operation of addition). Less standard examples include finite extensions of the field $\mathbb{Q}$ (e.g., the set of numbers of the form $x + y\sqrt{2}$, where $x, y \in \mathbb{Q}$, etc.). If we remove the requirement of being torsion-free, i.e., consider the theory DAG, then a standard model for such a theory would arguably be the group of rotations of a circle $U(1)$ (also known as $SO(2)$). Here, the operation $+$ is the addition of rotation angles. It is obvious that this is a divisible group, as any rotation can be divided into n equal rotations, and at the same time, this group has torsion elements, namely, a rotation by an angle of $360°/n$, added to itself n times, will result in a zero rotation.

In physics (optics, acoustics, quantum mechanics), the phase of a wave is a key characteristic. Phases are added modulo 2π, so their natural model is $U(1)$. Divisibility is important, for example, in the analysis of interference from n sources.

Divisible abelian groups arose as a natural and important object at the intersection of the study of infinite abelian groups, the search for structure theorems, and the development of homological algebra. Their complete and simple classification as $\mathbb{Q}$-vector spaces made them particularly useful and historically significant.

Let us note one very important aspect of comparing formal theories and general mathematical theories: a formal theory consists exactly of those theorems that can be derived from the axioms according to the inference rules, whereas a general mathematical theory studies not only the consequences within the formalism itself, but also all models of this formalism by means of set theory. And most of the famous and important theorems of mathematical group theory are theorems about the models of the formal theory of groups. Such theorems include, for example, Cayley's theorem, Lagrange's theorem, Sylow's theorems, the isomorphism theorems for groups, classification theorems, etc. All of them, in one way or another, use meta-theoretic constructions and definitions: functions between groups, the cardinality of a group (number of elements), factorization of a group, etc.

Therefore, if we open a textbook on group theory, we immediately see the definition of a group as an object of set theory (a structure), i.e., a set with an operation subject to the axioms of the formal theory of groups. The objects of general mathematical group theory are the groups themselves and their elements, while the objects of formal group theory are only the elements of a group.

B1.2.8 Some Concepts of Model Theory

Model theory allows us to give meaning to formalisms and to analyze them externally. Let us give a few definitions based on the possibility of constructing models of formal theories. A theory $\mathfrak{T} \subset \mathbb{F}$ is called satisfiable if it has a model. For example, the theory of strict linearorder and the theory of groups discussed above are satisfiable.

A formula $\varphi \in \mathbb{F}$ is called **valid** (notation: $\vDash \varphi$) if it is true in every model. Here are some examples of valid formulas:[10]

$$\neg(\forall x \varphi(x)) \leftrightarrow \exists x \neg\varphi(x), \quad \neg(\exists x \varphi(x)) \leftrightarrow \forall x \neg\varphi(x),$$

$$\exists x \forall y \varphi(x, y) \rightarrow \forall y \exists x \varphi(x, y),$$

where φ is an arbitrary formula of the language $\mathbb{F}$. All substitutions of formulas into tautologies of propositional logic are also valid formulas. A formula φ **logically (semantically) follows** from a set of formulas $\Gamma \subset \mathbb{F}$ (notation: $\Gamma \Vdash \varphi$) if for any model in which all formulas of Γ are identically true, the formula φ is also identically true. Usually, only non-logical formulas are considered as elements of Γ, while all logical ones are considered to be included by default. Therefore, if we say that the set of premises Γ is empty, it means that it consists

[10] The third formula will be derived in Part II, Theorem D2.7

of the pure predicate theory in the given language. Thus, the notation $\Vdash \varphi$ means that the formula φ logically follows from the pure predicate theory, which, as we will see later, is equivalent to its validity: $\Vdash \varphi \Leftrightarrow \vDash \varphi$.

Formulas are *equivalent* if they are simultaneously true or false in the same models (i.e., they logically follow from each other). To denote the equivalence of formulas, we use the same notation as in propositional logic: $\varphi \equiv \psi$. By the way, from the equivalence of propositional formulas automatically follows the equivalence of all substitutions of predicate calculus formulas into them. For example, we know that

$$\mathcal{A} \to \mathcal{B} \equiv \neg\mathcal{B} \to \neg\mathcal{A},$$

then, by making the substitutions $[\mathcal{A}/\varphi]$ and $[\mathcal{B}/\psi]$, regardless of the language and models, we will obtain the equivalence

$$\varphi \to \psi \equiv \neg\psi \to \neg\varphi.$$

B1.2.9 Basic Facts about Predicate Logic

Soundness and Completeness of Derivations

Let us now recall that for propositional logic, we mentioned such properties as soundness and completeness, and these properties were directly related to representing propositional logic formulas as binary functions. The set of such functions is nothing other than a model of propositional logic. Classical propositional logic has other models (for example, the power set of some non-empty set), but they are all in some (namely, algebraic) sense equivalent. The soundness and completeness of propositional logic meant two facts for us: "the turnstile doesn't lie" (everything derivable from tautologies is a tautology) and "the axiomatics are complete" (all tautologies are derivable from the axioms). Roughly the same can be said about predicate logic, with a reservation about the diversity of models.

Theorem B1.5 (Soundness Theorem). *If in a certain model all formulas of the set $\Gamma \subset \mathbb{F}$ are true and the formal derivation $\Gamma \vdash \varphi$ holds, then φ is also true in that model.*

In other words, if the axioms of a theory are true in a model, then all the theorems are also true ("the turnstile doesn't lie"). This theorem is proven by induction on the length of the derivation $\mathfrak{T} \vdash \varphi$ (i.e., the metatheory must include the axiom of induction). Indeed, the truth of the premises is already given to us (the set Γ), and all transitions to subsequent formulas in the deductive derivation are carried out strictly according to the inference rules. Thus, the inductive step is to verify that if the premises of an inference rule are formulas true in the model, then the conclusion is also a formula true in the model. This is fairly easy to see by checking each of the inference rules (MP and Bernays' rules). A theory $\mathfrak{T} \subset \mathbb{F}$ is called **consistent** if among its theorems there is not both some formula and its negation (i.e., one cannot derive $\varphi \wedge \neg\varphi$ or, more concisely, $\bot$).[11] Otherwise, the theory is *inconsistent*. For an inconsistent theory, the set of theorems consists of all closed formulas of the language $\mathbb{F}$, because once a contradiction is obtained in a derivation, any formula can be derived thereafter.

Theorem B1.6. *If a theory is satisfiable (has a model), then it is consistent.*

This theorem presupposes the consistency of the theory in which the model is constructed, i.e., set theory. Indeed, if the theory were inconsistent, then one could derive φ and $\neg\varphi$ in it. But then the interpretations of both these formulas would be true in the given model (by the Soundness Theorem), i.e., a falsehood would be true in the model. This means we would have managed to derive a contradiction in the metatheory.

Theorem B1.7 (Gödel's Completeness Theorem for Predicate Logic).

(1) *If a theory is consistent, then it has a model (is satisfiable).*
(2) *Equivalent formulation:*[12] *if $\Gamma \Vdash \varphi$, then $\Gamma \vdash \varphi$.*

The first part of the theorem, characteristically for mathematical logic, is proven essentially by constructing a model from... the formulas themselves. Indeed, if we have nothing but the language $\mathbb{F}$, why not build a universe from the letters we have? First, we include all constant symbols in the universe, as they must have a correspondence in the model. Second, from all true closed

[11] Recall that we defined the symbol $\bot$ as the propositional formula $\mathcal{A} \wedge \neg\mathcal{A}$, but all substitutions of predicate logic formulas into propositional formulas are also formulas of predicate logic, so the definitions of the symbols $\bot$ and $\top$ can be easily extended to predicate logic.

[12] Why these formulations are equivalent from the perspective of the metatheory will be discussed in Part II, see Section D2.2

formulas with an existential quantifier (*existential formulas*), we extract their witnesses, i.e., we use the special fact that any such formula has a witness expressible by a term in our language.[13] As a result, the carrier set of the model consists of constants and such terms that make the existential formulas true. The truth of all other axioms and theorems of the theory is then checked logically. The second part, equivalent to the first, states that if φ "actually" follows from Γ, then it is also formally derivable. Thus, derivability ($\vdash$) and logical (semantic) consequence ($\Vdash$) are equivalent in classical predicate logic, and the consistency of a theory is equivalent to its satisfiability (having a model). The proof of the completeness theorem for predicate logic by Kurt Gödel in 1929 was one of the greatest achievements of 20th-century logic. It showed that the syntactic inference rules of first-order logic are sufficient to derive all semantically true statements. The following theories are consistent (relative to the metatheory of sets):

- predicate logic with an empty set of non-logical axioms (pure predicate calculus);
- theory of partial order; theory of linear order;
- pure theory of equality;
- theory of semigroups; theory of groups;
- Tarski's elementary geometry;
- first-order Peano arithmetic.

Asserting the consistency of set theory itself is more difficult, as it would require presenting a model of set theory within set theory itself. Such models can only be constructed by strengthening Zermelo-Fraenkel set theory with special axioms of large sets (e.g., the axiom of a strongly inaccessible cardinal number). In other words, the consistency of set theory can only be justified in an even stronger theory. Nevertheless, the simplicity of the formulations of the ZF axioms, especially concerning hereditarily finite sets (i.e., without the axiom of infinity), and the possibility of their interpretation in recursive (read: computer) arithmetic, give us confidence in the consistency of set theory and confidence that it can be a reliable metatheory.

---

[13] This property of a theory is called the *Henkin condition*, and the possibility of extending any theory to one that satisfies the Henkin condition is proven separately, and in the case of uncountable languages, not without the help of a suitable version of the axiom of choice.

Completeness of a Theory

A closed formula φ is said to be **independent** of a set of formulas $\Gamma \subset \mathbb{F}$ if both φ and $\neg\varphi$ are not derivable from Γ in predicate logic ($\Gamma \nvdash \varphi$ and $\Gamma \nvdash \neg\varphi$).

Theorem B1.8. *If the theories $\mathfrak{T} \cup \{\varphi\}$ and $\mathfrak{T} \cup \{\neg\varphi\}$ both have models, then the formula φ is independent of $\mathfrak{T}$.*

This theorem is the main tool for proving the independence of mathematical statements from some axiomatic basis. Here are some of the most famous examples of independent formulas:

- Euclid's axiom (a formalization of the V postulate) relative to absolute geometry (postulates I–IV).
- Goodstein's theorem relative to first-order Peano arithmetic.
- The axiom of choice relative to ZF set theory.
- Cantor's continuum hypothesis relative to ZFC.

If one views the axiomatics of a theory as the basis of a vector space, the following analogies can be drawn:

- the axiomatics can be considered as a system of vectors (not necessarily independent);
- the set of all theorems with given axiomatics is like the linear span of the system of vectors;
- an independent formula is like an independent vector.

This analogy, as is usual, has its flaws. For example, the negation of a formula cannot be understood as the opposite vector, since the opposite vector adds nothing to the linear span of the vectors, whereas adding the negation of a formula to the original system gives a contradictory system from which the entire language $\mathbb{F}$ is derivable. Furthermore, the analogy reveals its limits in more complex situations.

In logic, it is possible for two mutually exclusive hypotheses, φ_1 and φ_2, to be independent of a theory Γ. Yet, they can both lead to the same new conclusion ψ which was not provable from Γ alone. That is, we can have $\Gamma, \varphi_1 \vdash \psi$ and $\Gamma, \varphi_2 \vdash \psi$ while $\Gamma \nvdash \psi$. This highlights a behavior distinct from the more straightforward geometry of vector spaces. In the case of derivability, the theories generated by these systems of axioms will be different, but in each of them, ψ will be a theorem, while at the same time it will not be in the theory generated by the axiomatics Γ. As an example, it is sufficient to take $\Gamma =$ ZF,

φ_1 as the axiom of choice AC, φ_2 as the axiom of determinacy AD, and ψ as the countable axiom of choice for sets of real numbers $\mathsf{AC}_\omega(\mathbb{R})$. Here $\mathsf{AC}_\omega(\mathbb{R})$ asserts that every countable sequence of non-empty sets of real numbers has a choice function.

If a *consistent* theory $\mathfrak{T} \subset \mathbb{F}$ is such that no closed formula $\varphi \in \mathbb{F}$ is independent, then the theory $\mathfrak{T}$ is called **complete** (*in the sense of derivability*).

Undoubtedly, the central and even epoch-making result about first-order formal theories is

Theorem B1.9 (Gödel's First Incompleteness Theorem). *If a theory is*

(G1) *consistent;*
(G2) *recursively enumerable (i.e., there exists an algorithm that can enumerate all its theorems);*
(G3) *capable of interpreting a sufficiently strong arithmetic (for example,* PA*),*[14]

then it is incomplete, i.e., there exists a sentence in the language of this theory that can neither be proved nor disproved within its system of axioms.

Kurt Gödel

Recursively enumerable theories include, for example, all theories with a finite or countable alphabet (including the signature), since the grammar and inference rules allow for the algorithmic "computation" of theorems, so this is not a very strong restriction on Gödelian theories. A stronger restriction is the ability to interpret arithmetic, i.e., to translate arithmetic into the language of formulas of this theory in such a way that the translated axioms of arithmetic become theorems of this theory. In the case of set theory, this means the ability to construct a model of natural numbers (e.g., von Neumann ordinals, as will be shown later). Embedding arithmetic into a theory means the ability to encode its entire language, including theorems and proofs of theorems, by internal means (the formulas of this theory). After which it becomes possible to write some statement about natural numbers that will be independent in this theory. There is also Gödel's second incompleteness theorem, which states that any Gödelian theory, i.e., one satisfying conditions (G1)–(G3), cannot prove its own consistency (or, more precisely, the statement about it encoded in arithmetic). After Gödel's

[14] More precisely, in standard formulations it is sufficient to interpret Robinson arithmetic Q (or an equivalent weak arithmetic sufficient for arithmetization of syntax).

incompleteness theorems, it became clear that any sufficiently strong formal theory will contain true but unprovable statements. The search for and classification of such statements (like the Paris-Harrington theorem or Goodstein's theorem) became an important direction in mathematical logic. Nevertheless, by weakening conditions (G1)–(G3), examples of complete theories appear. First of all, of course, theories in a countable language are of interest, meaning those that violate condition (G3), i.e., are unable to interpret first-order arithmetic within themselves. The following examples of complete theories can be given.

- The theory of dense linear order without first and last elements (DLO_∞). The axioms of this theory include the axioms of strict linear order, as well as:

$$(a < b) \to \exists x\ (a < x) \wedge (x < b), \quad \forall x \exists y\ (y < x), \quad \forall x \exists y\ (x < y),$$

  where the first asserts that between any two elements there is always a third, the second that there is no smallest element, and the third that there is no largest element.
- The theory of equality on an infinite set. Here, to the axioms of equality, one must add an infinite series of axioms forbidding finite carriers (see below).
- The theory of real closed fields (RCF, see below).
- Tarski's elementary geometry.
- Presburger arithmetic (arithmetic without the multiplication operation).

B1.2.10 More on the Theory of Equality

We have already defined the pure theory of equality as a theory with a single predicate = and three axioms: reflexivity, symmetry, and transitivity. In this theory, there is no need to add congruence axioms, as there are no other function or predicate symbols. However, it is easy to see that the pure theory of equality is not complete, as it is quite easy to provide examples of independent formulas. Let n be a positive integer. Consider the formula

$$\varphi_n \overset{\text{def}}{\leftrightarrow} \exists y_1 \exists y_2 \dots \exists y_n \forall x\ (x = y_1) \vee (x = y_2) \vee \dots \vee (x = y_n).$$

As one might guess, the formula tells us that there exist n objects such that any object coincides with one of them, i.e., that there can be no more than n objects in the universe of this theory. In mathematics, special abbreviations are often used to write homogeneous iterations with a parameter, otherwise known as bindings; one need only recall formulas with the summation $\sum$ or product $\prod$ symbols. Similarly, one can use the large disjunction and conjunction signs. Thus, the formula φ_n can be rewritten more concisely:

$$\varphi_n \overset{\text{def}}{\leftrightarrow} \exists y_1 \exists y_2 \dots \exists y_n \forall x \bigvee_{i=1}^{n} (x = y_i).$$

Moreover, when we write homogeneous quantifiers over numbered variables consecutively, we can also simplify our lives by using the notation $\exists \vec{y}$ instead of $\exists y_1 \exists y_2 \dots \exists y_n$, where the number of variables is restored from the subsequent part of the formula. Nevertheless, this is also convenient:

$$\varphi_n \overset{\text{def}}{\leftrightarrow} \exists \vec{y}\ \forall x \bigvee_{i=1}^{n} (x = y_i).$$

We do not consider notations of this kind to be part of the language $\mathbb{F}$, but rather as "syntactic sugar" at the metatheory level. However, they easily fit into the general approach of generating lexemes of the language $\mathfrak{Math}$ that we described in Section A1.2. For such binding lexemes, conventions are also used for what to consider them equal to in the case where the index set is empty (e.g., for $n = 0$ in our case). Algebraic considerations (associativity) are used for this. Thus, an empty conjunction is replaced by $\top$, and an empty disjunction by $\bot$. For the quantifier $\exists \vec{y}$ in the absence of indices, the convention should be adopted that it should be replaced by an empty string, i.e., removed from the text of the formula altogether, since the variables y_i do not exist for $n = 0$. If we now want to add the condition that all y_i are pairwise distinct, we will get a formula that says there must be exactly n objects in the universe:

$$\psi_n \overset{\text{def}}{\leftrightarrow} \exists \vec{y} \bigwedge_{i<j} (y_i \neq y_j) \wedge \forall x \bigvee_{i=1}^{n} (x = y_i),$$

where the symbol $\bigwedge_{i<j}$ denotes one long conjunction with the number of pairs y_i, y_j being compared $(i < j)$ equal to the number of combinations of n things taken 2 at a time. Thus, the pure theory of equality with the axiom ψ_n can only have finite sets with n elements as models. From this it is also clear that any two axioms ψ_n and ψ_m contradict each other for $n \neq m$, but since each one individually has its own model, the axioms ψ_n represent an infinite series of examples of statements independent of the axioms of the pure theory of equality. Nevertheless, if we consider the theory $\mathfrak{T}_n$, generated by axiomatics consisting of the axioms of equality and the axiom ψ_n, then such a theory will be complete. This is quite easy to explain, because any closed formula in such a poor language (where there is nothing but the equality predicate) is either true in all models with n elements or false in all such models, and therefore it either logically follows from the axioms or is refuted. But then, by virtue of the completeness of predicate logic, for any such formula or its negation, a proof exists, i.e., either the formula itself or its negation is a theorem in $\mathfrak{T}_n$. Thus, the theory $\mathfrak{T}_n$ is complete.

Finally, let us consider a theory of equality where instead of a single axiom ψ_n, the negations $\neg\psi_n$ are added for all natural numbers n. Such a theory can only have an infinite model. And such a theory will also be complete, but by virtue of a more complex theorem, namely, the Łoś-Vaught test for completeness, which will be discussed in Part II (see Theorem D2.17).

B1.3 Summary and Self-check Questions

At Level B1, we have made significant progress in the area of term and formula construction, introduced numerous concepts related to formalisms, as well as their semantics—models. Mastery of Level B1 in this book, as in the case of learning a spoken language, means the ability to read and understand the basic concepts and structures of the mathematical language that typically arise in various sources, the ability to express most simple facts with formulas, an understanding of the concept of proof, and the ability to reproduce it in the simplest cases. The key skills at this level are an understanding of the following concepts and their properties:

- the language of propositional formulas, including rules for parenthesis reduction;
- the language of first-order predicate logic with an arbitrary one-sorted signature;
- tautologies and valid formulas;
- derivability in both types of logic (propositional and predicate);

- soundness and completeness of derivations in both logics;
- the deduction theorem;
- axioms of equality and congruence;
- interpretation of a language and a theory in a model;
- logical (semantic) consequence;
- consistency and satisfiability of a theory;
- completeness of a theory and independent formulas.

We now have the opportunity to specify more precisely the recipe for creating a formal theory that describes a certain subject domain, following the same scheme that was presented in Table A2.1. For simplicity, we consider here only an unsorted (homogeneous) subject domain (Table B1.1).

Table B1.1. Recipe for Creating a Formal Theory

(1) Define the alphabet from which all lexemes of the language will be assembled.
(2) Specify the signature of the language (constant, function, and predicate symbols with their arities).
(3) Enumerate the non-logical axioms of the theory in accordance with the concept that we are defining with these axioms.

We have not included here the steps describing the construction of terms and formulas, nor the conventions for parenthesis reduction, as they are universal for all formal theories. We have also not included the step where the logical axioms and inference rules are defined, as we assume the use of classical predicate logic unless otherwise specified. In short, we have focused in this recipe on what truly defines the character of a formal theory. Following this recipe, we have already defined several specific examples of formal theories above:

- the theory of strict linear orders with complete modifications;
- the pure theory of equality with complete modifications;
- the theory of semigroups and the theory of groups.

Everything we have studied shows us the close connection between formalism and semantics, and also clearly demonstrates the fact that the rigor of mathematics rests on some simple but mandatory prerequisites, namely, on the ability to define recursively-natured objects (terms, formulas, proofs), the ability to "materialize" logic in set-theoretic semantics, and the understanding that one is inseparable from the other. To conclude Section B1, we suggest the reader answer the following self-check questions.

Q1 What are the main components (symbols) of the language of propositional logic ($\mathbb{L}$) and what are the rules for constructing its formulas? [p. 88]

Q2 What is the role of axioms and inference rules (in particular, Modus Ponens and the substitution rule (B1.6)) in the system of propositional logic? [p. 91]

Q3 Explain the concepts of soundness and completeness of derivations for propositional logic. [p. 88]

Q4 What is functional completeness in propositional logic and how is it related to truth tables and the equivalence of formulas? [p. 94]

Q5 What is the difference between terms and formulas in predicate logic? Describe the rules for constructing terms, including variables, constants, and function symbols. [p. 99]

Q6 What are the rules for constructing formulas in predicate logic, including the use of predicate symbols, logical connectives, and quantifiers? [p. 100]

Q7 List the main groups of logical axioms of predicate logic and the main inference rules. [p. 103]

Q8 What is an interpretation of a language and a model of a theory in predicate logic? When is a formula considered true in a model? [p. 109]

Q9 What is Gödel's completeness theorem for predicate logic? What is the connection between the consistency of a theory and the existence of a model? [p. 116]

Q10 What are a formal theory, non-logical axioms, and theorems in the context of predicate logic? Provide an example of a simple formal theory described in the text. [p. 105]

Q11 What is the difference between a formal theory and a general mathematical theory? Explain using the example of group theory. [p. 115]

Exercises

B1.1 Determine which of the following are tautologies (always true):

1. $(\mathcal{A} \to \mathcal{B}) \to (\neg\mathcal{B} \to \neg\mathcal{A})$;
2. $((\mathcal{A} \to \mathcal{B}) \to \mathcal{A}) \to \mathcal{A}$ (Peirce's Law);
3. $(\mathcal{A} \to \mathcal{B}) \vee (\mathcal{B} \to \mathcal{A})$ (Linearity of implication);

4. $(\mathcal{A} \wedge \mathcal{B}) \rightarrow (\mathcal{A} \vee \mathcal{B})$.

B1.2 Find a valuation (assignment of $0/1$ to $\mathcal{A}, \mathcal{B}, C$) that makes the following formulas false:

1. $(\mathcal{A} \vee \mathcal{B}) \rightarrow (\mathcal{A} \wedge \mathcal{B})$;
2. $(\mathcal{A} \rightarrow \mathcal{B}) \rightarrow \mathcal{A}$;
3. $(\mathcal{A} \vee \mathcal{B}) \wedge C \rightarrow \mathcal{A} \wedge (\mathcal{B} \vee C)$.

B1.3 Using De Morgan's laws and double negation, simplify:

1. $\neg(\mathcal{A} \vee \neg\mathcal{B})$;
2. $\neg(\mathcal{A} \rightarrow (\mathcal{B} \wedge \neg C))$;
3. $\neg(\forall x \exists y(\varphi(x, y) \rightarrow \neg\psi(x)))$.

B1.4 Following the convention of this book (Section B1.2.1), free variables (parameters) should be denoted by letters a, b, c, and bound variables by x, y, z. Rewrite the following "raw" formulas to comply with this convention (rename free variables to a, b and bound ones to x, y):

1. $\forall k(P(k) \rightarrow Q(m))$;
2. $(\forall x P(x)) \wedge Q(x)$ (Hint: The second x is free);
3. $\exists z(u < z \rightarrow \forall u(u = z))$ (Hint: Resolve the collision of u);
4. $\int_0^x t^2 dt$ (Interpret the upper limit as a parameter).

B1.5 Let $\varphi = \exists y(a < y)$ (where a is free).

1. Perform the substitution $\varphi[a/b]$.
2. Perform the substitution $\varphi[a/S(b)]$.
3. Explain why the substitution $\varphi[a/y]$ is forbidden (invalid) in our logic.

B1.6 Using the Deduction Theorem B1.1, prove the Hypothetical Syllogism rule:

$$\mathcal{A} \rightarrow \mathcal{B},\ \mathcal{B} \rightarrow C \vdash \mathcal{A} \rightarrow C.$$

B1.7 Prove the Law of Contraction:

$$\vdash (\mathcal{A} \rightarrow (\mathcal{A} \rightarrow \mathcal{B})) \rightarrow (\mathcal{A} \rightarrow \mathcal{B}).$$

(Hint: Use the axiom $(\mathcal{A} \rightarrow (\mathcal{B} \rightarrow C)) \rightarrow ((\mathcal{A} \rightarrow \mathcal{B}) \rightarrow (\mathcal{A} \rightarrow C))$).

B1.8 Using the axioms of Predicate Logic (PC2) and Bernays' rules, prove:

$$\forall x P(x) \vdash \exists x P(x).$$

(Hint: Use the term a and axioms ($\forall$) and ($\exists$)).

B1.9 The rule (R1) states: $\frac{\varphi \to \psi}{\varphi \to \forall x \psi}$. Why is the restriction "variable a does not occur in φ" necessary? Show that if we violate this rule (e.g., let $\varphi = P(a)$ and $\psi = P(a)$), we can derive a false statement like $P(a) \to \forall x P(x)$.

B1.10 Using the equivalence $\neg\forall x\varphi \leftrightarrow \exists x\neg\varphi$ and $\neg\exists x\varphi \leftrightarrow \forall x\neg\varphi$, move the negation inward:

1. $\neg\forall x(P(x) \to Q(x))$;
2. $\neg\exists x\forall y(R(x, y) \vee S(x))$;
3. $\neg(\exists x P(x) \to \forall y Q(y))$.

B1.11 Prove (semantically or syntactically) that if x does not occur free in P, then:

$$\forall x(P \vee Q(x)) \leftrightarrow P \vee \forall x Q(x).$$

B1.12 Disprove $\exists x(P(x) \wedge Q(x)) \leftrightarrow (\exists x P(x) \wedge \exists x Q(x))$.

B1.13 Construct a model (universe and predicates) where the following formulas are false:

1. $\exists x P(x) \to \forall x P(x)$;
2. $\forall x\exists y P(x, y) \to \exists y\forall x P(x, y)$;
3. $\forall x(P(x) \vee Q(x)) \to (\forall x P(x) \vee \forall x Q(x))$.

B1.14 Is the following set of formulas satisfiable?

$$\{\forall x\ (x < x),\ \forall x\forall y\forall z(x < y \wedge y < z \to x < z),\ \forall x\exists y\ \neg(x < y) \wedge \neg(y < x)\}$$

B1.15 Write down the specific instances of the Congruence Axiom Scheme $(a = b) \to (\varphi[x/a] \leftrightarrow \varphi[x/b])$ for:

1. $\varphi(x) \equiv P(x)$ (unary predicate);
2. $\varphi(x) \equiv (x < c)$ (binary predicate with parameter).

B1.16 Write a formula in first-order logic with equality expressing:

1. "There are at least 2 elements";
2. "There are at least 3 elements".

B1.17 Write a formula expressing:

1. "There is exactly 1 element" ($\exists!x$);
2. "There are exactly 2 elements".

B1.18 Express the following set-theoretic statements using logical connectives and the membership predicate:

1. $A \subseteq B \cup C$;
2. $A \cap B = \emptyset$;
3. $A \setminus B \subseteq C$.

B1.19 Let the universe U consist of three pencils:

$$U = \{p_1(\text{Red, Short}),\ p_2(\text{Red, Long}),\ p_3(\text{Blue, Long})\}$$

Let the predicate $R(x)$ mean "x is Red" and $L(x)$ mean "x is Long". Determine the truth value of the following formulas in this model:

1. $\forall x(R(x) \lor L(x))$;
2. $\exists x(R(x) \land \neg L(x))$;
3. $\forall x(L(x) \to \neg R(x))$;
4. $\forall x(R(x) \to \exists y(L(y) \land y \neq x))$.

B1.20 Consider the formula $\varphi = \forall x \exists y R(x, y)$. Determine its truth value in the following interpretations (models):

1. Model 1 (Arithmetic): Universe $\mathbb{N}$, $R(x, y)$ is "$x < y$".
2. Model 2 (Genealogy): Universe People, $R(x, y)$ is "x is the child of y".
3. Model 3 (Chess): Universe Chess pieces on a board, $R(x, y)$ is "x attacks y". (Assume standard setup for definiteness).

Level B2
The Foundations of Mathematics

God created the natural numbers, all else is the work of man.

— Leopold Kronecker

B2.1 Peano Arithmetic

In the late 19th century, mathematics faced the need for a more rigorous foundation for its branches, particularly for the arithmetic of natural numbers. The Italian mathematician Giuseppe Peano became one of the key figures in this process of formalization.

In 1889, Peano introduced his system of axioms, designed to provide a strict and formal definition of the natural numbers. His approach was based on a minimal set of primitive notions: the concept of zero (or one) and the successor operation. The central idea of Peano's system was the principle of mathematical induction, which was included in the axiomatics. This principle, along with the axioms for zero and the successor, laid the groundwork for the rigorous definition and proof of the properties of arithmetic operations. Within this axiomatic framework, it became possible to formally justify the recursive definitions of addition and multiplication and to prove their fundamental properties.

Giuseppe Peano

N. I. Kazimirov, *Mathematics as a Foreign Language*, Mathematics in Mind,
https://doi.org/10.1007/978-3-032-30721-7_4

Peano's system was a significant step towards the complete formalization of mathematics. With the subsequent development and systematization of first-order predicate logic in the early 20th century, Peano's axioms were reformulated within this powerful logical apparatus.

Today, first-order Peano Arithmetic (often denoted by PA) is understood to be precisely this formal theory. It consists of a finite number of axioms describing the basic properties of zero and the successor, and an axiom schema of induction. The latter means that for each formula in the language of arithmetic, there is a separate induction axiom corresponding to that formula. Despite the apparent simplicity of the initial concepts, PA is a sufficiently rich theory, capable of formalizing a significant portion of elementary number theory. However, according to Gödel's incompleteness theorems, it cannot be both complete and consistent.

But let us begin by describing the concept of a natural number. After all, to formalize its definition (to make it mathematical), we need to describe as precisely as possible what we expect from this concept.

So, natural numbers are abstract entities required to perform the following functions: counting (enumerating) arbitrary objects and comparing the sizes of collections of objects. Counting presupposes that natural numbers can be assigned to arbitrary objects, starting from a specific one, and then successively assigning the next natural number to each subsequent object. This gives rise to the idea that for each natural number n, there must be an immediate successor, denoted by $S(n)$, and also that every number (except for an initial one) is the successor of some other number. This creates the need for a function symbol for the successor, S, with specific properties.

Having the ability to enumerate objects, we can reason about the size of a collection based on which natural number exhausts the enumeration (if this is possible at all), taking this number as the measure of its size.

To compare the sizes of collections, we would like to be able to compare these measures, i.e., to determine which one is smaller. At this stage, the concept of an *order* on the natural numbers arises, specifically a linear order, so that we can confidently compare any two natural numbers. But from set-theoretic considerations, we also understand that besides answering the question "is collection A contained within collection B?", one can ask other questions, for example, "how large is the difference between these collections?" or "what needs to be added to collection A to obtain collection B?"

From here, we arrive at the concept of addition (and subtraction) on the natural numbers. We understand that adding a number n to a number m should mean for us performing the successor operation n times; that is, $m + n$ will be defined recursively through the basic operation S. From this, one can also note that comparison can be defined through addition: the smaller is the one to which something must be added to get the larger. Therefore, the comparison of natural numbers is often introduced after the formalization of arithmetic as a definition within the formal theory.

Finally, natural numbers can also be seen as a way to count the arithmetic operations themselves or, in other words, to define the multiplicity of these operations. Thus, the multiplicity of addition is multiplication. Therefore, we can (but do not have to) include the operation of multiplication in our concept. This theme can be developed further, moving to the operation of exponentiation, then to towers of powers, and so on, introducing the so-called hierarchy of fast-growing functions, but here we will limit ourselves to the first step—multiplication. This is quite sufficient to fundamentally strengthen the properties of arithmetic. So, let us proceed to the formalization of the concept of a natural number, following the scheme from Table B1.1.

B2.1.1 Language and Axioms

First, let's define the alphabet, in which we include the letters $a, \ldots, z$, as well as their modifications with primes and numerical (meaning digits $0, \ldots, 9$) indices as notations for variables. We will treat symbols from the beginning of the alphabet as free variables, and those from the end (usually u, v, w, x, y, z) as bound variables. The use of primes on variable symbols allows for the potential creation of arbitrarily many new variables, but we will strive for concise notation. Next, we include round and square brackets in the alphabet as interchangeable (to be able to highlight some brackets against others), as well as the signs for logical connectives and quantifiers.

Finally, let's define the signature. Constants: 0. Function symbols: S (unary), $+, \cdot$ (binary). Predicate symbols: $=, <$ (binary). We also stipulate that instead of $S(n)$ we will often write Sn if it does not cause confusion, instead of $+(a, b)$ we write $(a+b)$, instead of $\cdot(a, b)$ we write $a \cdot b$ or even shorter ab, instead of $= (a, b)$ we write $a = b$, instead of $\neg(a = b)$ we write $a \neq b$, and instead of $< (a, b)$ we write $a < b$ or $b > a$. This eliminates the need to include the comma in the language's alphabet. Furthermore, we immediately introduce two abbreviations for non-strict inequality: $(a \leqslant b) \overset{\text{def}}{\leftrightarrow} (a < b) \vee (a = b)$, $(a \geqslant b) \overset{\text{def}}{\leftrightarrow} (a > b) \vee (a = b)$.

Further, we adopt all the natural rules for reducing parentheses in accordance with the priorities of arithmetic operations to make the text of formulas and terms more readable.

Thus the language is defined; it remains to add the axioms. As noted, the axioms of classical first-order predicate logic and the rules of inference are assumed automatically. And, as usual, we use the non-terminal symbols φ, ψ (possibly with numerical indices) to denote arbitrary formulas, indicating in parentheses the variables that may appear in these formulas (here we will need a comma, but it will be an element of the meta-language).

Now let's write down the list of axioms of **first-order Peano Arithmetic**, after which we will discuss their meaning:

PA1 $a = a$; $(a = b) \to (b = a)$; $(a = b) \wedge (b = c) \to (a = c)$;

PA2 $a = b \to S(a) = S(b)$;

PA3 $S(a) = S(b) \to (a = b)$;

PA4 $S(a) \neq 0$;

PA5 $a + 0 = a$; $a + Sb = S(a + b)$;

PA6 $a \cdot 0 = 0$; $a \cdot Sb = ab + a$;

PA7 $\neg(a < 0)$; $(a < Sb) \leftrightarrow (a < b) \vee (a = b)$;

PA8 $\varphi[a/0] \wedge [\forall x\, (\varphi[a/x] \to \varphi[a/Sx])] \to \forall y\, \varphi[a/y]$.

The group of axioms PA1 are the standard axioms of equality and are included, as we have said, because they must always be included in a theory for the equality predicate (either as axioms or as theorems).

Axiom PA2 is the embodiment of the congruence axiom for terms. More precisely, from this axiom, with the help of the other axioms and by induction on the complexity of the formula, one can derive congruence in its general form (B1.9).

The remaining axioms are the proper axioms of arithmetic. PA3 tells us that the function S is injective, i.e., that the successors of different numbers are different numbers, which corresponds to our understanding of enumeration as a one-to-one correspondence.

Axiom PA4 states that zero is not the successor of any number, i.e., it is the first element among the numbers. With this, we implement the principle of the special nature of zero as the unique starting numeral. By the way, terms of the form $S \ldots S0$ are called *numerals*. Numerals are the notations for all natural numbers that can be constructed in a finite number of steps (one could also say that numerals are the Archimedean natural numbers).[1] We can devise more familiar notations for numerals by giving such definitions:

$$1 \stackrel{\text{def}}{\rightleftharpoons} S(0), \quad 2 \stackrel{\text{def}}{\rightleftharpoons} S(1), \quad 3 \stackrel{\text{def}}{\rightleftharpoons} S(2),$$
$$4 \stackrel{\text{def}}{\rightleftharpoons} S(3), \quad 5 \stackrel{\text{def}}{\rightleftharpoons} S(4), \quad \text{and so on.}$$

Incidentally, the symbols $\stackrel{\text{def}}{\rightleftharpoons}$ and $\stackrel{\text{def}}{\leftrightarrow}$ can be understood not only as operators from the meta-theory but can also be partially perceived as part of the language of arithmetic, i.e., as predicates subject to the axioms

$$(F \stackrel{\text{def}}{\rightleftharpoons} T) \to (F = T), \quad (P \stackrel{\text{def}}{\leftrightarrow} \varphi) \to (P \leftrightarrow \varphi), \tag{B2.1}$$

where F is a function (or constant) symbol not used in the axiomatics, T is an arbitrary term not containing the symbol F, P is a predicate symbol not used in the axiomatics, and φ is an arbitrary formula not containing the predicate P. The predicates $(a \neq b)$ and $(a \leqslant b)$ were defined in this way above. Thus, those terms (formulas) that are equal by definition are also equal (equivalent) in the sense of the original language (but not vice versa). However, not all of our past and future re-designations using the symbols $\stackrel{\text{def}}{\rightleftharpoons}$ and $\stackrel{\text{def}}{\leftrightarrow}$ reduce to these rules, so it is generally accepted to consider such abbreviations of formal texts as "syntactic sugar" at the meta-theory level and not to include this symbol in the formalism under investigation.

The group of axioms PA5 defines addition recursively via the successor operation. Firstly, it sets the base case $(a + 0 = a)$, and secondly, it defines the addition of a successor as the successor of the previous sum $(a + Sb = S(a + b))$. If one understands the operator S as adding one, these equalities become obvious. Moreover, using the already introduced notations, we get that $a + 1 = a + S0 = S(a + 0) = Sa$, where we have used axiom PA5, PA2, as well as the transitivity of equality and the Modus Ponens rule.

The group of axioms PA6 similarly defines the multiplication operation recursively. As noted, one can add similar axioms for other derived operations, such as exponentiation, to obtain more advanced arithmetic theories.[2]

---

[1] The irony is that the Archimedean property is itself defined via natural numbers. But in this definition, one means the natural numbers of the standard model. The construction of non-standard models is beyond the scope of this book.

[2] In fact, there is a significant difference between multiplication and all other recursive operations, namely: the correctness of recursive definitions cannot be proven without involving multiplication (since the coding of formulas uses division with remainder—see

The group of axioms PA7, firstly, states that zero is the minimal element (there is nothing less than zero). Secondly, it provides a recursive rule for comparing any two numbers. As an exercise, one can try to prove, for example, that $2 < 5$.

Finally, the most important axiom of Peano Arithmetic is the axiom of induction PA8. More precisely, it is a whole series (schema) of axioms, and each specific induction axiom is obtained by replacing φ with a specific formula of the language of arithmetic. We note, as usual, that any free variable can be substituted for the variable a, and any bound variables not present in the original formula φ can be substituted for x and y. The importance of the axiom of induction lies in the fact that it makes Peano Arithmetic a much stronger theory than a theory without induction. This is evident from the numerous properties of various arithmetics weaker than the one under consideration.

The theory generated by axioms PA1–8 in first-order predicate logic is called **Peano Arithmetic** and is denoted by PA.

Theorem B2.1. *The following theorems hold in the theory* PA *(with universal closure of the formulas understood):*

1° $a + b = b + a$ [commutativity of addition];

2° $a + (b + c) = (a + b) + c$ [associativity of addition];

3° $ab = ba$ [commutativity of multiplication];

4° $a(b + c) = ab + ac$ [distributivity];

5° $a(bc) = (ab)c$ [associativity of multiplication];

6° $(a < b) \leftrightarrow (Sa < Sb)$ [order is compatible with S];

7° $(a + c = b + c) \to (a = b)$ [cancellation law for +];

8° $(a < b) \leftrightarrow (a + c < b + c)$ [order is compatible with +];

9° $<$ is a linear order;

10° $(c \neq 0) \to [(a < b) \leftrightarrow (ac < bc)]$ [order is compatible with $\cdot$];

11° $(ac = bc) \wedge (c \neq 0) \to (a = b)$ [cancellation law for $\cdot$];

12° $(a \leqslant b) \wedge (b < Sa) \to (b = a)$ [discreteness of order];

13° $a \cdot S0 = a$ [identity element for multiplication];

section D3.4). Therefore, multiplication can be defined through addition, but the totality and correctness of such a definition cannot be proven, whereas exponentiation and all higher recursions are definably correct in the presence of multiplication. Therefore, multiplication is usually included in the axiomatics from the start.

14° $(a \cdot b = 0) \to (a = 0 \vee b = 0)$ [no zero divisors].

The proof of this theorem will be examined in Part II (see Theorem D3.1). Its meaning is that the natural numbers with the operations of addition and multiplication and the order $<$ form a structure that, in algebraic language, is called an *ordered discrete commutative semiring with unity*. It is worth noting that not all such semirings are models of PA, since the properties of such a ring do not guarantee that the axiom of induction holds. Nevertheless, as we can see, the axioms of PA are sufficient to obtain all the basic properties of numbers that we remember and love from childhood. Note also that all the listed properties are theorems of the formal theory PA. They are formulated and proven without resorting to external means (meta-theory). Let's demonstrate this with an example of proving the base case for the commutativity of addition, i.e., the formula $0 + a = a + 0$. We will conduct the proof by induction on a.

1. Base case $0 + 0 = 0 + 0$ follows from PA1 by substitution $[a/0 + 0]$;
2. Inductive step: $(0 + a = a + 0) \to (0 + Sa = Sa + 0)$.
   - 2.1. $0+a = a+0$ (hypothesis) and $(0+a = a+0) \to S(0+a) = S(a+0)$ (substitution $[a/a + 0; b/0 + a]$ into PA2);
   - 2.2. $S(0 + a) = S(a + 0)$ (hypothesis and Modus Ponens);
   - 2.3. $0 + Sa = S(0 + a)$ (substitution $[a/0; b/a]$ into PA5);
   - 2.4. $(0 + Sa = S(0 + a)) \wedge (S(0 + a) = S(a + 0))$ (tautology $\mathcal{A} \to (\mathcal{B} \to (\mathcal{A} \wedge \mathcal{B}))$ and twice MP, this rule is called "conjunction introduction"[3]);
   - 2.5. $0 + Sa = S(a + 0)$ (transitivity of equality, MP);
   - 2.6. $S(a + 0) = Sa$ (PA5, substitution $[a/a + 0]$ in PA2, MP);
   - 2.7. $0 + Sa = Sa$ (transitivity of equality, MP);
   - 2.8. $Sa = Sa + 0$ (PA5$[a/Sa]$, symmetry of equality, MP);
   - 2.9. $0 + Sa = Sa + 0$ (conjunction introduction, transitivity of equality, MP);
   - 2.10. $(0 + a = a + 0) \to (0 + Sa = Sa + 0)$ (deduction theorem).
3. $\forall x\ (0 + x = x + 0) \to (0 + Sx = Sx + 0)$ (quantifier introduction by rule (R1))[4]

[3] It has the following form: $\frac{\varphi \quad \psi}{\varphi \wedge \psi}$

[4] A simplified version of Bernays' rule is used, called the *generalization rule*: $\frac{\psi}{\forall x\ \psi[a/x]}$, where x does not occur in ψ. This rule is easily derived from Bernays' rule if one substitutes any valid formula for the premise φ.

4. $(0 + 0 = 0 + 0) \wedge [\forall x\ (0 + x = x + 0) \to (0 + Sx = Sx + 0)]$ (conjunction introduction);
5. $\forall y\ (0 + y = y + 0)$ (induction for the formula $\varphi \overset{\text{def}}{\leftrightarrow} (0 + a = a + 0)$).

Starting from the derived formula, one can then prove the commutativity of addition $a + b = b + a$ by induction on b. Before proceeding, we must address the issue of the congruence of equality in its broadest sense, namely, to obtain the properties (B1.9). The following is true:

Theorem B2.2 (Congruence). *Let T, T_1, T_2 be arbitrary terms of the language of* PA, *and φ be an arbitrary formula of the language of* PA. *Then*

$$(T_1 = T_2) \to T[a/T_1] = T[a/T_2]; \quad (T_1 = T_2) \to (\varphi[a/T_1] \leftrightarrow \varphi[a/T_2]),$$

where $[a/T_i]$ denotes the substitution of all occurrences of the free variable a with the term T_i, and these formulas are also valid for any other free variable (not just a).

The full proof of this theorem will be demonstrated in Part II, Theorem D3.2. Thus, we do not need to include congruence axiom schemas in the axiomatics of PA; instead, we add a basic congruence rule for the operator S and use the full power of arithmetic induction.[5]

B2.1.2 Order

In the previous section, when discussing the essence of the concept of "natural numbers", we said that a number a can be considered less than (or not greater than) a number b if b is obtained by adding some number to a. We then operated with the word "size", implying that numbers measure some sizes of concepts. However, later in the axioms, we introduced the definition of inequality recursively in the group of axioms PA7. But nothing prevents us from introducing another definition of inequality, by positing

$$\text{PA7'} \qquad (a <_1 b) \leftrightarrow \exists x\ (b = a + Sx).$$

[5] By "full power" here, one should understand the fact that many theorems are proven by so-called nested, or multi-parameter, induction, where the base case or step of one induction is obtained by applying a second induction, and so on.

It turns out that within the axiomatics of PA, these two definitions specify the same order. More precisely, the following is true:

Theorem B2.3. $(a < b) \leftrightarrow (a <_1 b)$.

This formula is already a proper theorem of PA (up to universal closure). The proof of the theorem will be given in Part II, Lemma D3.2. Related to inequality are two variants of simplified notation for quantifiers, the so-called *bounded quantifiers*. They are defined as follows:

$$\begin{aligned} \forall x < b : \ \varphi[a/x] &\overset{\text{def}}{\leftrightarrow} \forall x\ (x < b) \to \varphi[a/x], \\ \exists x < b : \ \varphi[a/x] &\overset{\text{def}}{\leftrightarrow} \exists x\ (x < b) \wedge \varphi[a/x], \end{aligned} \tag{B2.2}$$

where φ is an arithmetic formula, and a, b are free variables. The first formula, as is easy to see, returns a true value if and only if the formula φ is true for all values of the parameter a that are less than the parameter b. The second formula is true if and only if the formula φ is true for at least one value of the parameter a that is less than the parameter b. In predicate logic, it is not difficult to prove that

$$\neg(\forall x < b : \ \varphi[a/x]) \leftrightarrow (\exists x < b : \ \neg\varphi[a/x]),$$
$$\neg(\exists x < b : \ \varphi[a/x]) \leftrightarrow (\forall x < b : \ \neg\varphi[a/x]),$$

i.e., these quantifiers behave exactly like the standard ones, which corresponds to common sense. Bounded quantifiers are useful because determining their truth value requires only a finite number of evaluations of the formula φ. In essence, they embody the implementation of a for loop in arithmetic. The presence of such quantifiers greatly simplifies the computability of formulas with quantifiers, so formulas in which all quantifiers are bounded have even been placed in a separate class of formulas, called Δ_0-formulas. [6]

It is worth noting that if we refrain from introducing the relation $<$ into the axiomatics but wish to use bounded quantifiers, we can, of course, define them by expanding the definition of inequality in its strong form. However, such a definition would itself cease to be a Δ_0-formula, as it would use unbounded quantifiers. Thus, introducing inequality into the axiomatics (even in its weak form) makes the arithmetic stronger in the sense that it allows for the definition

[6] The class of Δ_0-formulas is only the lowest level of the so-called arithmetical hierarchy of formulas, in which all formulas are divided into classes depending on the order and number of alternating types of quantifiers in the formula—see, for example, [41].

of the class of Δ_0-formulas, although from the point of view of proof strength, these theories are equivalent: the theory PA without inequality and the theory PA with inequality contain the same theorems (up to the expansion of the < sign).

B2.1.3 Variants of Induction

It has already been noted that there are several statements equivalent to arithmetical induction. Here we will consider only a few. Let φ be a formula in the language of PA, and let a be a free variable occurring in this formula. We introduce the following notations using a definition schema:

$$Ind_a[\varphi] \overset{\text{def}}{\leftrightarrow} \varphi[a/0] \wedge \forall x\, (\varphi[a/x] \rightarrow \varphi[a/Sx])$$
$$Prog_a[\varphi] \overset{\text{def}}{\leftrightarrow} \forall x\, ((\forall y < x :\ \varphi[a/y]) \rightarrow \varphi[a/x])$$
$$Tot_a[\varphi] \overset{\text{def}}{\leftrightarrow} \forall y\, \varphi[a/y]$$

The first schema expresses the *inductiveness* of the formula φ, the second its *progressiveness*, and the third its *totality*. The axiom of induction PA8 is obviously written using the inductiveness schema as follows:

$$Ind_a[\varphi] \rightarrow Tot_a[\varphi], \tag{B2.3}$$

i.e., if a formula φ is inductive, then it is total. This corresponds well with the usual notion of mathematical induction when we talk about sets, i.e., essentially about the extensions of concepts. In an arithmetic where we allow the use of the concept of a "set" (second-order arithmetic), induction is formulated as: if a set is inductive, then it is total (i.e., it contains all natural numbers). And inductiveness means that the set contains the number 0, and also, for every number in it, its successor is also in it. In our case, the same idea is simply written with formulas. Using the progressiveness formula, one can write down what is known as **complete** (or *course-of-values*) induction:

$$Prog_a[\varphi] \rightarrow Tot_a[\varphi]. \tag{B2.4}$$

Induction in the form (B2.3), i.e., PA8, is then called *local*. The property of progressiveness seems weaker, as in it we say the following: if *all* elements smaller than x satisfy φ, then x also satisfies φ. For sets, this can be formulated as: if all elements less than x belong to a given set, then x itself belongs to it. The premise of progressiveness imposes more conditions on φ than the premise of inductiveness, which means that progressiveness is not stronger than inductiveness. And indeed, it is not difficult to verify that $Ind_a[\varphi] \to Prog_a[\varphi]$ always holds. From this, it follows that complete induction implies local induction, i.e., formally, complete induction appears stronger than its local variant. However, in PA arithmetic, both these forms of induction are equivalent.

Theorem B2.4. *The schema of complete induction is equivalent to the schema of local induction, modulo the other axioms of* PA.

For the proof, we refer, as usual, to Part II (Theorem D3.3), but we emphasize that this theorem is a meta-theorem; it does not at all mean the derivation of the equivalence of formulas (B2.3) and (B2.4) for *each specific* formula φ. It states that if we accept the *entire* schema of one induction, we can derive (by means of PA) the *entire* schema of the other induction, and vice versa. Let us now perform the following purely formal manipulations with formulas, using only predicate calculus:

$$\begin{gathered} Prog_a[\varphi] \to Tot_a[\varphi] \equiv \neg Tot_a[\varphi] \to \neg Prog_a[\varphi] \equiv \\ \equiv (\exists y\ \neg\varphi[a/y]) \to \exists x\ \neg((\forall y < x\ \varphi[a/y]) \to \varphi[a/x]) \equiv \\ \equiv (\exists y\ \neg\varphi[a/y]) \to \exists x\ ((\forall y < x\ \varphi[a/y]) \land \neg\varphi[a/x]). \end{gathered}$$

Now let's denote $\psi \overset{\text{def}}{\leftrightarrow} \neg\varphi$ and we get:

$$Prog_a[\varphi] \to Tot_a[\varphi] \equiv (\exists y\ \psi(y)) \to \exists x\ (\psi(x) \land \forall y < x :\ \neg\psi(y)),$$

where we have removed the now tiresome substitution of a bound variable for a free one, simplifying the notation at the meta-level. Thus, we see that if the schema of complete induction is valid (and it is in PA), then the following schema is also valid:

$$(\exists y\ \psi(y)) \to \exists x\ (\psi(x) \land \forall y < x :\ \neg\psi(y)), \tag{B2.5}$$

since the formula ψ here can be chosen arbitrarily (as the complete induction schema has an arbitrary formula φ). The schema (B2.5) is called the **well-foundedness principle** of the natural numbers.

Why so? The reason is that among partial orders (and even preorders), it is customary to single out those that have the property that every non-empty set has a *minimal element* (i.e., an element such that there is no smaller element in the set). For sets (extensions of concepts), this property is formally written as:

$$(A \neq \emptyset) \to \exists (x \in A)\ \forall y < x :\ (y \notin A).$$

But now, let's imagine that the set A is the extension of a concept defined by the formula ψ, i.e., instead of $x \in A$, we can write $\psi(x)$. Then our set-theoretic definition of well-foundedness turns into formula (B2.5). Thus, through purely predicate logic derivations, we have shown that the principle of well-foundedness in the language of PA (i.e., even without the involvement of any PA axioms) is equivalent to the principle of complete induction. And by virtue of Theorem B2.4, it turns out to be equivalent to ordinary induction as well, but this time within the framework of the PA axiomatics. But that's not all. Let's now transform the principle of induction slightly differently, again in pure predicate calculus:

$$\begin{aligned} Prog_a[\varphi] \to Tot_a[\varphi] &\equiv \neg Tot_a[\varphi] \to \neg Prog_a[\varphi] \equiv \\ &\equiv (\exists y\ \neg\varphi(y)) \to \neg(\forall x\ ((\forall y < x :\ \varphi(y)) \to \varphi(x))) \equiv \\ &\equiv (\exists y\ \neg\varphi(y)) \to \neg(\forall x\ (\neg\varphi(x) \to \exists y < x :\ \neg\varphi(y))) \end{aligned}$$

where we have used the equivalence $\mathcal{A} \to \mathcal{B} \equiv \neg\mathcal{B} \to \neg\mathcal{A}$. Now, let's again replace $\neg\varphi$ with ψ (the choice of formula remains arbitrary), and we obtain the following schema, equivalent to complete induction:

$$(\exists y\ \psi(y)) \to \neg(\forall x\ (\psi(x) \to \exists y < x :\ \psi(y))). \tag{B2.6}$$

Let's consider this formula, imagining that ψ defines some extension of a concept A, and substituting $(x \in A)$ for $\psi(x)$. Then this principle can be understood as: if A is not empty, then it is not true that for any element of A, there is a smaller element in A. In other words, within the set A, it is impossible to descend down the $<$ inequality infinitely (for some x, it will be impossible to find a smaller element in A). Thus, (B2.6) implements another important principle in Peano Arithmetic—the **principle of infinite descent**, used by Fermat back in the 17th century to prove his famous theorem for the case $n = 4$.

So, within the axiomatics of PA, the following four principles are equivalent (as schemas):

- local induction;
- complete induction;
- well-foundedness;
- the principle of infinite descent,

with the last three being equivalent in pure predicate calculus (i.e., the definition of the symbol < is irrelevant in this case).

B2.1.4 Provability and Expressibility

To emphasize once again the significance of induction for arithmetic, let's look at it from the perspective of the meta-theory. After all, induction (in its local form) essentially states that if something is true at zero, and the transition from x to $x + 1$ (which is Sx) is also true, then it is true for all natural numbers. But now, let's imagine that we can prove individual theorems of the form $\varphi(\underline{n}) \rightarrow \varphi(\underline{n+1})$ for each specific n, i.e., for each numeral $\underline{n} \overset{\text{def}}{\rightleftharpoons} \underbrace{S \ldots S}_{n} 0$.

Would the totality formula for the property φ then be provable? It turns out, no. And in this lies a very subtle property of the universal quantifier, embedded in the formula $Tot_a[\varphi]$. This subtle property is that if the formula $\forall n\ \varphi(\underline{n})$ is true from the meta-theory's perspective, the totality formula $\forall n\ \varphi(n)$ is not necessarily provable within the theory PA itself (a reminder: the red quantifier is written in the language of the meta-theory). There are numerous examples of this, stemming from the theory of recursive functions.

Let's consider this with an example. Let $G(n)$ be Goodstein's function.[7] For each specific numeral $\underline{n}$, we can prove in PA that the value of Goodstein's function at the point $\underline{n}$ is defined:

$$\forall n\ \mathsf{PA} \vdash \exists y\ (G(\underline{n}) = y),$$

i.e., there exists a finite chain of formulas (very long for large n) that constitutes a proof of the formula $\exists y\ (G(\underline{n}) = y)$ for each specific numeral representing the number n. However, the totality formula for the function G has no proof (nor refutation) in PA:

[7] It is not important what this is right now; what is important is to understand that the equality $y = G(x)$ can be written as an arithmetical formula.

$$\mathsf{PA} \nvdash \forall x\ \exists y\ (G(x) = y).$$

In other words, as soon as we wanted to bring the universal quantifier into the notation of the PA theorem, it ceased to be a theorem.

Therefore, despite the apparent simplicity of the principle of local induction, which we often take for granted, it makes a substantial contribution to the proof strength of the theory. Moreover, the principle of induction can be weakened in various ways (by forbidding nested inductions or by applying the induction schema not to the entire class of formulas, but only to a certain subclass, for example, only to Δ_0-formulas), thereby obtaining a great variety of arithmetical theories in the same language.

Finally, there is a possibility to classify arithmetical theories by their proof strength by finding the class of recursive functions for which that theory can prove their totality. The example of Goodstein's function given above shows that there are recursive functions whose totality cannot be proven even by PA with its induction schema without restrictions on parameters and classes of formulas.

It is worth adding that arithmetic can, on the contrary, be strengthened by adding previously unprovable principles to it (for example, Goodstein's theorem), or by strengthening its language by transitioning to so-called second-order arithmetic, which allows for the existence of arbitrary arithmetical sets. In such an arithmetic, induction takes on the familiar features from school, as it is formulated as a property of sets, not formulas. And in such an arithmetic, theorems like Goodstein's theorem, as well as many theorems of real analysis, become provable.

Along with the provability of any mathematical fact, one must also speak of its expressibility in one or another arithmetic. Indeed, the ability to write an arbitrary abstract fact as a formula is not always available in every specific language.

For example, in the case of the aforementioned Goodstein's function, we are able to write the formula for its totality $\forall x\ \exists y\ (G(x) = y)$, because the computation of Goodstein's function is a clear algorithmic computation procedure (in general, any function programmable on a computer can be defined by a sufficiently simple (from the point of view of quantifier structure) formula of the language of PA, see Theorem D3.8).

However, if we look at the natural numbers from a general mathematical point of view, we can notice that there is a more than countable (continuum) set of subsets of the set of natural numbers, while in the language of PA there is only a countable set of formulas. Consequently, not every subset of the set of natural numbers can be described by a formula.

For example, we can in a certain way encode all sentences of the language of PA, assigning them so-called Gödel numbers. This is done quite simply, much like any text in a computer is represented by a sequence of bits, i.e., by some huge number. Then we can single out among the Gödel numbers only those that correspond to true (from the point of view of general mathematics, i.e., the meta-theory) sentences. The set of all such numbers is obviously a subset of the set of natural numbers, but it is known that in the language of PA there is no formula $True(x)$ that would be true for all numbers of true sentences, and only for them. This statement is the content of *Tarski's undefinability of truth theorem.*

Arithmetic, nevertheless, is a very powerful language in terms of what mathematical facts can be expressed in it. This concerns not only Goodstein's theorem or theorems such as Fermat's Last Theorem or the Chinese Remainder Theorem.

Therefore, it is easiest to provide examples of inexpressibility for weaker languages. Thus, if we take the theory of torsion-free abelian groups (for example, the integers with the operation of addition), then in such a theory one cannot express the relation < (although, as we saw above, in the natural numbers this is possible). But if we add multiplication, then in the ring of integers (but not in any ring!) the order can already be expressed in the following way:

$$a < b \overset{\text{def}}{\leftrightarrow} \exists x_1, x_2, x_3, x_4 : \; Sa + x_1 \cdot x_1 + x_2 \cdot x_2 + x_3 \cdot x_3 + x_4 \cdot x_4 = b,$$

which is a consequence of Lagrange's four-square theorem. Another example (without detail): in the signature of the language of first-order geometry, such objects as a unit segment, any figure other than the entire plane, a direction along some axis, etc., are inexpressible.

Thus, we see that formal languages and theories have significant limitations compared to general mathematics (naïve set theory):

- not every concept from the model of a language/theory can be expressed (defined) in that language;
- even if some fact true in the model has been successfully expressed by a formula of the language, it is not always possible to prove or disprove it in the corresponding formal theory.

All these questions are dealt with by a complex of disciplines in the foundations of mathematics under the general name of **Proof Theory**. This is not just the art of constructing proofs, but a whole science that studies proofs themselves as mathematical objects. Just as linguistics studies the structure of language, proof theory, starting with the works of Gerhard Gentzen, analyzes the "anatomy" of formal deductions, their complexity, and expressive power. It is within the framework of proof theory that a rigorous assessment of the "strength" of a particular axiomatic system is obtained. For example, to prove the consistency of Peano Arithmetic, methods are required that go beyond the scope of this arithmetic itself. One of the central directions here is *ordinal analysis*, which assigns to each theory a certain transfinite ordinal that serves as a measure of its "proof-theoretic strength". Thus, the strength of PA corresponds to the ordinal $\varepsilon_0 = \omega^{\omega^{\omega^{\cdots}}}$, which clearly shows how far beyond "ordinary" induction one must go to grasp it from a meta-mathematical height. Thus, having encountered the limits of formalism, mathematicians have turned the study of these limits into a new, deep, and actively developing area of research.

B2.2 Prolegomena to Set Theory

Georg Cantor

Before we move on to a rigorous axiomatic theory, it is necessary to pay tribute to its creator—**Georg Cantor**. It was he who, at the end of the 19th century, brought about a revolution by proposing to consider infinite collections not as a potential, unfinished process, but as fully-fledged mathematical objects—**sets** (German: *Mengen*). His approach, which today is called *naive set theory*, was based on an intuitively clear principle: *for any property $\mathcal{A}(a)$, there corresponds a set of all objects a that possess this property.*

Armed with this powerful principle, Cantor developed tools for comparing infinite sets. He proposed that two sets (finite or infinite) be considered equinumerous if a one-to-one correspondence (a bijection) could be established between their elements. This led him to a stunning discovery: there are different "sizes" of infinity. Using his famous *diagonal argument*, he proved that the set of all subsets of the natural numbers (the continuum) is "more powerful" than the set of natural numbers itself. Thus was born the theory of transfinite cardinal and ordinal numbers—the arithmetic of the infinite.

It is this initial, intuitive, and extremely fruitful approach that constitutes the "paradise" which, in Hilbert's words, Cantor created for mathematicians. However, it soon became clear that this paradise had its flaws. The unrestricted application of the comprehension principle (any property defines a set) led to the discovery of logical paradoxes, the most famous of which is Russell's paradox (see below). It became evident that to save set theory from contradictions, it was necessary to restrict the ways of constructing sets by explicitly listing the permissible operations in the form of axioms. Zermelo took on this task.

In Section A2.3, we already introduced a number of set-theoretic notations for operations and predicates related to the concept of a set. In Section A2.7, we slightly expanded this apparatus by adding such important concepts as relations and functions. Now, our task is to provide, in the simplest possible form, a definition of the mathematical concept of a "set" and then to formulate a language and a theory that would correspond to this concept.

So, by a set in a broad sense, we understand the extension of a concept, i.e., any definable collection of entities. To write a set as a collection defined by some property $\mathcal{A}(a)$, the following comprehension notation is usually used:

$$\{x \mid \mathcal{A}[a/x]\},$$

which is read as: "the set of all x corresponding to the property $\mathcal{A}$". Logically, this comprehension construct has the defining property:

$$a \in \{x \mid \mathcal{A}[a/x]\} \leftrightarrow \mathcal{A}(a). \tag{B2.7}$$

In other words, the lexeme $\{x \mid \mathcal{A}[a/x]\}$ is a special kind of term constructed from a formula, which is not allowed in first-order languages. But on the other hand, the term $\{x \mid \mathcal{A}[a/x]\}$ itself can be regarded as syntactic sugar, whose definition is given at the meta-theory level. After all, in all formulas about sets, the term as such participates only as an operand of a predicate. Consequently, if we figure out how such a term can be expanded within the predicates of set theory, we will eliminate it from the language as a formally unnecessary lexeme.

What predicates are needed in set theory? In Section A2.3, several predicates were defined: $\in, =, \subseteq$. In fact, there are others, but here one can notice that they can all be reduced to a single basic predicate—that of *membership*. Indeed, $A \subseteq B$ is an abbreviation for the formula $\forall x\ ((x \in A) \to (x \in B))$ (every element of A is an element of B), and the equality of sets $A = B$ means that every element of A is an element of B and vice versa, i.e.,

$\forall x\ ((x \in A) \leftrightarrow (x \in B))$. It is most convenient to choose membership as the primitive predicate, since all others can be easily expressed through it. Thus, we arrive at the point that in the language of the minimalist set theory we want to construct, we can leave just one (binary) predicate: $\in$. By the way, we immediately add the definition $(a \notin b) \overset{\text{def}}{\leftrightarrow} \neg(a \in b)$. But in that case, if a formula $a \in \{x \mid \mathcal{A}[a/x]\}$ is written somewhere, we can write it much more concisely and, importantly, without using second-order terms, namely as $\mathcal{A}(a)$. If we encounter a formula of the form $\{x \mid \mathcal{A}[a/x]\} \in b$, we can replace it with the formula

$$\exists z((z \in b) \wedge \forall x\ (x \in z \leftrightarrow \mathcal{A}(x)))$$

i.e., instead of a single atomic formula, we write a conjunction of an atomic formula and a formula with a quantifier (instead of a term-based comprehension, a formula-based comprehension appears, which is allowed in predicate logic). This complication of the text can be seen as the price for getting rid of second-order constructions. After all, although the formula has become longer, it is now written in a simpler language. And the simpler the language, the easier it is to analyze its capabilities.

Getting rid of the term $\{x \mid \mathcal{A}[a/x]\}$ does not, however, save us from the paradoxes of set theory, in particular, Russell's paradox, which is as follows. Consider the set R defined by the formula $x \notin x$, i.e., $R \overset{\text{def}}{\rightleftharpoons} \{x \mid x \notin x\}$. In other words, the elements of the set R are those and only those sets x that are not elements of themselves (it is not crucial here whether cycle-sets, i.e., sets that are their own elements, exist at all). The question is—is it true or false that $R \in R$? As is easy to see, either assumption logically leads to its opposite. Indeed, let's try to eliminate the comprehension here according to the rules described above: since $R = \{x \mid x \notin x\}$, then $R \in R$ is equivalent to $R \notin R$ by rule (B2.7), which already yields a contradiction, since the formula $\mathcal{A} \leftrightarrow \neg\mathcal{A}$ is a tautological falsehood. In doing so, we have not eliminated the second-order term R. This happens because, since we consider R to be a set, we are entitled to substitute it for any variable ranging over sets, including in the formula $x \notin x$ which defines it. This gives rise to a so-called *self-referential definition*, where an object is defined, in part, through itself. It is this possibility for the variable x in the formula $x \notin x$ to range over the value of the set R itself, which is defined by this very formula, that creates a vicious circle.

This example is illustrative not only because of the contradiction, but also because we cannot correctly eliminate the comprehension term and switch to a first-order language due to the circularity of the definition. If we did not assume that $\{x \mid x \notin x\}$ is a correct term and, therefore, cannot be substituted for the variable in the formula $a \in a$, then there would be no paradox. Thus, it turns out that our rather liberal definition of a set as any definable collection leads to obvious problems. What is to be done?

At the beginning of the 20th century, several solutions were proposed, one of which belongs to Russell and is a theory of types, where each term of the form $\{x \mid \mathcal{A}[a/x]\}$ is a set of a higher type than the sets x used in its definition. Thus, the very notation $R \in R$ is impossible, since the elements of sets of a certain type cannot be sets of the same or a higher type. This construction turned out to be quite complex, and it also implicitly appealed to the concept of a natural number, which is required to compare the ranks of sets.

In the end, everyone agreed that instead of granting the right to be a set to any definable collection of objects, it is simpler to write a series of rules by which certain sets are allowed to be constructed, starting from some atomic sets that have no elements.[8] From here, we arrive at the main thesis of set theory: "Everything must be a set!". This should be understood to mean that in this theory, we have no other objects besides sets. Consequently, the formal theory is untyped, and in the atomic predicate $(a \in b)$, both operands are necessarily sets.

This is a strong commitment, as it now becomes difficult to form, for example, a set of pencils, birds, people, etc. However, from a physics perspective, all things are ultimately sets of elementary particles, albeit bound by some additional relations. So the question boils down to how to consider these elementary particles as sets. And here, a certain convention comes into play. Just as modern computers have character tables (mappings of letters to certain numbers), created arbitrarily once and for all, we can similarly consider that elementary particles are *encoded* by some arbitrarily chosen, fixed sets.

Moreover, we only need to take one atomic set—the empty set. More precisely, all atomic sets are empty by definition (they have no elements), but by virtue of the definition of equality, they all turn out to be pairwise equal, i.e., from the point of view of the set equality we discussed above, there is only one empty (atomic) set.

There are modifications of set theory with many atoms, whose equality is excluded by definition, since atoms are forbidden to appear on the right side of the membership predicate. However, such a theory is unnecessarily complex and brings nothing significantly new to the science, as it can be easily interpreted in classical set theory with a single atom.

[8] Much like in physics, everything is built from elementary particles in the Standard Model.

The existence of a single atomic set, however, does not relieve us of the need to associate different elementary particles with different sets. Therefore, elementary particles will have to be modeled by non-elementary sets, as long as we have a sufficient supply of them.

So, in the final analysis, we have the following requirements for the language and theory of sets:

1. a single sort of object—sets;
2. a single atomic predicate ($a \in b$), where both operands are any sets without restriction;
3. a single atomic set—the empty set;
4. all sets are defined according to a finite list of rules that prohibit the idea of self-reference;
5. there must be a potential infinity of sets.

To fulfill all these requirements, let's use the following conceptual model. We will imagine every set as a tree (or, if you like, as a folder in a file system that contains other folders, but not files!). The set itself is the root of the tree, and its elements are the immediate descendants (child nodes) of the root, which are also represented as trees. At the leaves of such a tree, there is always the empty set (see Fig. B2.1). Under this scheme, it seems there are many empty sets—as many as there are leaves. However, nothing prevents us from considering that "physically" there is only one leaf, and the tree is an expansion (or parse tree) of the set along the membership relation. Thus, different leaves are actually pointers or references to the same unique object—the empty set. The same can be said about other possible duplicate sets in our scheme. Note that if we consider the edges of the tree not as a direct embodiment of the $\in$ relation, but as references to the corresponding elements (similar to shortcuts to folders in a file system), then no problems with the uniqueness of each set will arise, despite the tree-like expansion suggesting otherwise.

Within this concept, all described objects are homogeneous—they are trees (branches of trees). There is also a single atomic object—a tree consisting of a single root vertex, and a single relation: edges-as-references. Problems of self-reference are also excluded here, because if we assume a situation like $x \in x$ or a more complex $x \in y \in z \in \cdots \in x$ (a membership cycle), then the tree of the set x would contain infinite paths. In principle, one could imagine this, but we will proceed from a finitary concept, and infinities will appear only as superstructures over finite structures. It is easy to see that the principles for constructing sets using trees are also limited to a certain finite set of

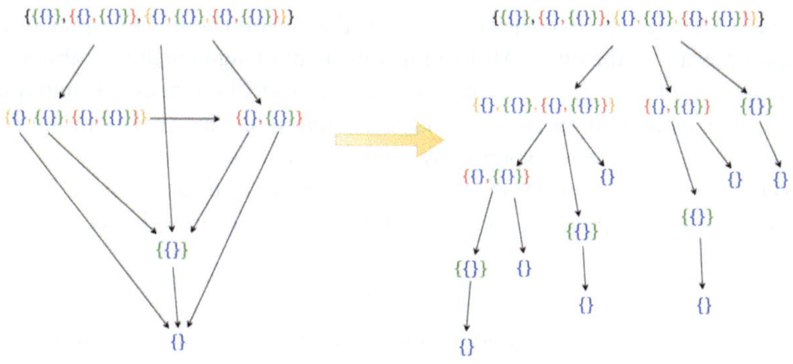

Fig. B2.1. A set and its tree expansion.

rules, which essentially boil down to a single one: the ability to link already prepared trees by connecting the leaves of some trees to the roots of others. Of course, such a graph-theoretic approach is far from the usual operations on sets we are accustomed to, but in this finitary concept of a "set," they can all be expressed through this single linking operation. Indeed, the union of two trees simply means identifying their roots, resulting in all elements of the unified sets becoming descendants of one common root instead of many. Intersection means selecting only the common root branches, and the set of all subsets (the power set) means, on the contrary, that instead of one root we add many roots, grouping the root branches in various combinations, and then link these new roots to one common root. And so on.

Note that tree-like objects have been following us throughout the book, as this concept underlies all regular constructions: lexemes, terms and formulas, grammar, logical inference. Trees also follow us in life whenever we use a file system, directories and hierarchies, study family trees, construct scenarios for the development of events, etc.

Now, keeping the tree-like concept of sets in mind, we can more easily imagine how the idea of a set should be formalized in a formal language. We will consider the tree-like concept to be an informal description of the notion of a "set".

Let's move on to the formal part. As for the language, it turns out to be extremely simple: it has no constants, no function symbols, but only one binary predicate $\in$ and two classes of variables—free and bound (here we will use both uppercase and lowercase Latin letters, possibly with numerical indices, still considering that free variables are at the beginning of the alphabet and bound ones are at the end). All formulas in such a language are combinations of variables (of a single type), the predicate $\in$, logical connectives, and quantifiers. It is astonishing that in such a primitive language (but with the right axioms), it is possible to express any mathematical theorem!

What rules for constructing sets might be useful to us as axioms? For this, let's recall the set operations we have already defined earlier in sections A2.3 and A2.7:

$$\cap, \cup, \setminus, \vartriangle, \times,$$

in addition, we need to be able to form sets and ordered tuples from given terms by listing them:

$$\{a, b, \dots, c\}, \quad \langle a, b, \dots, c\rangle.$$

Directly formalizing the informal rule of linking trees is quite problematic, but we can formalize some consequences of such linking, which were discussed just above. Furthermore, we will also need to add a definition of the predicate of set equality, as mathematics is unimaginable without equality.

It turns out that to construct almost all (finite) mathematical objects, it is sufficient to postulate a few simple things:

1. two sets are equal if and only if they consist of the same elements;
2. from any set, a definable part can be separated;
3. there must exist an empty set;
4. for every set, there must exist the set of all its subsets;
5. for every set, there must exist the union of all its element-sets;
6. from any two sets, a pair can be formed, both ordered and unordered.

When we write here that "can be formed" or "there must exist", we mean that there exists a set that is the result of such an operation.

Let's consider these ideas in more detail. Point 2 tells us that if we already have some set A, we can define a subset of it using a formula. In other words, one can form the set $\{x \mid (x \in A) \wedge \varphi(x)\}$, using the comprehension notation. There is one significant restriction in this comprehension: x are taken only from those already in A, not arbitrary sets as in Russell's paradox. Thus, we are, in a way, implementing the ancient Roman legal maxim: "He who is permitted to do the greater (A) is certainly permitted to do the lesser (a subset of A)". Formally, this idea is expressed by the **Axiom Schema of Separation**:

$$\forall X\ \exists Y\ \forall z((z \in Y) \leftrightarrow (z \in X) \wedge \varphi[a/z]), \tag{B2.8}$$

where, as usual, φ is an arbitrary formula of the language under consideration, not containing the variable Y. By the way, if the variable a is not in the formula φ, then the substitution $\varphi[a/z]$ does not change the formula, and in this case, we either get $X = Y$ when φ is true, or $Y = \emptyset$ when φ is false.

Point 3. Why must the empty set exist? Recall, the empty set is a set $\emptyset$ that has the property $\forall x\ (x \notin \emptyset)$. One could even introduce the constant $\emptyset$ into our language and add the definition of the empty set as an axiom (many authors do this). But we do not want to burden ourselves with even such a trifle.

The existence of the empty set is quite easy to explain (and in some sense, even to prove without direct postulation). The fact is, if we consider set theory to be consistent, then it must have a model (yes, a non-empty set, but in the meta-theory), which means the formula $\exists x\ \ (x = x)$ is true. But from this formula, along with the Axiom of Separation, the formula $\exists y\ \ (\forall z\ \ (z \notin y))$ is derivable by the rules of predicate logic, i.e., the existence of the empty set is derived. For this, it is sufficient to take y defined as a part of x using any false formula, for example, $z \neq z$, and then the existence of at least some set inevitably implies the existence of the empty set.

Point 4 is something like a declaration of equal opportunities. More precisely, if we have some set A, a huge number of possibilities open up before us to cut out a part of it (a subset), to reduce or collapse it to something smaller (much like the reduction of a wave function after a measurement is made). One can also think of it this way: you have 100 dollars, and with this money, you can potentially buy many different things, the total value of which is measured in trillions of dollars, but as soon as you buy something (make a measurement), your potential possibilities are realized into a single one. So, the set of all subsets is the collection of all potential possibilities, all such sets into which the original set A can collapse. Why should we deny this choice the right to exist? The choice itself, not just the results of each specific choice. It even turns out to be something like the Many-Worlds Interpretation of Everett's Quantum Mechanics, where all possibilities exist a priori, but in parallel worlds. And for us, they will exist in the set of all subsets, also called the **power set** of A. Using the previously defined predicate of set inclusion, the power set axiom is easy to formulate:

$$\forall X\ \exists Y\ \forall z\ ((z \in Y) \leftrightarrow (z \subseteq X)). \tag{B2.9}$$

Note that if z is a part of the set A, then in its power set, z is an element (or, as they also say, a point).

If point 4 generates sets in an upward direction with respect to the $\in$ relation (what was a subset becomes an element), then point 5 offers the reverse, downward process (what was an element becomes a subset's content), since, given a set A, we construct a new set in which we collect all the elements of all the elements of A.

If a set is thought of as a tree, where the root is the set itself and the elements are the immediate descendants of the root, then one can unwind this definition further, moving to the next layer of the tree (the elements of the elements), and so on. Principle number 5 allows us to discard the first layer (the direct descendants) in this tree and move directly from the root to the second layer.

It is difficult to come up with any physical-legal analogies here, except perhaps to recall some right of inheritance. But the persuasiveness of the graph theory narrative seems sufficient motivation for the **Axiom of Union**. Its formulation, by the way, is:

$$\forall X\ \exists Y\ \forall x\ ((x \in Y) \leftrightarrow \exists y\ ((x \in y) \wedge (y \in X))). \tag{B2.10}$$

Finally, point 6 implements the idea that from any two sets A and B, one can form the pair $\{A, B\}$, i.e., a set that can be defined by the comprehension lexeme $\{x \mid (x = A) \vee (x = B)\}$ or the **Axiom of Pairing**:

$$\forall x \forall y \exists Z\ \forall z\ ((z \in Z) \leftrightarrow (z = x) \vee (z = y)). \tag{B2.11}$$

Here we already have a fair suspicion: will we not run into some paradox like Russell's again? After all, we are simply forming an arbitrary collection, not cutting out a part of something already known.

But, firstly, as in the world of humans, we do not deny any individuals the right to form pairs (and larger coalitions, which will follow from the ability to form a pair), and secondly, the resulting set is not as all-encompassing as in the known paradoxes: we can not only specify its elements precisely but even list them, so to speak, by name. And everything that can be written in a finite number of steps, we accept as existing without question, otherwise, we could not even work with the simplest formal languages.

As for the ordered pair, we will simply express it through unordered pairs in a special (but permitted by the theory) way, and through it, we can then implement the direct product of sets. So the unordered pair of sets can be considered something more fundamental than these constructions.

However, as we will see later, the Axiom of Pairing and the Axiom of Separation can be derived from another, more general axiom, the motivation for which is also not unfounded, and in the presence of such an axiom, even the unordered pair of sets will become a secondary object.

Nevertheless, the question of whether self-reference will arise in the definition of a pair remains open. Indeed, nothing prevents us from imagining the equation $A = \{A, B\}$ with a circular relation $A \in A$. We do not approve of such things, and therefore we enlist a special axiom to combat them, one that postulates the well-foundedness of the $\in$ relation. It is called the **Axiom of Regularity** and it states that in every non-empty set X, there is a minimal element with respect to membership (an element $x \in X$ such that any y "smaller" than it with respect to membership is not an element of X). On diagram B2.1 this property is expressed by the fact that not all two-step root paths can be shortened (on the picture this is the path $\to \{\{\}\} \to \{\}$).

The Axiom of Regularity prohibits cycles of the form $a \in b \in c \in \cdots \in d \in a$, which is in perfect agreement with the tree-like concept presented above, from which the prohibition of self-reference obviously follows, since all higher nodes of the tree cannot coincide with lower ones if we do not allow infinite looped trees (fractal structures). Formally, the Axiom of Regularity postulates that any descending chain with respect to $\in$ must terminate, which corresponds exactly to the requirement of the finiteness of all paths in the tree-like expansion from the root to the leaves. In fact, the essence and purpose of the Axiom of Regularity is to ensure that the entire universe of sets is very well-ordered, i.e., that not only does each set have the form of some regular tree, but the entire universe is essentially such a single tree, growing from the empty set. The Axiom of Regularity will later lead us to the definition of the von Neumann cumulative universes, which show the strict hierarchical nature of the class of all sets. But we will deal with this at level C1 (see section C1.3).

Thus, we have almost completely formulated and explained the axioms of set theory, although we still leave out the possibilities of forming the set-theoretic constructions familiar to us: $\cup$, $\cap$, $\setminus$, $\times$, etc. Let's now move directly to the practice of using the aforementioned axioms.

B2.3 Hereditarily Finite Sets

Here we will construct a formal theory of sets that describes so-called finite mathematics, or the mathematics of the finite. In this theory, we essentially implement precisely the idea of representing sets as finite graphs.[9]

B2.3.1 Language and Axioms

The language of the theory, as already noted, is built upon a single predicate symbol $\in$. However, we consider set theory as a theory with equality (and, consequently, we include the standard axioms of equality from predicate logic), so the predicate symbol $=$ is also present here. It can, however, be considered secondary to $\in$ by virtue of the axiom of extensionality. It can be replaced in all formulas using this axiom, but then both the formulation and the very presence of the axiom of congruence would cease to be obvious. So, for the convenience of notation and consistency with the ideas of predicate logic with equality, we will consider that the language of set theory is constructed from two atomic binary predicates: $\in$ and $=$. As variables, we allow the use of lowercase and uppercase Latin letters (as usual, separating free and bound variables). Furthermore, we use all the previously introduced abbreviations for first-order formulas, namely:

$$\begin{gathered} A \subseteq B \overset{\text{def}}{\leftrightarrow} \forall x\ ((x \in A) \to (x \in B)), \\ (a \notin b) \overset{\text{def}}{\leftrightarrow} \neg(a \in b), \quad (a \neq b) \overset{\text{def}}{\leftrightarrow} \neg(a = b), \\ \exists x \in A :\ \varphi(x) \overset{\text{def}}{\leftrightarrow} \exists x\ ((x \in A) \wedge \varphi(x)), \\ \forall x \in A :\ \varphi(x) \overset{\text{def}}{\leftrightarrow} \forall x\ ((x \in A) \to \varphi(x)). \end{gathered} \tag{B2.12}$$

The axioms:

HF0 Negation of the Axiom of Infinity.

HF1 Congruence: $(a = b) \to ((a \in M) \leftrightarrow (b \in M))$.

HF2 Extensionality: $(A = B) \leftrightarrow (\forall x\ ((x \in A) \leftrightarrow (x \in B)))$.

HF3 Separation [for a formula φ]: $\exists X\ \forall x\ ((x \in X) \leftrightarrow (x \in A) \wedge \varphi(x, \vec{p}))$.

---

[9] The notation for the theory, HF, comes from the English term *hereditarily finite sets*.

HF4 Power Set: $\exists P\ \forall x\ ((x \in P) \leftrightarrow (x \subseteq A))$.

HF5 Union: $\exists U\ \forall x\ ((x \in U) \leftrightarrow (\exists a \in A :\ x \in a))$.

HF6 Elementary Sets: $\exists X\ \forall y\ (y \notin X)$ (empty set);
$\exists X\ \forall x\ ((x \in X) \leftrightarrow (x = a) \vee (x = b))$ (pair).

HF7 Regularity: $(\exists x \in A) \rightarrow \exists X \in A :\ \forall y \in X :\ (y \notin A)$.

Here, as usual, we present the formulas of the axioms without the enclosing universal quantifiers for ease of perception. We omit the formulation of the axiom of infinity (and its negation) for now, as we will dedicate a separate section to it. Moreover, it could be omitted from the list of axioms for the theory HF altogether, but then this theory would admit more complex models than we wish to imagine in the context of considering finite mathematics. Furthermore, HF with the negation of the axiom of infinity is mutually interpretable with the arithmetic PA, which is also important for a better understanding of the foundations of finite mathematics.

On the other hand, in addition to the previously mentioned axioms, we have introduced here the axiom of congruence of equality, which is usually implied for the equality predicate automatically. However, it is necessary to single it out, as it has the same meaning as the axiom PA2 in arithmetic, which we discussed earlier. Axiom HF1 expresses only congruence for the left argument of the predicate $\in$, since congruence for the right argument is already embedded in the axiom of extensionality, and general congruence (B1.9) can be proven at the meta-theory level by induction on the complexity of formulas in the language of set theory.

Following the canons of constructing a theory in a first-order language, we could split the axiom of extensionality into two separate axioms:

$$(A = B) \rightarrow (\forall x\ ((x \in A) \leftrightarrow (x \in B))),$$
$$(\forall x\ ((x \in A) \leftrightarrow (x \in B))) \rightarrow (A = B),$$

then the first of them would be the missing part of the axiom of congruence (for the second argument of the predicate $\in$), and the second would be the axiom proper, linking equality and membership. But our formulation clearly demonstrates how the equality of sets is expressed through membership (which is essentially the definition of equality in set theory).

Next follows the axiom schema of separation, which depends on the formula φ, where $\vec{p}$ denotes a set of parameters, i.e., free variables $p_1, \ldots, p_n$, since the separation of a part of a set can be parameterized. Then come the familiar axioms of power set and union, then the axiom postulating the existence of the empty set and the pair, and finally, the axiom of regularity, which states that every non-empty set has a minimal element with respect to membership.

As already noted, the existence of the empty set follows directly from the assumption of the theory's consistency and the axiom of separation, so this requirement can be omitted. But, again, for historical continuity, we retain this axiom. The axiom of pairing is a direct constructive way to create new sets not as parts of already defined ones, i.e., partially (in a finitary context) realizing the principle of an arbitrary collection. As in the case of Peano arithmetic, for the system HF, one needs to derive the basic set-theoretic relations familiar to us from childhood.

Theorem B2.5. *The following theorems are derivable in the theory* HF *(with universal closure of the formulas understood):*

1° *equality is reflexive, symmetric, and transitive;*

2° $\exists X\, \forall x\, ((x \in X) \leftrightarrow (x \in A) \wedge (x \in B))$;

3° $\exists X\, \forall x\, ((x \in X) \leftrightarrow (x \in A) \vee (x \in B))$;

4° $\exists X\, \forall x\, ((x \in X) \leftrightarrow (x \in A) \wedge (x \notin B))$;

5° $\exists X\, \forall x\, ((x \in X) \leftrightarrow ((x \in A) \wedge (x \notin B)) \vee ((x \notin A) \wedge (x \in B)))$;

6° $\exists X\, \forall x\, ((x \in X) \leftrightarrow (x = a_1) \vee (x = a_2) \vee \cdots \vee (x = a_n))$;

7° *there exists an ordered pair and the Cartesian product of sets.*

Proof. The first property states that our equality satisfies the standard axioms of equality:

$$a = a, \quad (a = b) \to (b = a), \quad (a = b) \wedge (b = c) \to (a = c).$$

This follows directly from extensionality in pure predicate logic.

The second property is obtained directly from the axiom of separation by substituting $\varphi \equiv (x \in B)$, where B is a parameter. The next property is more difficult; it requires the use of two axioms at once—pairing and union. To make their use clear, it is better to resort to an extended signature of the language, enriching it with new binary function symbols $\{\}, \cup, \cap, \setminus, \triangle, \langle\rangle, \times$, unary function symbols $\mathcal{P}, \bigcup$, and the constant $\emptyset$.

A language with such an extended signature will require new axioms that define the rules for using the new symbols. However, all theorems of the extended language can be translated back into the original language, where they will also be theorems. Such an extension of a language and theory is called *conservative*. This technique is often used to improve the visual perception of the language and theorems without changing the substance. But for this, a certain amount of syntactic work must be done, which we will now undertake.

So, first, we will introduce re-designations for the function symbols to move away from the standard Polish notation of predicate calculus, such as $F(a, b)$, namely, we set:

$$\{a, b\} \overset{\text{def}}{\rightleftharpoons} \{\}(a, b), \quad a \cup b \overset{\text{def}}{\rightleftharpoons} \cup(a, b), \quad a \cap b \overset{\text{def}}{\rightleftharpoons} \cap(a, b), \tag{B2.13}$$

$$a \setminus b \overset{\text{def}}{\rightleftharpoons} \setminus(a, b), \quad a \triangle b \overset{\text{def}}{\rightleftharpoons} \triangle(a, b), \tag{B2.14}$$

$$\langle a, b\rangle \overset{\text{def}}{\rightleftharpoons} \langle\rangle(a, b), \quad a \times b \overset{\text{def}}{\rightleftharpoons} \times(a, b). \tag{B2.15}$$

Let us emphasize once again that these are not axioms, but merely syntactic re-designations at the meta-theory level; they are still not present in the formal language itself.

HF8 Axioms for the symbols $\{\}, \cup, \cap, \setminus, \triangle, \langle\rangle, \times, \mathcal{P}, \bigcup, \emptyset$:

$$\begin{aligned}
c \in \{a, b\} &\overset{\text{def}}{\leftrightarrow} (c = a \vee c = b),\\
c \in a \setminus b &\overset{\text{def}}{\leftrightarrow} (c \in a \wedge c \notin b),\\
c \in a \cup b &\overset{\text{def}}{\leftrightarrow} (c \in a \vee c \in b),\\
c \in a \triangle b &\overset{\text{def}}{\leftrightarrow} (c \in a \setminus b) \vee (c \in b \setminus a),\\
c \in a \cap b &\overset{\text{def}}{\leftrightarrow} (c \in a \wedge c \in b),\\
c \in \langle a, b\rangle &\overset{\text{def}}{\leftrightarrow} (c = \{a, a\} \vee c = \{a, b\}).
\end{aligned}$$

The remaining symbols are governed by the following axioms:

$$\begin{aligned}
c \in \mathcal{P}(a) &\overset{\text{def}}{\leftrightarrow} c \subseteq a,\\
c \in \bigcup a &\overset{\text{def}}{\leftrightarrow} \exists x \in a :\ c \in x,\\
c &\notin \emptyset,\\
c \in A \times B &\overset{\text{def}}{\leftrightarrow} \exists x \in A : \exists y \in B :\ c = \langle x, y\rangle.
\end{aligned}$$

Note that here we have used the previously undefined predicates $c = \{a, b\}$ and $c = \langle x, y\rangle$. Syntactically, they are, of course, well-formed, but how do we determine their truth or falsity? It turns out this is easy to do. From the existing axioms, we can derive properties for the new terms when they are substituted into the equality predicate and into the membership predicate on the left (and not on the right, as in the definition).

Theorem B2.6. *Let the term $t(\vec{p})$ be defined by the formula $\varphi(a, \vec{p})$, i.e.,*

$$a \in t(\vec{p}) \overset{def}{\leftrightarrow} \varphi(a, \vec{p}), \tag{B2.16}$$

where $\vec{p}$ is a set of parameters (free variables). Then the following are theorems of the theory HF*:*

$$C = t(\vec{p}) \leftrightarrow \forall x\ ((x \in C) \leftrightarrow \varphi(x, \vec{p})). \tag{B2.17}$$

$$(t(\vec{p}) \in A) \leftrightarrow \exists x \in A:\ (x = t(\vec{p})), \tag{B2.18}$$

Proof. To prove the first equivalence, we first use extensionality

$$C = t(\vec{p}) \leftrightarrow \forall x\ ((x \in C) \leftrightarrow (x \in t(\vec{p}))),$$

and then replace the formula under the quantifier with an equivalent one, using (B2.16). In fact, the reverse transition from the first theorem to the definition is also valid. To do this, it is sufficient to substitute the term $t(\vec{p})$ for the variable C in (B2.17) and use the reflexivity of equality, as well as the axiom of predicate logic for the universal quantifier.

To prove the second equivalence, we will use Theorem D2.6 (Part II), where we substitute the formula $(a \in A)$ for $\varphi(a)$, and the term $t(\vec{p})$ for t. □

From theorem (B2.17), we can notice that if we substitute the comprehension lexeme $\{x \mid \varphi(x, \vec{p})\}$ for C, then on the left we get the familiar *direct definition* of the term $t(\vec{p})$ through this comprehension, and on the right—exactly the definition of the comprehension term (B2.7), which, in general, allows us to introduce definitions of new function symbols using comprehension. But the problem with such a definition is precisely that it is not a first-order formula, and therefore cannot be perceived as an axiom in the language of set theory (neither in the original nor in the extended one).

Therefore, if somewhere in the text we encounter a direct definition of the form $t(\vec{p}) \overset{\text{def}}{\rightleftharpoons} \{x \mid \varphi(x, \vec{p})\}$ with a first-order formula φ, we must immediately mentally convert it into the first-order axiom $a \in t(\vec{p}) \overset{\text{def}}{\leftrightarrow} \varphi(a, \vec{p})$, thereby legitimizing this direct definition. In the group of axioms HF8, the axiom for the empty set symbol stands apart, as it does not have the form (B2.16), but we can easily bring it to this form. The axiom $c \notin \emptyset$ means that the formula $c \in \emptyset$ must be false, hence it can be rewritten as

$$c \in \emptyset \overset{\text{def}}{\leftrightarrow} \varphi(c) \wedge \neg\varphi(c),$$

where φ is any formula of the original language of the theory HF. And from this, we obtain a direct definition of the empty set through comprehension: $\emptyset \overset{\text{def}}{\rightleftharpoons} \{x \mid \varphi(x) \wedge \neg\varphi(x)\}$. Note also that this definition does not depend on the choice of the formula φ due to the laws of propositional logic. If we want to use logical constants in our language, the definition can be simplified to $\emptyset \overset{\text{def}}{\rightleftharpoons} \{x \mid \bot\}$.

It is worth noting that when extending the signature of the language, we not only introduced definitions for new symbols but also extended the action of the congruence axioms (B1.9) to them. It is always important to remember this when extending the language of a theory with equality!

We can apply another abbreviation, setting $\{a\} \overset{\text{def}}{\rightleftharpoons} \{a, a\}$, so that with this abbreviation, the presented axioms, and the transitivity of equality, we can obtain a direct definition of the ordered pair according to Kuratowski:

$$\langle a, b\rangle = \{\{a\}, \{a, b\}\}.$$

Note also that often $\{\}$ is written instead of $\emptyset$, effectively treating this function symbol as 0-ary.

Since we have extended the language with new function symbols, terms are now not only variables but also all possible compositions of these new symbols and variables. Therefore, it would be necessary to stipulate the behavior of the newly defined function symbols when different terms are substituted into them. Fortunately, the given axioms are sufficient, as we have the axioms of predicate calculus, which allow any free variables to be replaced by any terms. So, for example, having the formula $c \in \{a, b\}$ and two arbitrary terms t_1, t_2, we can substitute them into the axiom and get a definition for $c \in \{t_1, t_2\}$, which reduces the use of the more complex term $\{t_1, t_2\}$ to equalities with two less complex terms: $c = t_1$ and $c = t_2$. And since the parse tree of any term is finite, it is clear that we can unwind the original formula completely, i.e., down to formulas with variables, for which our axioms HF8 work.

Thus, it turns out that all formulas in the richer language of set theory with a signature including the symbols $\{\}, \cup, \cap, \setminus, \vartriangle, \langle\rangle, \times, \mathcal{P}, \bigcup, \emptyset$ can be uniquely translated into the poorer language without these symbols.

Using such richness, we can rewrite some axioms of the theory HF in a more pleasant way:

HF4 Power Set: $\exists P\ (P = \mathcal{P}(a))$.

HF5 Union: $\exists U\ (U = \bigcup a)$.

HF6 Elementary Sets: $\exists X\ (X = \emptyset)$; $\exists X\ (X = \{a, b\})$.

By the way, a formula of the form $\exists x\ (x = t)$, where t is an arbitrary term, is usually abbreviated as $\exists t$. This will make the notation of the axioms even simpler.

We can generalize the definition of the pair $\{a, b\}$ to the definition of an arbitrary finite set $\{a, b, \ldots, c\}$, where the ellipsis is a meta-symbol and means that an arbitrary sequence of variables of finite length is located here. In other words, we would need to introduce separate function symbols $\{\}(\ldots)$ for each finite arity and write axioms for each of them, similar to those presented above. But we will not do this, as such terms are more conveniently defined using pairs and unions through direct definition, for example, like this:

$$\{a, b, c\} \overset{\text{def}}{\rightleftharpoons} \{a, b\} \cup \{c\}, \quad \{a, b, c, d, e\} = \bigcup \{\{a, b, c\}, \{d, e\}\}. \quad \text{(B2.19)}$$

Simultaneously, this allows us to establish the existence of such sets by referring to the axioms.

Finally, we can return to the theorem stated above and rewrite some of its points more simply:

2° $\exists(A \cap B)$;

3° $\exists(A \cup B)$;

4° $\exists(A \setminus B)$;

5° $\exists(A \triangle B)$;

6° $\exists\{a_1, \ldots, a_n\}$;

7° $\exists\langle a, b\rangle, \exists(A \times B)$.

The existence of $A \cap B$ follows from the axiom of separation, as already noted.

The existence of $A \cup B$ follows from the fact that $A \cup B = \bigcup\{A, B\}$, i.e., the set $A \cup B$ is a composite term of those that we have already defined and whose existence is guaranteed by the axioms of pairing and union.

The existence of $A \setminus B$ is a direct consequence of the axiom of separation with the formula $\varphi(x, B) \equiv (x \notin B)$.

The existence of $A \triangle B$ follows from the fact that $A \triangle B = (A \setminus B) \cup (B \setminus A)$.

The existence of an arbitrary finite set $\{a_1, \ldots, a_n\}$ follows from its construction, which was mentioned above, and the other axioms.

The existence of the ordered pair $\langle a, b\rangle$ follows from its Kuratowski definition.

We leave the question of why it satisfies the property of an ordered pair

$$(\langle a, b\rangle = \langle c, d\rangle) \leftrightarrow (a = c) \wedge (b = d),$$

as an exercise for the reader. As for the Cartesian product $A \times B$, its existence follows from the fact that it can be defined as a part of the set $\mathcal{P}(\mathcal{P}(A \cup B))$, which exists by virtue of the axioms. □

B2.3.2 The Bracket Model

In the extended language of set theory (taking into account the use of expressions like $\{a, b, \ldots, c\}$), which was defined above, it is permissible to describe sets with terms whose notation contains only brackets and commas:

$$\{\}, \quad \{\{\}\}, \quad \{\{\}, \{\{\}\}\}, \quad \{\{\{\{\}\}\}, \{\{\}\}, \{\}\}$$

Sets that can be described by such terms are called **hereditarily finite**. Their general mathematical (recursive) definition is as follows: hereditarily finite sets are finite sets whose elements are also hereditarily finite.

The main idea of these sets is that they are finite objects defined recursively. Moreover, it is not difficult to write a decision algorithm in any universal programming language that, given any bracket notation, can determine in a finite number of steps whether the given notation is a correct representation of a set or not.

The rules for constructing bracket notations in Backus-Naur Form are as follows. We use two non-terminal symbols, *list* and *set*, and the terminal symbols $\{, \}, \Lambda$. Then

P1 *list* $\rightarrow$ $\Lambda \mid$ *list set*;

P2 *set* $\rightarrow$ $\{$*list*$\}$,

where the symbol Λ denotes the empty string. Thus, a list is either an empty string or a sequence of sets, and a set is a list enclosed in curly brackets. This results in set notations like

$$\{\Lambda\}, \quad \{\Lambda\{\}\}, \quad \{\Lambda\{\Lambda\{\}\}\}, \quad \{\Lambda\{\Lambda\{\}\}\Lambda\{\}\Lambda\{\}\},$$

which can be easily converted to a more familiar form by removing Λ and replacing all occurrences of the character pair "}{" with the three-character string "},{":

$$\{\},\quad \{\{\}\},\quad \{\{\{\}\}\},\quad \{\{\{\}\},\{\},\{\}\}.$$

Having defined the universe of bracket notations, we interpret them as sets. For any notation s, let $s_1, \dots, s_n$ be the list of elements (substrings) contained within the outer brackets of s. Similarly, let t contain elements $t_1, \dots, t_m$. Then we define the predicates $\in$ and $=$ by simultaneous recursion on the nesting depth (or length of the string):

$$s = t \iff (\forall i \leqslant n\ \exists j \leqslant m : s_i = t_j) \wedge (\forall j \leqslant m\ \exists i \leqslant n : t_j = s_i), \tag{B2.20}$$

$$s \in t \iff \exists j \leqslant m : s = t_j. \tag{B2.21}$$

In this definition, the quantifiers range over the finite set of indices of the immediate sub-elements. This definition reduces the equality of sets to the equality of their elements, which are strictly smaller strings, ensuring the termination of the recursion.

Why do the axioms HF0–7 hold under such definitions of $\in$, $=$? The absence of infinite sets follows automatically from the finitary nature of the bracket notation construction itself. Extensionality holds obviously, as it is embedded in the definition of equality (B2.20). From this, in particular, it follows that equality is reflexive, symmetric, and transitive. Congruence follows obviously from the fact that if $a = b$ and $a \in M$, then the notation M contains a substring t equal to a, and it will also be equal to b, hence $b \in M$ by virtue of (B2.21).

The axioms of power set and separation hold because, in the bracket model, all subsets of any set M exist, regardless of whether they are defined by formulas or not. We can write an algorithm that lists all possible subsets of M, collects them into a list, and wraps them in curly brackets, thereby constructing $\mathcal{P}(M)$. After that, it will not be difficult to find among them the subset corresponding to the formula φ, i.e., to perform separation. The construction of the union $\bigcup M$ is even simpler: one just needs to remove the curly brackets that form the sets in the list of set M according to rule P2 (i.e., the second-level nested brackets), which will result in a set whose list is the concatenation of the lists of the elements of set M.

Note that our bracket model perfectly embodies the tree-like principle of constructing sets mentioned in the previous section. In fact, the syntactic parse trees of the bracket notations realize exactly the same set trees. The empty set, as already noted, is written as the string {} (in our original formalism, it was $\{\Lambda\}$). The pair $\{A, B\}$ exists directly by virtue of the construction of bracket notations. And the axiom of regularity is checked by induction on the nesting depth of the brackets.

Thus, the bracket notations described above represent a *model* of sets that obey the axioms of the theory HF in its original language. Such a model is called the *standard model of hereditarily finite sets.*[10]

Moreover, among all bracket notations, we can distinguish *canonical* ones, which will have certain uniqueness properties, namely: each canonical notation will have its own code (a natural number), and the elements of the generating list of a canonical notation will be arranged in descending order of this code and will not be duplicated. Suppose that a notation S has the hereditary property of uniqueness of elements. This means that in its list of elements $L(S) = \{s_1, \ldots, s_n\}$, all s_i are pairwise distinct (if $i \neq j$, then $s_i \neq s_j$ according to formula (B2.20)), and this property holds recursively for all elements s_i. Then, following Ackermann, we set:

$$\mathrm{Code}(S) = \sum_{s \in S} 2^{\mathrm{Code}(s)}, \quad \mathrm{Code}(\{\}) = 0. \tag{B2.22}$$

With such a coding, it turns out that $a \in b$ if and only if the bit at the position $\mathrm{Code}(a)$ in the binary representation of the code $\mathrm{Code}(b)$ is 1.

It is not difficult to prove (by induction on the construction of the set) that different sets correspond to different codes, and by virtue of the property of uniqueness of elements, all terms in this sum will be different. Moreover, given an arbitrary natural number N, we can find a set with a code equal to N. To do this, the number N must be represented as a sum of powers of two (written in the binary system), and then the same must be done with the powers of two, and so on, until we reach zero powers. After which, we write {} instead of zeros, s instead of 2^s, and put commas instead of plus signs and wrap the terms in curly brackets (see Fig. B2.2).

These algorithms (for coding and decoding notations) converge for purely arithmetical reasons. Finally, to decide in what order to arrange the elements of a set in such a notation, we agree that they are listed in descending order of their codes. Thus, we define the canonical notations of hereditarily finite sets and establish a one-to-one correspondence between them and the natural

[10] More precisely, the standard model is the universe V_ω, but its sets are exactly those that correspond to our bracket notations.

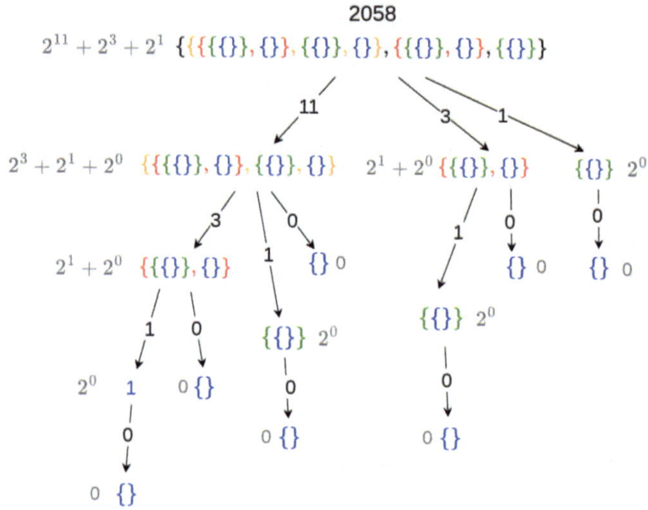

Fig. B2.2. Coding of a canonical set notation.

numbers. The possibility of mutual interpretation of the theory of hereditarily finite sets and Peano arithmetic is based on this fact (see Part II, section D4.5).

It is interesting to note that there exists a completely different, algebraic approach to encoding finite sets of natural numbers. Any finite set $S \subseteq \mathbb{N}$ can be represented by its characteristic function, which is essentially a finite binary sequence (a word). To ensure unambiguous encoding, such a sequence must end with a one, corresponding to the maximal element of the set; for example, the set $\{2, 5\}$ is represented by the word 00101, and not 001010, which contains the same information but would lead to a different matrix. On the other hand, it is well known that the matrix monoid $SL_2(\mathbb{N})$ with non-negative integer elements and a determinant equal to 1 is a free semigroup with two generators:

$$A = \begin{pmatrix} 1 & 1 \\ 0 & 1 \end{pmatrix}, \quad B = \begin{pmatrix} 1 & 0 \\ 1 & 1 \end{pmatrix}.$$

This means that any matrix in $SL_2(\mathbb{N})$ is uniquely represented as a product of these two matrices. Thus, a bijection is established between binary words ending with a one (and the empty word) and the matrices, which allows one to encode any finite set of natural numbers in the form of a single 2×2

matrix. This example vividly illustrates how deep results from group theory and number theory can provide unexpected tools for solving problems in the foundations of mathematics.

The bracket model we have constructed, or the standard model of the theory of hereditarily finite sets in the language of bracket notations, illustrates the possibility of implementation and thus the consistency of the HF axioms. Although HF itself only talks about finite sets, its standard model relies on a stronger theory that accepts actual infinity (since the model itself is infinite). This points to the subtle interaction between finitary reasoning (which HF seems to embody) and the infinitary assumptions necessary to study its metatheory.

Finally, it is worth noting that in the universe of hereditarily finite sets, analogues of the axiom of replacement and the axiom of choice are provable for the finite case, but their full power is manifested only in theories with infinite sets. This fact indicates that in such a rather primitive (primitive recursive) construction, we are able to model almost the entire arsenal of ordinary ZFC theory, with the exception of the axiom of infinity. But since these models themselves are obviously infinite, or more precisely, countable, there is practically no room for doubt about the consistency of at least the countable fragment of ZFC (where the action of the power set axiom is limited only to finite sets). In fact, all our reasoning about hereditarily finite sets and the detailed analysis of the theory HF are aimed precisely at ensuring the sufficient reliability of this theory, so that it can later be considered our meta-theory, within which Peano arithmetic, recursive functions, and, consequently, the coding of formal languages and derivability can be realized.

Thus, the circle closes, leading from the first timid attempts to describe naive set theory through the thickets of formalizing first-order logic to the definition of already quite precise and mathematically rich constructions, such as Peano arithmetic PA and the theory HF.

From this moment on, we can consider that we are standing firmly on the foundation of the theory of formal systems, and we can give precise mathematical definitions to all those objects and inferences that we will encounter in the future.

Note additionally that axioms HF1–6 (excluding regularity), which we have cited above, were invented and published by Ernst Zermelo as early as 1908. He immediately added the axiom of infinity and the axiom of choice to them. Such a theory is usually denoted by Z after the first letter of the author's surname. Abraham Fraenkel and Thoralf Skolem introduced the axiom of regularity and the axiom of replacement in the early 1920s, and now Zermelo's theory without the axiom of choice but with the axioms of regularity and replacement is denoted by ZF, and if we also keep the axiom of choice in it, then ZFC.

B2.4 Self-Check Questions

Q1 What is the main goal and what are the key ideas in the formal axiomatization of natural numbers? [p. 129]

Q2 Briefly describe the language of PA. What is the conceptual meaning of the axioms defining the successor operation, zero, addition, and multiplication? [p. 131]

Q3 Explain the essence of the principle of mathematical induction in PA. Why is it represented as an axiom schema? How are other forms of induction, such as complete induction or the well-foundedness principle, conceptually related to it? [p. 138]

Q4 How is the order relation usually defined in Peano arithmetic? What is the conceptual difference and significance between the weak and strong definitions of inequality? [p. 136]

Q5 Explain the difference between the truth of a statement for all natural numbers (in the metatheory) and its formal provability as a universal statement ($\forall x\ \varphi(x)$) within PA. What conclusions about the proof strength of PA can be drawn from this? [p. 141]

Q6 Why does the intuitive understanding of the term "set" lead to problems (e.g., Russell's paradox)? What general principles underlie the axiomatic approach to defining sets? [p. 144]

Q7 Explain the conceptual purpose of the main axioms of the theory HF that allow for the construction of sets. How do they contribute to building the universe of hereditarily finite sets? [p. 154]

Q8 Describe the conceptual idea of "hereditarily finite sets." What does it mean for a set to be hereditarily finite, and how is this related to the tree-like representation of sets? [p. 161]

Q9 What is the conceptual connection or mutual interpretability between HF and PA? [p. 164]

Exercises

B2.1 Express the statement "Every natural number is either zero or a successor" in the language of PA.

B2.2 Formalize the definition of a "square number" using only multiplication.

B2.3 Express the following arithmetic concepts in the language of PA:

- "x is an even number".
- "x is a divisor of y" ($x \mid y$).
- "x is a prime number".
- The Goldbach conjecture: "Every even integer greater than 2 is the sum of two primes".

B2.4 Write a definition of a prime number using *only* the divisibility predicate ($x \mid y$) and the predicate of being invertible (a unit) $Inv(x) \overset{\text{def}}{\leftrightarrow} \exists y(xy = 1)$.

B2.5 Express the "Twin Prime Conjecture": there are infinitely many pairs of primes p, q such that $q = S(S(p))$.

B2.6 Formalize the statement: "For any two numbers, there exists a common multiple".

B2.7 Write a formula for the "Greatest Common Divisor" relation $G(a, b, d)$ (d is the gcd of a and b).

B2.8 Using the axioms of PA, prove the transitivity of the divisibility relation:

$$\forall x, y, z\ ((x \mid y) \wedge (y \mid z) \rightarrow (x \mid z)).$$

B2.9 Using the induction schema, prove that $\sum_{i=0}^{n} i = \frac{n(n+1)}{2}$.

B2.10 Prove by induction that for every $n \geqslant 0$, the number $n^3 - n$ is divisible by 3.

B2.11 Prove Bernoulli's inequality $(1 + x)^n \geqslant 1 + nx$ for natural n and real $x > -1$.

B2.12 Prove by induction that the sum of the first n odd numbers equals n^2.

B2.13 Using complete induction, prove that every natural number $n > 1$ can be represented as a product of primes.

B2.14 Using the principle of infinite descent, prove that $\sqrt{2}$ is irrational (assume $a^2 = 2b^2$ leads to a smaller solution).

B2.15 Draw the parse tree for the bracket notation of the von Neumann ordinal $3 = \{\emptyset, \{\emptyset\}, \{\emptyset, \{\emptyset\}\}\}$.

B2.16 Give the recursive definition of the rank of a hereditarily finite set.

B2.17 Let $\sim$ be an equivalence relation on set A. Express the formula defining the predicate $(x \in A/_{\sim})$ for a factor.

B2.18 In the theory HF, write a formula expressing the fact that a set A is an ordered pair $\langle x, y\rangle$ (using Kuratowski's definition).

B2.19 Let $n \in \mathbb{N}, n > 1$. Prove that the relation $a \equiv b \pmod{n}$ is a congruence with respect to addition.

B2.20 Prove that the relation $a \equiv b \pmod{n}$ is a congruence with respect to multiplication.

B2.21 Calculate the Ackermann code for

1. the empty set $\emptyset$.
2. the singleton $\{\emptyset\}$.
3. the set $\{\{\emptyset\}\}$.
4. the set $\{\emptyset, \{\emptyset\}\}$.
5. the ordered pair $\langle\emptyset, \emptyset\rangle$ (using Kuratowski's definition).

B2.22 Determine the hereditarily finite set encoded by the number a) 11; b) 15.

Level C1
A Mathematician's Paradise

No one shall expel us from the Paradise that Cantor has created for us.

— David Hilbert

C1.1 Zermelo-Fraenkel Set Theory

C1.1.1 The ZFC Axiom System

Ernst Zermelo

So, we proceed to the construction of the most fundamental mathematical theory—Zermelo-Fraenkel set theory. In its standard version, the language of this theory includes only variables (free and bound) and two binary predicates, $\in$ and $=$. However, we will immediately extend it, just as we did for the theory HF, by adding the binary function symbols $\cup, \cap, \setminus, \vartriangle, \langle\rangle, \times$, the unary function symbols $\mathcal{P}, \bigcup$, the constant $\emptyset$, and the schema of function symbols $\{\dots\}$ with finite arities for writing down finite sets.

Naturally, we adopt the same abbreviations for formulas as in (B2.12) and for function symbols as in (B2.13)–(B2.15), as well as the definitions HF8 and (B2.19) for the function symbols. Furthermore, let us introduce several new quantifiers:

N. I. Kazimirov, *Mathematics as a Foreign Language*, Mathematics in Mind,
https://doi.org/10.1007/978-3-032-30721-7_5

$$\begin{aligned}
\exists! x\ \varphi[a/x] &\overset{\text{def}}{\leftrightarrow} \exists x\ \varphi[a/x] \wedge \forall y\ (\varphi[a/y] \to (y = x)),\\
\exists! x \in A :\ \varphi[a/x] &\overset{\text{def}}{\leftrightarrow} \exists x \in A :\ \varphi[a/x] \wedge \forall y \in A :\ (\varphi[a/y] \to (y = x)),\\
\exists x \subseteq A :\ \varphi[a/x] &\overset{\text{def}}{\leftrightarrow} \exists x :\ (x \subseteq A) \wedge \varphi[a/x],\\
\forall x \subseteq A :\ \varphi[a/x] &\overset{\text{def}}{\leftrightarrow} \forall x :\ (x \subseteq A) \to \varphi[a/x],
\end{aligned}$$

where the formula φ may have other free variables (parameters) besides a. The first two quantifiers express the existence and uniqueness of a witness for the formula φ (either globally or only within the set A). This syntactic shorthand is as useful as the bounded quantifiers defined in (B2.12). The third and fourth quantifiers are variations of bounded quantifiers.

Now we are ready to formulate the full axiom system of ZFC:

ZF1 Congruence: $(a = b) \to ((a \in M) \leftrightarrow (b \in M))$.

ZF2 Extensionality: $(A = B) \leftrightarrow (\forall x\ (x \in A) \leftrightarrow (x \in B))$.

ZF3 Power Set: $\exists \mathcal{P}(A)$.

ZF4 Union: $\exists \bigcup A$.

ZF5 Regularity: $(A \neq \emptyset) \to (\exists x \in A : \forall y \in x :\ y \notin A)$.

ZF6 Replacement $[\varphi]$:

$$(\forall x \in A : \exists! y : \varphi(x, y, \vec{p})) \to (\exists Y\ \forall y\ (y \in Y) \leftrightarrow \exists x \in A :\ \varphi(x, y, \vec{p})).$$

ZF7 Infinity: $\exists X\ (\emptyset \in X) \wedge (\forall x \in X :\ x \cup \{x\} \in X)$.

ZF8 Choice (AC):

$$(\emptyset \notin A) \to \exists f \subseteq A \times \bigcup A :\ (\forall x \in A :\ \exists! y \in x :\ \langle x, y\rangle \in f).$$

As one can easily see, the Axiom of Separation HF3 and the Axiom of Elementary Sets (Empty Set and Pair) HF6 have disappeared from our list. This is because we can now prove them. The Axiom of Infinity, which has appeared here, directly postulates the existence of a set X, one of whose elements is the empty set; therefore, the empty set exists.

Now, let there be a formula $\varphi(a, \vec{p})$ and a set A, and we want to construct its definable part $\{x \mid (x \in A) \wedge \varphi(x, \vec{p})\}$. Firstly, we note that if $\varphi(x, \vec{p})$ is false for all $x \in A$, or if $A = \emptyset$, then we already know the answer—the empty set. Suppose that A is non-empty and there exists an $s \in A$ such that $\varphi(s, \vec{p})$. Consider the following formula:

$$\psi(a, b, s, \vec{p}) \stackrel{\text{def}}{\leftrightarrow} (\varphi(a, \vec{p}) \to (b = a)) \wedge (\neg\varphi(a, \vec{p}) \to (b = s)).$$

It is not difficult to see that the formula ψ satisfies the premise of the Axiom of Replacement ZF6. This formula defines what is called a *definable mapping* of type $a \mapsto b$ from the set A into itself, which is the identity where $\varphi(a, \vec{p})$ is true, and is constant and equal to s at the other points of A. In this way, it maps A to the truth domain of the formula φ on A. In ordinary mathematics, we would write such a mapping with the following lexeme:

$$f(a) = \begin{cases} a, & \varphi(a, \vec{p}), \\ s, & \neg\varphi(a, \vec{p}). \end{cases}$$

Now, by the Axiom of Replacement, we obtain that there exists a set Y, which can be written as a binding-lexeme as $\{y \mid \exists x \in A : \psi(x, y, s, \vec{p})\}$, and this is precisely the set $\{x \mid (x \in A) \wedge \varphi(x, \vec{p})\}$, since

$$(a \in A) \wedge \varphi(a, \vec{p}) \leftrightarrow \exists x \in A : \psi(x, a, s, \vec{p}).$$

Therefore, the Axiom Schema of Separation is derivable in our system of axioms.[1]

To justify the Axiom of Pairing, we first define a unique pair: $\{\emptyset, \{\emptyset\}\}$. It exists as the set $\mathcal{P}(\mathcal{P}(\emptyset))$. After that, we need to devise a definable mapping that will uniquely map the elements of this pair to arbitrary sets B and C. The formula for such a mapping is:

$$\varphi(g, f, B, C) \stackrel{\text{def}}{\leftrightarrow} (g = \emptyset) \wedge (f = B) \vee (g = \{\emptyset\}) \wedge (f = C).$$

Note that $\emptyset \neq \{\emptyset\}$, so the formula φ will satisfy the premise of the Axiom of Replacement ZF6, where $A = \{\emptyset, \{\emptyset\}\}$. Then the set Y whose existence is asserted by this axiom will be identical to the pair $\{B, C\}$.

Thus, the axiom system ZFC extends the axiom system HF without axiom HF0, which is the negation of axiom ZF7. It should also be noted that in the study of set theory, the axiom system without the Axiom of Choice ZF8, denoted AC, is often used. This version of the set theory axiom system is denoted ZF. If we exclude the Axiom of Regularity and the Axiom of Replacement from ZFC, but retain its consequences—the Axiom of Separation and the Axiom of Pairing—then the resulting theory is called Zermelo's

[1] It is additionally worth noting that it is derivable directly from the Axiom of Replacement, the existence of $\emptyset$, and the axioms of equality; that is, the other axioms are not used here.

axiomatics and is denoted by Z. The relationship between the theories can be expressed in Table C1.1.

Table C1.1. Comparison of set theory axiom systems

Axiom	HF	Z	ZF	ZFC
Congruence	✓	✓	✓	✓
Extensionality	✓	✓	✓	✓
Pairing	✓	✓	theorem	theorem
Power Set	✓	✓	✓	✓
Union	✓	✓	✓	✓
Separation	✓	✓	theorem	theorem
Empty Set*	✓	theorem	theorem	theorem
Regularity	✓	–***	✓	✓
Replacement	–	–	✓	✓
Infinity	x**	✓	✓	✓
Choice (AC)	–	–	–	✓

* The Axiom of Empty Set is usually derived from Separation or Infinity.
** Negation of the Axiom of Infinity.
*** A dash indicates that the axiom is absent from the theory's axiom list.

We have distinguished the first 7 axioms as a kind of *absolute set theory*, which one would always like to have at hand, but which does not have a special designation. Its models would include, for example, hereditarily finite sets and bracket notations, but at the same time, other, more complex universes of sets. We provide this definition by analogy with absolute geometry, which includes only the first four postulates of Euclid. In all these set theories, including absolute set theory, the existence of the sets $a \cap b, a \cup b, a \setminus b, a \triangle b, \langle a, b \rangle, a \times b$ is proven in exactly the same way as before, relying on the Schema of Separation, Union, and Pairing.

Let's now see what the Axiom of Infinity gives us. To begin, let us introduce the notation $S(a) \overset{\text{def}}{\rightleftharpoons} a \cup \{a\}$, which is analogous to the arithmetic successor operation that expresses the addition of one. Here, too, we are in a sense adding one, namely, one new point a to the set a. By the Axiom of Regularity, as we have already seen in the theory HF, $a \notin a$, so it is indeed a new point relative to a as a set. And now, just as in arithmetic, we can provide a series of definitions for the *numerals*:

$$0 \overset{\text{def}}{\rightleftharpoons} \emptyset, \quad 1 \overset{\text{def}}{\rightleftharpoons} S(0), \quad 2 \overset{\text{def}}{\rightleftharpoons} S(1), \quad 3 \overset{\text{def}}{\rightleftharpoons} S(2), \ldots$$

These are the so-called von Neumann natural numbers, which are a characteristic feature of the von Neumann universe of sets that satisfies the Axiom of Regularity.

The Axiom of Infinity tells us that there exists some set X whose elements are the von Neumann numbers. In essence, this axiom asserts the existence of the set of natural numbers as an object of the theory. In other words, it postulates actual infinity. A set a that satisfies the property $Ind(a) \overset{\text{def}}{\leftrightarrow} (0 \in a) \wedge (\forall x \in a : S(x) \in a)$ is called **inductive**. By taking an arbitrary inductive set a (which exists by the Axiom of Infinity), we can separate a part of it according to the rule:

$$\omega \overset{\text{def}}{\rightleftharpoons} \{x \mid (x \in a) \wedge (\forall y \, Ind(y) \rightarrow (x \in y))\}.$$

The set ω exists by the Axiom of Separation and is nothing other than the intersection of all inductive sets, so it can also be written as:

$$\omega = \bigcap_{Ind(x)} x.$$

The last expression is a binding-lexeme, for which one can introduce the following first-order definition-axiom

$$a \in \bigcap_{\varphi(y,\vec{p})} y \overset{\text{def}}{\leftrightarrow} \exists x \, \varphi(x, \vec{p}) \wedge \forall y \, (\varphi(y, \vec{p}) \rightarrow (a \in y))$$

with a first-order formula φ. If the formula φ is identically false (for the given values of the parameters $\vec{p}$), then we get that $\bigcap_{\varphi(y,\vec{p})} y = \emptyset$. If the formula φ has at least one witness, then, as stated above, the lexeme $\bigcap_{\varphi(y,\vec{p})} y$ will denote a set whose existence is guaranteed by the Axiom of Separation. The addendum $\exists x \, \varphi(x, \vec{p})$ in this definition of the binding is needed precisely for the case of an identically false formula $\varphi(x, \vec{p})$, because if it were removed, the implication $\varphi(y, \vec{p}) \rightarrow (a \in y)$ would be true for all a in that case, and our binding-lexeme would cease to denote a set (it would be the class of all sets).

Thus, the set ω is the minimal (in the sense of the relation $\subseteq$) inductive set. This set consists of precisely the von Neumann natural numbers.

C1.1.2 Relations, Functions, and Classes

Note that the definition of a mathematical concept, given on page 42, can now be reformulated in the language of ZF as follows:

- The definition of a concept (*intension*) is a formula $\varphi(a, \vec{p})$ in the language of ZF that expresses the collection of properties (attributes) of sets named by the variable a and dependent on parameters $\vec{p}$.
- The *extension* of a concept is a class of all sets a that satisfy the formula $\varphi(a, \vec{p})$, and only those. The extension of the concept depends on the parameters $\vec{p}$.
- The designation of a concept is its established name in natural language (it may include a reference to the parameters).
- The notation for a concept is a lexeme of the Set Theory language (not necessarily the original language of ZF), introduced by a definition-axiom as a shorthand for either the defining formula $\varphi(a, \vec{p})$, or the term denoting the extension of the concept, or it is a variable that ranges only over the elements of the extension of the concept.

For example, the concept "von Neumann natural number" is defined by the formula $\forall y\ (Ind(y) \to (a \in y))$ without parameters, its extension contains all von Neumann natural numbers and only them, and its notation is the constant ω.[2]

Another example: the concept "Cartesian product of sets A and B" is defined by the formula $\exists x \in A : \exists y \in B : (a = \langle x, y\rangle)$ with parameters A and B. In strict accordance with the definition of a mathematical concept given above, the extension of this concept is the set of all Kuratowski ordered pairs $\langle x, y\rangle$ in which $x \in A$ and $y \in B$, and only them, and the corresponding notation is the term $A \times B$.

Informally, the notion of a class is used in ZF set theory. At the meta-theory level, a **class** is a mathematical concept in the language of ZF. Most often, a class is identified with its extension. Thus, if we have a mathematical concept defined by a formula $\varphi(a, \vec{p})$, then the class is identified with the term $\{a \mid \varphi(a, \vec{p})\}$. If, upon instantiation of the parameters $\vec{p}$, the class does not coincide with any set, it is called a *proper class*.

[2] Here one must distinguish between the notation for the extension of the concept and the general notation for its elements. Usually, elements of ω are denoted by Latin letters from the middle of the alphabet: l, m, n, k, etc.

As we have just noted, the extension of the concept "Cartesian product of A and B" upon instantiation of the parameters is a specific set $A \times B$ and is not a proper class. However, if we look at the concept "Cartesian product of two sets" in general, without specific parameters in mind, we are dealing with a proper class consisting of all Cartesian products of arbitrary sets. Such a class will no longer be a set, but a proper class. The Russell class $R = \{x \mid x \notin x\}$ is also a proper class, since otherwise we would obtain the well-known contradiction.

For the formal legitimization of the concept of a *class of sets*, a formal system with two sorts of objects is used—sets and classes, where every set is a class, but not vice versa. This is Gödel–Bernays set theory (GB), which is a two-sorted first-order theory of sets. It is characterized by the fact that its list of axioms is finite (there are no axiom schemata), quantifiers over classes are allowed, but the theory itself is a conservative extension of ZF (ZFC).[3]

Henceforth, we will not be so meticulous in showing all the necessary components of a mathematical concept, but will introduce them as is customary in mathematics, highlighting the designation in italics or bold, understanding by this concept its corresponding class or set of objects, depending on parameters if any are present in the definition.

Now it is time to finally formalize the set-theoretic concepts introduced in section A2.7. A *relation* between sets A and B is any subset $R \subseteq A \times B$. If R is a relation, it is customary to write aRb instead of the formula $\langle a, b\rangle \in R$. If $A = B$, the relation between A and B is called a relation on A. For example, we can define the relation $<$ on the elements of ω as follows:

$$a < b \overset{\text{def}}{\leftrightarrow} a \in b.$$

In Part II, we will prove that this relation $<$ is a strict linear ordering (and even a well-ordering) on the von Neumann natural numbers (see Theorem D4.4). In accordance with the adopted notation, we can now write, for example, $3 < 5$ or $n < S(n)$ for $n \in \omega$.

Accordingly, we can reproduce the types of a relation R on a set A:

- *reflexive* — $\forall x \in A : xRx$;
- *irreflexive* — $\forall x \in A : \neg(xRx)$;
- *symmetric* — $\forall x, y \in A : xRy \to yRx$;
- *the inverse of relation R* — $\forall x, y\ \langle x, y\rangle \in R \leftrightarrow \langle y, x\rangle \in R^{-1}$

[3] This means that any theorem formulated exclusively in terms of sets is provable in GB if and only if it is provable in ZF.

- *antisymmetric* — $\forall x, y \in A : xRy \land yRx \to x = y$;
- *transitive* — $\forall x, y, z \in A : xRy \land yRz \to xRz$;
- *connex* — $\forall x, y \in A : (xRy) \lor (x = y) \lor (yRx)$;[4]
- *preorder* — reflexive and transitive;
- *equivalence relation* — reflexive, symmetric and transitive;
- *partial (non-strict) order* — an antisymmetric preorder;
- *strict partial order* — irreflexive and transitive;
- *linear order* — a connex partial order.
- a well-ordering on A — a linear order for which

$$\forall X \subseteq A : (X \neq \emptyset) \to \exists x \in X : \forall y \in X : (xRy)$$

(such an element x is called the minimum of the set X and is denoted $\min_R X$).

Accordingly, a set together with a (partial/linear/well-order) relation defined on it is called a (partially/linearly/well-) *ordered set*. Here, the word "together" implies that we are considering a pair that includes the set and the relation on it.

And also the types of a relation $R \subseteq A \times B$:

- *total* — $\forall x \in A : \exists y \in B : xRy$;
- *functional* — $\forall x \in A : \forall y, z \in B : xRy \land xRz \to y = z$;
- a **function from** A **to** B — total and functional.

The fact that a relation f is a *function* from A to B is written with the shorthand formula $f : A \to B$. Note that this is a formula with three parameters, not a term. If we have a function $f : A \to B$, then A is called its domain, and the set $\{y \mid (y \in B) \land \exists x \in A : \langle x, y\rangle \in f\}$ is called the range of the function f and is denoted by the term $\mathrm{ran}(f)$.

These definitions can be extended by assuming that instead of sets A, B, R, first-order formulas $\varphi_A, \varphi_B, \varphi_R$ are used, and instead of the membership predicate, the formulas $\varphi_A(x), \varphi_B(y), \varphi_R(x, y)$ are used throughout the definitions. This allows us to speak of a *definable relation on a class* (between classes).

---

[4] In Order Theory, this property is sometimes called *total* or *linear*. However, since "total" is used for the property of the domain of definition (being everywhere defined), we use "connexity" to avoid confusion. This terminology is standard in general algebra and logic (see, e.g., [37]).

Thus, for example, we can introduce the class $\mathfrak{V}$ of all sets, whose defining formula would be $(x = x)$. Moreover, using a binding-lexeme, we could say that $\mathfrak{V} = \{x \mid x = x\}$, but here we should recall Russell's paradox and the fact that we have no axioms that would allow us to prove that $\mathfrak{V}$ is a set. Therefore, although the concept of a class is often used, one must always understand that a class $\mathfrak{C}$, defined by a formula $\varphi(a, \vec{p})$, can only be placed on the right side of the membership predicate, $a \in \mathfrak{C}$, and this formula should be understood every time as nothing other than the formula $\varphi(a, \vec{p})$.

Now, if we have a formula $\varphi(x, y, \vec{p})$ that defines a functional relation on the universe of all sets $\mathfrak{V}$, and if it is total on some class $\mathfrak{C}$, then we say that the formula φ defines a **definable mapping** on $\mathfrak{C}$, which can be written abbreviatedly as $\varphi : \mathfrak{C} \to \mathfrak{V}$. But in this case, we do not use the designation *function*, as it is reserved strictly for functional relations between sets!

However, if we have a functional relation that is total on a set, then by the Axiom of Replacement, it is indeed a function, since the Axiom of Replacement guarantees that the range of such a mapping is a set. Finally, let us define the *value of a function at a point a* for a function $f : A \to B$:

$$f(a) \overset{\text{def}}{\rightleftharpoons} \bigcup\{y \mid (y \in B) \wedge \langle a, y\rangle \in f\}.$$

Since such a y exists and is unique by virtue of f being total and functional, the set $\{y \mid (y \in B) \wedge \langle a, y\rangle \in f\}$ consists of a single point, which is denoted by the term $f(a)$. Thus,

$$b = f(a) \leftrightarrow \langle a, b\rangle \in f.$$

Let there be a function $f : A \to B$ and two subsets $X \subseteq A$ and $Y \subseteq B$. The **image** of the set X (under f) is the set $f[X] \overset{\text{def}}{\rightleftharpoons} \{b \mid \exists a \in X : (b = f(a))\}$, the **preimage** of the set Y (under f) is the set $f^{-1}[Y] \overset{\text{def}}{\rightleftharpoons} \{a \mid (a \in A) \wedge (f(a) \in Y)\}$.

To distinguish the value of a function at a point from the image of a set, we use round parentheses in the first case and square brackets in the second. From the given definitions, it follows that for any function $f : A \to B$

$$\operatorname{ran}(f) = f[A], \quad f^{-1}[B] = A.$$

C1.1.3 Why Functions are Needed

The concept of a function is, perhaps, one of the central mathematical concepts, as it is used everywhere in various mathematical sciences. Therefore, the terminological apparatus surrounding functions is extremely developed, and we consider here at least the main types of functions:

- $f : A \to B$ is **constant** if there exists $b \in B$ such that $\forall x \in A : f(x) = b$;
- $f : A \to A$ is the **identity** on A if $\forall x \in A : f(x) = x$, in which case f is denoted as id_A;
- $f : A \to \{0, 1\}$ is the **indicator function** of a set $S \subseteq A$ if $\forall x \in A : f(x) = 1 \leftrightarrow (x \in S)$, in which case f is denoted χ_S;
- $f : A \to B$ is an **injection** (or *embedding*) if $f(a) = f(b) \to a = b$;
- $f : A \to B$ is a **surjection** (or *mapping "onto"*) if $\forall y \in B : \exists x \in A : y = f(x)$;
- $f : A \to B$ is a **bijection** (or *one-to-one correspondence*) if it is both an injection and a surjection, in which case the predicate for the function is written as: $f : A \leftrightarrow B$;
- if $f : A \to B$ and $g : B \to C$, then the composition $g \circ f$ is the function $h : A \to C$ such that $h(a) = g(f(a))$ (in terms of relations: $g \circ f = \{\langle x, z\rangle \mid \exists y \in B : \langle x, y\rangle \in f \wedge \langle y, z\rangle \in g\}$);
- a function $g : B \to A$ is called the **inverse** of $f : A \to B$ if $\forall x \in A : g(f(x)) = x$ and $\forall y \in B : f(g(y)) = y$, in which case the inverse function is denoted by f^{-1}.

The existence of a bijection $f : A \leftrightarrow B$ indicates that there is a one-to-one correspondence between these sets (much like a zipper), i.e., they have the "same number" of elements, so to speak. Of course, this notion is difficult to extend to infinite sets, but without some additional structure on the sets, it is hard to imagine how else one could say anything about their "sameness" other than by presenting a bijection. Therefore, the following definition exists: two sets A and B are **equinumerous** ($A \simeq B$) if there exists a bijection $f : A \leftrightarrow B$. From this arise secondary definitions: a set is **finite** if it is equinumerous with some $n \in \omega$, and **infinite** otherwise;[5] **countable** if it is equinumerous with ω, **at most countable** if it is finite or countable.

[5] It should be noted that there are several variants of the definition of these designations, for example, Tarski-finite (our definition) and Dedekind-finite, which are equivalent within ZFC but may differ in weaker theories.

The relation of equinumerosity $\simeq$ is an equivalence relation on the class of all sets, i.e., on the universe $\mathfrak{V}$; it is a definable class-relation. Furthermore, the class of all sets equinumerous with a set A is often denoted as $\mathrm{Card}(A)$ and is called the cardinality of the set A.[6]

Let us provide several examples of the use of the function concept in mathematics.

- Vectors, understood as tuples of elements, are functions defined on a finite support (usually on the set $\{1, \ldots, n\}$).
- A function defined on the product of von Neumann natural numbers $n \times m$ can be understood as a matrix of size $n \times m$.
- The set of all bijections of a set H forms a group with respect to the composition operation $\circ$. The identity of this operation is the function id_H. In particular, if H is finite, such a group of bijections is called the symmetric group, or the group of permutations.
- Sequences, series, and integrals fundamentally use the concept of a function. In particular, concepts such as Hilbert spaces (also used in quantum mechanics) are sets of functions with specific properties.
- Norm, metric, functional, linear operator, measure, random variable, tensor, scalar product—all these are different types of functions.

C1.2 The Axiom of Choice

Let us now turn to the discussion of the last axiom on our list—the Axiom of Choice,—and state its formulation once more:

$$(\emptyset \notin A) \to \exists f \subseteq A \times \bigcup A : \ (\forall x \in A : \ \exists! y \in x : \ \langle x, y\rangle \in f).$$

The first thing it asserts is that A contains no empty element. The second is that f is a relation between the sets A and $\bigcup A$. The third is that this relation is total and functional. Finally, the fourth is that $f(x) \in x$. And the main thesis of the axiom is that such an f exists. Using functional notation, the Axiom of Choice can be reformulated as:

[6] Cantor defined cardinality as follows: it is that which is common to all sets equinumerous with a given one.

$$(\emptyset \notin A) \to \exists f\ (f : A \to \bigcup A) \wedge \forall x \in A : f(x) \in x.$$

In other words, whatever the set A without an empty element may be, there always exists a function f that chooses one element from each of its elements. This formulation is intuitively clear: if you have a "bag of bags," and each inner bag is non-empty, then you can choose one item from each bag. Such a function f is called a **choice function** on the set A. Note that the only requirement for the set A is the absence of the empty element. This requirement is not too significant, because if some set B contains the empty element, we can always move to the set $A = B \setminus \{\emptyset\}$ and apply the axiom to it. Nevertheless, one must remember that when we speak of a function $f : A \to \bigcup A$, it must be total on A. This means that on a set containing the empty element, a choice function does not exist (since it is impossible to satisfy the requirement $f(\emptyset) \in \emptyset$).

Logic and common sense also require us to check whether any contradiction arises if the set A is empty. Firstly, let us note that

$$c \in \bigcup \emptyset \leftrightarrow \exists x \in \emptyset : c \in x,$$

where the right-hand side is a false formula, and thus $\bigcup \emptyset = \emptyset$. Secondly,

$$c \in \emptyset \times \emptyset \leftrightarrow \exists x \in \emptyset : \exists y \in \emptyset : c = \langle x, y \rangle,$$

where the right-hand side is again a false formula, and thus $\emptyset \times \emptyset = \emptyset$. Moreover, the equalities $A \times \emptyset = \emptyset$ and $\emptyset \times B = \emptyset$ are also true, so the empty set is the zero element of the Cartesian product operation on sets. It is obvious that if $f \subseteq \emptyset \times \emptyset$, then $f = \emptyset$, and consequently, the predicate $\langle x, y \rangle \in f$ is false. But the axiom contains the quantifier $\forall x \in A$, which expands by definition as:

$$\forall x\ (x \in A) \to \exists! y \in x : \langle x, y \rangle \in f.$$

In the case $A = \emptyset$, the premise of the implication will be false, and thus the entire implication will be true, and no contradiction will arise. That is, in this case, the choice function will be empty, which does not contradict its main property. In general, one can say that for the empty set, any property beginning with a universal quantifier holds ("all elements of the empty set are empty," "all elements of the empty set are infinite," etc.). This is a direct consequence of the use of *material implication*. Examples: on the set $\{\emptyset\}$, no choice function exists, while on the set $\emptyset$, a choice function exists and is equal to $\emptyset$.

The Axiom of Choice has another, often more convenient, formulation:

$$(I : A \to B) \wedge (\forall x \in A : I_x \neq \emptyset) \to \\ \exists f : A \to \bigcup_{x \in A} I_x : \forall x \in A : f(x) \in I_x. \quad \text{(C1.1)}$$

First, let us comment on the liberties we took when writing this formula. Firstly, we denoted the value of the function I at point x as I_x instead of $I(x)$. This is a common way in mathematics to write the value of a function using an index. Secondly, we slightly abbreviated the syntactically correct notation $\exists f\ f : A \to \bigcup_{x \in A} I_x$ by removing the consecutive repetition of the letter f. Such abbreviations are also permissible (at the meta-level) as they do not cause ambiguity. Thirdly, we introduced a binding-lexeme that can be given a direct definition:

$$\bigcup_{x \in A} I_x \overset{\text{def}}{\rightleftharpoons} \bigcup \operatorname{ran}(I),$$

since the range of a function is a set by the Axiom of Replacement. The formulation of AC as (C1.1) says that if we have an indexed family of sets (which is a set by virtue of the Axiom of Replacement), then we can choose an indexed family of elements of these sets, taking one from each. Intuitively, if we have a series of numbered boxes, we can take one item from each, attach labels with the box numbers to them, and put them all in one bag. The function f, whose existence is stated in (C1.1), is also called a *choice function*, but not on the set A, but *for the family of sets* I_x.

Theorem C1.1. *The formulations of the Axiom of Choice in the form* ZF8 *and in the form* (C1.1) *are equivalent in* ZF.

Proof. The axiom in the form ZF8 follows trivially from (C1.1): simply set $I(x) = x$. Conversely, let $I : A \to B$ and every I_x be non-empty for $x \in A$. Apply the Axiom of Choice in the form ZF8 to the set $\operatorname{ran}(I)$: let $f : \operatorname{ran}(I) \to \bigcup \operatorname{ran}(I)$ be a choice function. Then the function $g : A \to \bigcup_{x \in A} I_x$, defined by the composition rule $g(x) = f(I_x)$, is the desired choice function in (C1.1). □

The Axiom of Choice AC was first explicitly formulated by Ernst Zermelo in 1904 in an attempt to prove the theorem that any set can be well-ordered. However, it was used intuitively by mathematicians long before its formal introduction, as it seemed to be an obvious fact. Immediately after its explicit formulation, AC sparked considerable controversy due to its non-constructive nature: it postulates the existence of a choice function without providing a specific method for its construction. In 1938, Kurt Gödel proved that if the other

Zermelo-Fraenkel axioms are consistent, then adding AC to them does not lead to a contradiction (i.e., AC is relatively consistent with ZF). In 1963, Paul Cohen proved that AC cannot be derived from the ZF axioms (i.e., AC is independent of ZF), using the method of forcing. This means that there are models of set theory where AC is true, and models where it is false. AC is of enormous importance in many areas of mathematics, allowing for the proof of important theorems that are impossible or extremely difficult to prove without it within ZF. It is equivalent to several other important statements, such as Zermelo's well-ordering theorem and Zorn's lemma.

Theorem C1.2 (Equivalents of the Axiom of Choice). *The following statements are equivalent in the theory* ZF*:*

(1) *the Axiom of Choice;*
(2) *Zermelo's theorem: any set can be well-ordered;*
(3) *Zorn's lemma: if in a partially ordered set every chain has an upper bound, then the set has a maximal element.*
(4) *Hausdorff's maximal principle: in any partially ordered set, there exists a maximal chain with respect to inclusion.*

From the perspective of the language of mathematics, it is important for us to be able to convert words into formulas and vice versa. Therefore, instead of proving this theorem, for which we again refer to Part II (see Theorem D4.13), we will show how these three theorem formulations are translated into the (enriched) language of ZF.

With Zermelo's theorem, there are no particular problems, since we have already introduced the definition of a well-ordered set. This is a linearly ordered set in which every non-empty subset has a minimum. Now for the translation:

$$\begin{aligned}\exists R \subseteq A \times A : [\forall x, y, z \in A : (xRx) \wedge (xRy \wedge yRz \longrightarrow xRz) \wedge \\ (xRy \wedge yRx \longrightarrow x = y) \wedge (xRy \vee x = y \vee yRx)] \wedge \\ (\forall X \subseteq A : (X \neq \emptyset) \longrightarrow \exists y \in X : \forall x \in X : yRx)\end{aligned}$$

Let us analyze this formula. Firstly, it has one free variable A, which corresponds to "any set". It then asserts the existence of a certain relation R on A ($R \subseteq A \times A$) that has a number of properties. The properties listed in the square brackets are those of a non-strict linear order: reflexivity, transitivity, antisymmetry, and connexity. To better understand the meaning of the formula, one should mentally replace R with $\leqslant$. Finally, the last condition on R tells us that in any non-empty subset $X \subseteq A$, there exists an element y that is less than or equal to any $x \in X$, i.e., it is a minimum.

By the way, let us give a definition. If $\leqslant$ is a partial (non-strict) order on X and $A \subseteq X$, then a *minimal* (*maximal*) element of the set A is an element $m \in A$ such that there are no elements in A smaller (greater) than it, and a *minimum* (*maximum*) is an element $M \in A$ such that $M \leqslant x$ ($x \leqslant M$) for all $x \in A$. Note: "no elements smaller" does not mean it is smaller than all others. There is a fundamental difference between the concepts of a minimal element and a minimum. For example, if we consider the divisibility relation on the natural numbers starting from 2 (excluding 1 and 0), then all prime numbers will be minimal elements, as there are no numbers that divide them other than themselves. However, if we return 1 to this set, it will be the minimum, as it divides all positive natural numbers. In a linearly ordered set, where all elements are pairwise comparable, the difference between these concepts disappears.

Let us also add the definitions of bounds and suprema/infima. An element $b \in X$ is called an upper (lower) **bound** of a set A if $\forall x \in A : x \leqslant b$ ($\forall x \in A : b \leqslant x$). An upper (lower) bound is called the least upper bound (greatest lower bound) if it is the minimum (maximum) of the set of all upper (lower) bounds of the set A. The least upper bound of a set A is denoted $\sup A$, the greatest lower bound is $\inf A$. It is clear from the definition that they may not always exist. Note also that if $A = \emptyset$, then all elements of the set X will be its upper and lower bounds, which is a bit strange, since some lower bounds can be greater than some upper bounds. Therefore, it is usually assumed that the concepts of bounds are defined only for a non-empty set A. These definitions allow us to introduce new notations and new quantifiers:

$$m = \min_{\leqslant} A \overset{\text{def}}{\leftrightarrow} (m \in A) \wedge \forall x \in A : m \leqslant x,$$

$$\exists \min_{\leqslant} A \overset{\text{def}}{\leftrightarrow} \exists m \in A : m = \min_{\leqslant} A,$$

$$M = \max_{\leqslant} A \overset{\text{def}}{\leftrightarrow} (M \in A) \wedge \forall x \in A : x \leqslant M,$$

$$\exists \max_{\leqslant} A \overset{\text{def}}{\leftrightarrow} \exists M \in A : M = \max_{\leqslant} A.$$

In this notation, the last condition on R in Zermelo's theorem can be rewritten more concisely: $\forall X \subseteq A : (X \neq \emptyset) \to \exists \min_R X$. And if we add a special predicate

$$
\begin{aligned}
Lin(A, R) \stackrel{\text{def}}{\leftrightarrow} \forall x, y, z \in A : (xRx) \wedge (xRy \wedge yRz \rightarrow xRz) \wedge \\
(xRy \wedge yRx \rightarrow x = y) \wedge (xRy \vee x = y \vee yRx),
\end{aligned}
$$

which means that R is a linear order on A, then Zermelo's theorem takes on a quite respectable form:

$$
\exists R \subseteq A \times A : Lin(A, R) \wedge \forall X \subseteq A : (X \neq \emptyset) \rightarrow \exists \min_R X.
$$

It is also worth noting that if the set A is empty, then it is well-ordered by the only possible relation on it, $\emptyset$.

Zorn's lemma will require more effort to translate into the language of ZF, as we first need to provide definitions. Firstly, we say that a relation Q is *induced* (*restricted*) on a set B by a relation $R \subseteq A \times A$ (notation: $R \restriction B$) if $Q = R \cap (B \times B)$. Secondly, a *chain* in a partially ordered set is a subset such that the partial order induced on it is linear.[7] If one imagines a partial order as a tree, then a chain would be any path (and this path can skip some vertices, using transitivity). Finally, we say that a chain $C \subseteq A$ is *bounded above* (below) if there exists an upper (lower) bound for C in A.

Let us introduce another predicate for partial order:

$$
\begin{aligned}
Part(A, R) \stackrel{\text{def}}{\leftrightarrow} \forall x, y, z \in A : \\
(xRx) \wedge (xRy \wedge yRz \rightarrow xRz) \wedge (xRy \wedge yRx \rightarrow x = y).
\end{aligned}
$$

Now, let us write down Zorn's lemma:

$$
\begin{aligned}
\big(Part(A, R) \wedge (\forall C \subseteq A : (Lin(C, R \restriction C) \rightarrow \exists b \in A : \max_R(C \cup \{b\}))))\big) \\
\rightarrow \exists m \in A : \forall x \in A : (mRx) \rightarrow (m = x).
\end{aligned}
$$

The last subformula $\forall x \in A : (mRx) \rightarrow (m = x)$ means that m is a maximal element, i.e., there are no elements greater than it (this is written as: if there is something greater than or equal to it, then it must be equal to it, since we are using a non-strict partial order here).

As we can see, the formulations of any reasonably substantial theorems in the language of ZF, even with a bunch of abbreviations, take on the form of

[7] The dual concept to a chain is an antichain: a subset whose elements are pairwise incomparable under the induced order. The maximal antichain principle is also equivalent to the axiom of choice.

a long poem. It's hard to imagine what these formulations would look like in the "assembly language" of mathematics—the original language of set theory based only on the single predicate $\in$. Nevertheless, all our previous constructions unequivocally testify that such a translation is possible. We suggest that the reader write down Hausdorff's maximal principle on their own.

The Axiom of Choice has numerous useful consequences in a wide variety of mathematical fields, for example:

- Tychonoff's theorem* on arbitrary products of compact spaces,
- The theorem* that every vector space has a basis,
- The theorem* that every subspace of a vector space has a direct complement,
- Krull's theorem* that every nonzero ring with unity has a maximal proper ideal,
- The theorem* on the existence of a maximal proper ideal in a nontrivial distributive lattice with a maximum element,
- The Boolean Prime Ideal Theorem** (BPI),
- Tarski's ultrafilter theorem**,
- Gödel's completeness theorem** for predicate logic (for uncountable languages),
- Stone's representation theorem for Boolean algebras**,
- The Artin-Schreier theorem** on the ordering of a formally real field.
- The Hahn-Banach theorem on the extension of a linear functional,
- The theorem on the algebraic closure of a field,
- The Nielsen-Schreier theorem that a subgroup of a free group is free,
- Szpilrajn's extension theorem (any partial order can be extended to a linear order).

The theorems marked with * are equivalent to the Axiom of Choice, those marked with ** are equivalent to each other and are strictly weaker than AC. The Nielsen-Schreier theorem implies the axiom of choice for families of finite sets, but it is not known whether it is equivalent to the full axiom of choice. The Hahn-Banach theorem is strictly weaker than BPI. Szpilrajn's theorem (also known as the order extension principle) is also strictly weaker than BPI, but if the cofinality principle (every linear order contains a cofinal well-ordered subset) is added to it, then together they imply AC. The theorem on the algebraic closure of a field follows from BPI; whether it is equivalent to BPI is not known. To observe this magnificent picture of dependencies between various mathematical statements related in one way or another to the axiom of choice, we recommend consulting the book [38].

Some consequences of AC may seem counter-intuitive. The most famous example is the Banach-Tarski paradox, which states that a three-dimensional ball can be divided into a finite number of parts from which two balls of the same size can be assembled (the concept of a ball implies that it has no voids). Of course, these parts of the ball will be non-measurable in the Lebesgue sense.

Despite controversies and counter-intuitive consequences, the Axiom of Choice is widely accepted by the majority of mathematicians as a standard tool and is part of the standard axiomatics. Nevertheless, the study of mathematics without AC is also an active area of research. Here, various options for weakening the general Axiom of Choice are proposed, for example, to its countable version AC_ω, which asserts the existence of a choice function only for countable families of non-empty sets, as well as alternative axioms: the Axiom of Dependent Choice, the Axiom of Determinacy, etc.

It should be noted that AC_ω has a good supply of useful consequences, primarily in theories that do not go beyond the continuum (such as real analysis and second-order arithmetic), and at the same time, it has no known paradoxical consequences like the Banach-Tarski theorem or the Vitali set. Within the framework of the countable axiom of choice, one can prove, for example, the following consequences:

- any infinite set contains a countably infinite subset (see Theorem D4.11);
- every infinite set is Dedekind-infinite (see Theorem D4.11);
- a countable union of countable sets is countable;
- Ramsey-type principles for arbitrary infinite sets;
- the set of all real numbers is not a countable union of countable sets;
- any subspace of a separable metric space is separable;
- the Lebesgue measure is countably additive on sequences of measurable sets;
- König's infinity lemma: in any infinite tree with finite branching, there exists an infinite path.

All the stated propositions are weaker than the axiom AC_ω, but may be equivalent to some weaker versions of the axiom of choice. For example,

König's lemma in this formulation (without specifying the countability of the tree and the numbering of vertices) is equivalent to the axiom $\mathsf{AC}_{\omega}^{fin}$ (countable choice from finite sets), Ramsey-type principles correspond to weaker forms of countable choice depending on their precise formulation, and the theorem that a countable union of countable sets is countable is equivalent to countable choice from countable sets.

After these examples about AC and AC_{ω}, one might get the impression that all known theorems are either equivalent to one of the forms of the axiom of choice, or weaker than them, or are even provable in ZF. However, this is not the case. One can provide an "ascending list" of principles (theorems, axioms) from which AC follows, but which are not equivalent to it. These include, for example, the Generalized Continuum Hypothesis GCH (which is strictly stronger than AC), and Gödel's Axiom of Constructibility $V = L$ (which is strictly stronger than GCH and, therefore, than AC). There are also a number of alternatives to the axiom of choice. It is sufficient to mention, for example, the *Axiom of Determinacy* AD, which has many useful consequences, but does not have known paradoxes.

Thus, from the point AC in the language of ZF, one can move both along the deductive flow ($\vdash$) and up against it, as well as sideways. And each time, different versions of mathematics will arise, varying in their proof-theoretic strength and diversity of facts.

The choice function allows for a generalization of the Cartesian product of sets: $\prod_{i \in I} A_i$ is the set of all choice functions for the sets A_i, $i \in I$. For example, if $I = \{1, 2, 3\}$, then the elements of the Cartesian product of the sets A_1, A_2, A_3 will be sequences $\langle a_1, a_2, a_3 \rangle$, where $a_i \in A_i$, each of which we consider a shorthand notation for the set of the choice function itself $f = \{\langle 1, a_1 \rangle, \langle 2, a_2 \rangle, \langle 3, a_3 \rangle\}$.

Thus, a choice function is a generalization of the concept of an ordered tuple: it can be of any "length," which is determined only by the size of the indexing set, and with components of any nature, which is determined by the specific sets A_i. Therefore, finite ordered tuples are usually defined as a special case of a choice function with a finite index set, which is usually a segment of the natural numbers (von Neumann numbers). Of course, the question immediately arises—what about the ordered pair $\langle a, b \rangle$? We defined it according to Kuratowski, but it can also be defined as a function from the set $\{0, 1\}$. The problem is that the very definition of a function is already based on the ordered pair, and therefore we cannot discard the Kuratowski definition (unless we replace it with some other one not related to functions). So for ordered pairs, we have a certain dichotomy in the choice of their definition where we want to use them. It seems correct to assume that when dealing with some fundamental structures that do not imply multi-valued relations or multi-dimensional vectors, with the analysis of the foundations of set theory, with the expressibility of certain predicates in set theory, etc., it is better to operate with native pairs (according to Kuratowski) and, moreover, to generate finite tuples as compositions of binary ones, assuming, for example, $\langle a, b, c\rangle = \langle\langle a, b\rangle, c\rangle$, etc.

In areas more distant from the Foundations, say, in linear algebra, topology, analysis, geometry, etc., it is easiest to use homogeneous ordered tuples defined by choice functions, and in the case of finite tuples, not to think about their definition method at all. After all, we can enrich the language of set theory with n-ary operators $\langle\rangle$ and for each of them provide a corresponding axiom for an ordered tuple of length n. In this case, we will not have to think about their internal structure and the indexing set, and their existence can be proven using the axiom of replacement and the Kuratowski ordered pair.

The concept of a *sequence* is also often used, as a function of the form $f : \omega \to A$, which in this case is written with the binding-lexeme $\{f_k\}_{k=0}^{\infty}$. Such a term is easily legitimized in ZF by the axiom

$$a \in \{f_k\}_{k=0}^{\infty} \overset{\text{def}}{\leftrightarrow} \exists k \in \omega : a = f(k),$$

which essentially coincides with the definition of the set $\operatorname{ran}(f)$, but graphically indicates its nature. As an alternative, terms like $(f_k)_{k=0}^{\infty}$ or $\langle f_k\rangle_{k=0}^{\infty}$ are also used, which, through the use of different brackets, indicate the "tuple-like" nature of this notation and simply replace the term f and its definition as a function of a natural number (without specifying the range).

In the case of a finite indexing set, the term $\prod\limits_{i\in I} A_i$ is also usually written in a more expanded form, for example:

$$A \times B \times C, \quad A_1 \times A_2 \times \cdots \times A_n, \quad A^n.$$

The latter means that all A_i are equal to the same set A, and the indexing set is $\{1, 2, \ldots, n\}$. The lexeme A^n leads us to another common notation: A^B, which denotes the set of all functions from B to A. But in fact, this is a special case of $\prod\limits_{i\in B} A_i$, because if all A_i are equal to A, then the choice functions become simply functions from B to A.

Another interesting use of the power construction is noteworthy: 2^A often denotes the same thing as $\mathcal{P}(A)$. The idea is that $2 = \{0, 1\}$ in accordance with the definitions of von Neumann numbers, and therefore, every element $f \in 2^A$ is an *indicator function* of a subset of A. Indeed, to each function f corresponds a unique set $\{x \mid f(x) = 1\}$ and vice versa, for every subset $B \subseteq A$, one can define its indicator function by the rule

$$f(x) = \begin{cases} 1, & x \in B, \\ 0, & x \in A \setminus B. \end{cases}$$

In other words, the set 2^A is the set of indicator functions of the elements of $\mathcal{P}(A)$. But they are not quite the same thing.

Finally, it is interesting to look at the algebraic properties of the operation A^B. Let us list without proof the following properties[8] of set exponentiation, using the equinumerosity relation $\simeq$:

1. $2^A \simeq \mathcal{P}(A)$;
2. $(A^B)^C \simeq A^{B \times C}$;
3. $A^{B \cup C} \simeq A^B \times A^C$, if $B \cap C = \emptyset$;
4. $(A \times B)^C \simeq A^C \times B^C$;
5. $\emptyset^B = \emptyset$, if $B \neq \emptyset$;
6. $A^\emptyset = \{\emptyset\}$, which is equinumerous to 1, for any A;
7. $\{a\}^B \simeq \{a\}$ (and also $1^B \simeq 1$);

We have devoted such a detailed review to the Axiom of Choice for several reasons. Firstly, it is indeed a very significant axiom lying at the foundations of mathematics, and if the reader wants to better grasp not only the language of mathematics but its very essence, then without the Axiom of Choice this understanding will be frankly incomplete. The Axiom of Choice in logic and mathematics plays a role similar to Euclid's fifth postulate in geometry: it is also independent of the basic axioms, it also hid from mathematicians for a long time under the guise of obviousness, and without it or its alternatives, pure ZF theory becomes as pale and timid as geometry without some form of the parallel axiom.

Secondly, the formulation of the Axiom of Choice and its dependent principles is itself an instructive example for mastering Lingua Mathematica, for

[8] All of them easily follow from Theorem D4.10, proven in Part II.

the ability to convert abstract images into precise formulas, and ultimately, for understanding the process of mathematical thinking.

Thirdly, if we still look at mathematics as a foreign language, we must not only learn the vocabulary and grammar but also delve into the relevant cultural and idiomatic context of this language. And the Axiom of Choice represents a phenomenon of our mind as profound as universal grammar in linguistics, highlighting the very essence, and sometimes the conceptual fragility, of the foundations of mathematics.

Fourthly, this section has a secondary goal of conveying to a wide range of readers the most accurate definition of the Axiom of Choice and its impact on mathematics, in order to neutralize the speculations on this topic present in popular science circles, as well as various kinds of incorrect interpretations of the principle of choice.

Finally, fifthly, we want to emphasize separately that even when it is possible to prove a theorem without using the Axiom of Choice (say, Hilbert's Nullstellensatz), it usually turns out that with the Axiom of Choice, the proof becomes much shorter and simpler. That is, the Axiom of Choice also provides us with a certain deductive sugar (by analogy with syntactic sugar), allowing us to express the complex simply and beautifully, and this property can be considered a manifestation of the expressiveness of the mathematical language.

C1.3 The von Neumann Universe

John von Neumann

When we discussed the general concept of defining sets in Section B2.2, we imagined a set as a directed graph, its tree-like expansion, or a file system directory. This view aligns well with the implementation of the finitary part of set theory, namely, hereditarily finite sets and their bracket notations. But then we added the Axiom of Infinity, as a result of which the set ω became available to us, whose tree is a root from which edges go to all the roots of the trees of all natural numbers. In other words, it is a tree with countable branching at the root, in which every proper branch is finite. Moreover, each subsequent branch is the union of all previous branches, since $n + 1 = \{0, \dots, n\}$. At the same time, it is difficult to call such a graph fractal, since it has no infinite chains of self-repetition, such as those that arise if one abandons the Axiom of Regularity. In the tree of the set ω, all paths are finite!

Thus, the introduction of infinite sets leads to the fact that for the graphical representation of sets, we are forced to use infinite branching of the corresponding trees, but the paths descending along the membership relation in such a tree still remain finite. At the same time, it must be understood that the length of these paths is also not limited; in the same set ω, although all paths are finite, we can find a path of any arbitrarily large length. These considerations are intended to lead us to the idea that all sets can be classified by their "depth," i.e., by the potential possible length of paths from the root to the leaves in the tree-expansion of the membership relation. Such a concept indeed exists in set theory, and it is called the *rank of a set*. However, in order to introduce it, it is first necessary to define a scale from which we will choose these ranks. The simplest approach is to take the scale of natural numbers, i.e., the set ω. Indeed, since all paths are finite, either there is a maximum path length, and we will call it the rank of the set, or there is no maximum length, and then the rank of the set will be infinite.

However, this approach works well only for hereditarily finite sets and infinite sets whose elements are hereditarily finite (for example, ω). But consider, for example, the set $\{\omega\}$. In its membership graph, a new vertex appears, standing one step above the previous root. How will the paths in such a graph change? We will simply get paths that are one unit longer than we had before. But the trick is that if you add one to the natural numbers, they still remain natural numbers (which is what the axiom of infinity tells us), i.e., the limit of the lengths of such paths will not differ from the limit of the lengths of the paths in the set ω. At the same time, for hereditarily finite sets, the construction $\{a\}$ clearly increases the rank of the set by 1. Why, in the case of some sets, do we see a difference in ranks when adding brackets, while in the case of others, we do not? From this arises the need for a more precise scale, capable of distinguishing the ranks of infinite sets.

Such a scale exists—it is the class of ordinals.

The concept of an ordinal is to extend the idea of well-ordering from the natural numbers to all infinite sets, or at least to some of them. The well-ordering of a set implies that in this set, every element is always followed by a next one (if there are any other elements there at all), i.e., a successor. Therefore, without delving into the theory of ordered sets, let us move directly to its main result: the construction of standard well-orderings, i.e., ordinals.

From ordinals, we want the property of inductiveness to hold, with the ability to overcome limit points. Recall that an inductive set is closed under the successor operation $S(x) = x \cup \{x\}$. Such is, for example, the set ω. However, it itself is not obtained as a result of this operation, but is the limit of an increasing chain of von Neumann natural numbers: $\omega = \bigcup\{n \mid n \in \omega\}$. Having obtained ω, we can continue the process of adding new points using the operator S: $S(\omega)$, $S(S(\omega))$, etc. After performing these steps ω times, we will again have to take the limit to get the set

$$\omega + \omega = \bigcup\{\omega, S(\omega), S(S(\omega)), \dots\}.$$

We do not stop here and continue to use the operator S again. *Et cetera ad infinitum*. The speculative constructions presented, however, are not formal from the point of view of the language of ZF. Fortunately, we have an alternative approach that will give us all the ordinals in one fell swoop. So, let us say that a set a is *transitive* if the following property holds for it:

$$Tr(a) \overset{\text{def}}{\leftrightarrow} \forall x \in a : x \subseteq a.$$

After which, we say that a set a is an **ordinal** if it is transitive and the relation $\in$ induces a well-ordering on it:

$$\text{Ord}(a) \overset{\text{def}}{\leftrightarrow} Tr(a) \wedge Lin(a, \in) \wedge \forall x \subseteq a : (x \neq \emptyset) \to \exists \min_{\in} x,$$

where $\text{Ord}(a)$ is a predicate defining the class of ordinals, which is also denoted by Ord. It turns out that ordinals defined in this way have all the properties we need.

Before proceeding to the formulation of theorems, let us enrich our ZF language with new variables—Greek letters, possibly with the addition of numerical indices and accents. We will assign the property of being an ordinal to these variables by introducing two new quantifiers according to the following rules:

$$\exists\alpha : \varphi(\alpha) \overset{\text{def}}{\leftrightarrow} \exists\alpha\ \text{Ord}(\alpha) \wedge \varphi(\alpha), \quad \forall\alpha : \varphi(\alpha) \overset{\text{def}}{\leftrightarrow} \forall\alpha\ \text{Ord}(\alpha) \to \varphi(\alpha).$$

Of course, it would be more correct to give new notations to these quantifiers, for example, $\forall^{\text{ord}}$ and $\exists^{\text{ord}}$, but in our case, their logical novelty is already evident from the fact that we place Greek letters under them, and also use a colon, as in bounded quantifiers (which these quantifiers essentially are). By the way, it is not difficult to verify that the new quantifiers satisfy analogous axioms and rules of inference for the quantifiers of standard predicate logic.

Note further that we formally consider all Greek variables to be bound, since we are interested in theorems about ordinals, i.e., formulas without free variables. However, as in the case of writing axioms, we reserve the right to remove the closing universal quantifier (both ordinary and ordinal) from theorems, leaving the bound Greek variable in the formula where it becomes syntactically free, but by convention is understood as universally quantified over ordinals. Let us explain with an example how this works. Suppose we have proven that for all ordinals α, the formula $\varphi(\alpha)$ holds. This fact can be written in a closed form as: $\forall\alpha : \varphi(\alpha)$, which by definition expands to $\forall\alpha \; \mathrm{Ord}(\alpha) \to \varphi(\alpha)$. We can omit the closing quantifier in two ways: either simply write $\varphi(\alpha)$, assuming the premise $\mathrm{Ord}(\alpha)$ since α is a Greek variable, or explicitly write the formula $\mathrm{Ord}(\alpha) \to \varphi(\alpha)$. Naturally, the first method is preferable, although it is syntactic sugar and not a standard ZF formula. In the second method of writing, we can replace α with a, thereby returning to the standard language of *ZF*.

Theorem C1.3. *The following theorems are provable in* ZF*:*

1° *an element of an ordinal is an ordinal;*

2° *for any ordinals* α, β*:* $\alpha \in \beta \leftrightarrow (\alpha \subsetneq \beta)$*;*

3° *for any ordinals* α, β*:* $\alpha \subseteq \beta$ *or* $\beta \subseteq \alpha$*;*

4° *the union and intersection of any set of ordinals is an ordinal;*

5° *the class of all ordinals is well-ordered by the relation* $\in$ *and is a proper class;*

6° *for any ordinal* α*:* $S(\alpha)$ *is an ordinal;*

7° *the von Neumann natural numbers and* ω *are ordinals;*

8° *induction* $[\varphi]$*:* $[\forall\alpha : (\forall x \in \alpha : \varphi(x, \vec{p})) \to \varphi(\alpha, \vec{p})] \to \forall\alpha : \varphi(\alpha, \vec{p})$*;*

9° *definition by recursion* $[\varphi]$*: let* $\varphi(g, b, \vec{p})$ *define a mapping of type* $g \mapsto b$*, i.e.* $\forall x \exists! y : \varphi(x, y, \vec{p})$*, then for any ordinal* α *there exists a function* f *defined on* α *and such that for any* $\beta < \alpha$*:*

$$\varphi(f \restriction \beta, f(\beta), \vec{p}). \tag{C1.2}$$

Property 5 tells us that membership on ordinals behaves just like on the von Neumann natural numbers: firstly, it linearly orders them, and secondly, in any non-empty subclass of the class of ordinals, there exists a least element. Note that we are talking about a subclass here, i.e., a definable part of the class of all ordinals, which is essentially any property of ordinals expressible in ZF.

Property 2 demonstrates the equivalence of membership and proper inclusion of ordinals as sets. This means that ordinals form a chain of nested sets in accordance with the order on them. And since membership plays the role of order, one usually writes $\alpha < \beta$ instead of $\alpha \in \beta$, except in cases where the ordinal needs to be perceived as a set rather than a number.

Property 4 (taking into account properties 2 and 3) makes it possible to find the minimum and supremum of a set of ordinals by performing basic set theory operations—intersection and union:

$$\min A \overset{\text{def}}{\rightleftharpoons} \bigcap A, \quad \sup A \overset{\text{def}}{\rightleftharpoons} \bigcup A,$$

if A is a non-empty set of ordinals. The appearance of min here is directly related to the fact that the membership relation on ordinals is well-founded.[9]

Here we have acquired another useful predicate, $x \subsetneq y$, which is a shorthand for the formula $(x \subseteq y) \wedge (x \neq y)$. Sometimes the symbol $\subset$ is used instead, but in some sources, it is often understood as non-strict inclusion, which makes its use undesirable. Ordinals are usually divided into three subclasses: the zero ordinal, a successor ordinal, and a limit ordinal. A successor ordinal has the form $S(\alpha)$, and limit ordinals include all ordinals that are not zero and not successor ordinals. The class of all successors is denoted by Succ, the class of all limit ordinals by Lim.

Property 8 is called **transfinite induction** and generalizes complete induction from Peano arithmetic. It states that if the truth of a formula passes from the elements of an ordinal to the ordinal itself, then such a formula is true for the entire class of ordinals.

The last point of the theorem gives us a powerful tool for constructing sets and classes using recursion. This theorem generalizes arithmetic recursion to the class of ordinals. To understand the complex formulation, let us add a little more syntactic sugar to our language.

Firstly, if we have a definable mapping $g \mapsto b$ given by the formula φ, then instead of $\varphi(g, b, \vec{p})$ we will write $b = \Phi_{\vec{p}}(g)$ (it is clear that the correspondence between the letters φ and Φ is conventional here, since these are metatheoretical notations, and in each specific case one must know exactly which formula φ was used). Such functional notation is much easier to perceive. The mapping $\Phi_{\vec{p}}$ is then called the *recursion generator*.[10]

---

[9] Here, the axiom of regularity is not even needed, since well-foundedness is built right into the definition of an ordinal.

[10] This is not a standard designation, but I find it fitting.

Secondly, recall that if we have a function $f : A \longrightarrow B$, then by $f \restriction S$ we denote the **restriction** of the function to the set $S \subseteq A$, which is the set of pairs $\{\langle x, f(x)\rangle \mid x \in S\}$.

Then the formula (C1.2) can be written in the following way:

$$f(\beta) = \Phi_{\vec{p}}\,(f \restriction \beta)\,, \tag{C1.3}$$

i.e., the value of f at the point β is the value of the mapping defined by formula φ applied to the restriction of f to β (the history of values). The recursion here consists precisely in the fact that the values of the function f on subsequent ordinals are determined by all its values on the preceding ordinals. Considering the adopted abbreviations, we can rewrite the entire last point of the theorem with a single formula:

$$(\forall x \exists! y :\ \varphi(x, y, \vec{p})) \longrightarrow \forall \alpha \exists f : \alpha \longrightarrow \mathfrak{V} :\ \forall \beta < \alpha :\ f(\beta) = \Phi_{\vec{p}}(f \restriction \beta).$$

In fact, it can be shown that all the functions f, whose existence is asserted here, are restrictions of a single class-mapping F with the same property $F(\alpha) = \Phi_{\vec{p}}\,(F \restriction \alpha)$, and in a set theory with classes, the theorem on recursive definitions would be formulated more simply:

$$(\Phi_{\vec{p}} : \mathfrak{V} \longrightarrow \mathfrak{V}) \longrightarrow \exists F : \mathrm{Ord} \longrightarrow \mathfrak{V} :\ \forall \alpha : F(\alpha) = \Phi_{\vec{p}}(F \restriction \alpha),$$

where Ord is the class of all ordinals. In the theory ZF, one has to say that the first-order formula

$$\psi(\alpha, d, \vec{p}) \overset{\text{def}}{\leftrightarrow} \exists f : S(\alpha) \longrightarrow \mathfrak{V} :$$
$$\forall \beta < S(\alpha) :\ f(\beta) = \Phi_{\vec{p}}(f \restriction \beta) \wedge (d = f(\alpha))$$

defines a mapping of type $\alpha \mapsto d$ with this property. For the proof of the given theorem, we, as usual, refer to Part II (see Theorem D4.6), and now we will begin to apply it diligently.

Firstly, let us note that, as a rule, the action of the generator $\Phi_{\vec{p}}$ is defined differently for all three types of ordinals: zero, successor, and limit ordinals. This makes the formulas simpler and more intuitive. One might ask here: how does the generator recognize which ordinal it is working on? Since the input to the generator is the function $g = F \restriction \alpha$ (the history of the process), its domain is exactly the ordinal α ($\mathrm{dom}(g) = \alpha$). Thus, the generator can easily determine the type of α (zero, successor, or limit) by inspecting the domain of its input argument.

Therefore, a typical recursive definition looks like this:

$$F(0) = f_0, \quad F(S(\alpha)) = \Psi_{\vec{p}}(F(\alpha)), \quad F(\lambda) = \bigcup\{F(\gamma) \mid \gamma < \lambda\},$$

where λ is a limit ordinal. By the way, in such definitions, it is always implied that λ denotes a limit ordinal. And again, we use a simplification of syntax, typical for mathematicians, but not entirely formal from the point of view of the ZF language. Namely, we write the binding-lexeme $\{F(\gamma) \mid \gamma < \lambda\}$ in a non-standard way, placing the value to which it is equated in the subsequent formula directly in place of the bound variable. Compare:

$$\{F(\gamma) \mid \gamma < \lambda\} \quad \text{v.s.} \quad \{x \mid \exists\gamma < \lambda : x = F(\gamma)\}.$$

However, at the C1 level of the language of mathematics, we should already be well able to read such "reworkings" of the formal construction, understanding how it is transformed into the original language of set theory (or any other formal theory). As we can see, we have moved away from substituting the full history (restriction $F \upharpoonright \alpha$), replacing it first with the image $F[\alpha]$ (which is the set of values), and then with the union of all point-wise values. This is a *special*, but very common, type of recursive definition, used mainly for defining monotonic mappings from Ord to Ord or cumulative hierarchies. And we will use it repeatedly in the future.

As an exercise, the reader is invited to consider how such a particular definition is formally reduced to (C1.3), utilizing the fact that the domain of the input function uniquely identifies the step of the recursion. Secondly, let us define addition and multiplication on ordinals recursively:

$$\begin{aligned} \alpha + 0 &= \alpha, & \alpha \cdot 0 &= 0, \\ \alpha + S(\beta) &= S(\alpha + \beta), & \alpha \cdot S(\beta) &= \alpha \cdot \beta + \alpha, \\ \alpha + \lambda &= \sup\{\alpha + \gamma \mid \gamma < \lambda\}, & \alpha \cdot \lambda &= \sup\{\alpha \cdot \gamma \mid \gamma < \lambda\}, \end{aligned}$$

where α is a parameter of the recursive definition (not to be confused with the argument of the function f). And here is how the recursion generator $\Phi(g)$ (where the input g is the restriction $f \upharpoonright \beta$) can be defined to cover all cases in a single formula, using the values $g(\gamma)$ accumulated in the function:

$$\alpha + \beta = \alpha \cup \sup\{S(\alpha + \gamma) \mid \gamma < \beta\}, \quad \alpha \cdot \beta = \sup\{(\alpha \cdot \gamma) + \alpha \mid \gamma < \beta\}.$$

To understand how the first definition of multiplication reduces to the second, one first substitutes $S(\gamma)$ for β in the second definition, so that in the successor case the supremum reduces to evaluation at the maximal point γ. For a limit ordinal λ, we use

$$\sup\{(\alpha \cdot \gamma) + \alpha \mid \gamma < \lambda\} = \sup\{\alpha \cdot S(\gamma) \mid \gamma < \lambda\},$$

and then note that for every $\delta < \lambda$ we have $\delta \leqslant S(\delta) < \lambda$ (since λ is limit), so taking the supremum over successors below λ gives the same value as taking it over all ordinals below λ. Hence

$$\sup\{\alpha \cdot S(\gamma) \mid \gamma < \lambda\} = \sup\{\alpha \cdot \delta \mid \delta < \lambda\},$$

which is exactly the limit clause in the first definition. Similar reasoning applies to addition. Furthermore, when $\beta = 0$, the set for which we take the sup is empty, and therefore its formal supremum is 0. Thus, in the case of addition, we also add α to the formula. Defining arithmetic operations by recursion may not fully reflect their essence, but it is very convenient for calculations and refers us back to Peano arithmetic, where addition and multiplication are defined in the same recursive way. From this, by the way, it follows that the addition and multiplication of ordinals on ω is precisely the addition and multiplication of natural numbers. Thus, we have practically proven that Peano arithmetic is modeled on the ordinal ω. Let us list some properties of ordinal arithmetic.

1° $S(\alpha) = \alpha + 1, 0 + \alpha = \alpha, 0 \cdot \alpha = 0, \alpha \cdot 1 = \alpha = 1 \cdot \alpha$.

2° Non-commutativity: $\omega + 2 \neq 2 + \omega$, $\omega \cdot 2 \neq 2 \cdot \omega$.

3° Right distributivity fails: $(1+1)\cdot\omega = 2\cdot\omega = \omega$. But $(1\cdot\omega)+(1\cdot\omega) = \omega+\omega$.

4° Associativity: $(\alpha + \beta) + \gamma = \alpha + (\beta + \gamma)$, $(\alpha \cdot \beta) \cdot \gamma = \alpha \cdot (\beta \cdot \gamma)$.

5° Left distributivity: $\alpha \cdot (\beta + \gamma) = (\alpha \cdot \beta) + (\alpha \cdot \gamma)$.

6° Right strict monotonicity of addition: $\beta < \gamma \Leftrightarrow \alpha + \beta < \alpha + \gamma$.

7° Right strict monotonicity of multiplication: if $\alpha > 0$, then $\beta < \gamma \Leftrightarrow \alpha \cdot \beta < \alpha \cdot \gamma$.

8° Left non-strict monotonicity: if $\alpha < \beta$, then $\alpha+\gamma \leqslant \beta+\gamma$ and $\alpha\cdot\gamma \leqslant \beta\cdot\gamma$.

9° No zero divisors: if $\alpha \cdot \beta = 0$, then $\alpha = 0$ or $\beta = 0$.

10° Uniqueness of subtraction: if $\beta \leqslant \alpha$, then there exists a unique ordinal γ such that $\beta + \gamma = \alpha$.

11° Division with remainder: for any α and $\beta > 0$, there exists a unique pair of ordinals γ, δ such that $\alpha = \beta \cdot \gamma + \delta$, where $\delta < \beta$.

Let us return to the stated topic of defining the rank of a set. Using recursive definitions, we can construct the hierarchy of von Neumann universes V_α according to the following rules:

$$V_0 = \emptyset, \quad V_{S(\beta)} = \mathcal{P}(V_\beta), \quad V_\lambda = \bigcup\{V_\alpha \mid \alpha < \lambda\}, \tag{C1.4}$$

or, within the framework of the original definition (C1.3), utilizing a generator that takes the union of power sets of the range of the history function, they are defined as follows:

$$V_\alpha = \bigcup\{\mathcal{P}(V_\beta) \mid \beta < \alpha\}.$$

Next, it is not difficult to show that if $\beta < \alpha$, then $V_\beta \subsetneq V_\alpha$, i.e., the sequence of universes we have constructed is strictly monotonic with respect to set inclusion. Indeed, let $\beta < \alpha$. We can immediately assume that $0 < \beta$, since for $\beta = 0$ we get $V_0 = \emptyset \subseteq V_\alpha$. For $\beta \neq 0$, the set V_β is not empty. Let $x \in V_\beta$. Then, following the definition (C1.4), there exists a $\gamma < \beta$ such that $x \in \mathcal{P}(V_\gamma)$. But then, since $\gamma < \alpha$, we have that $x \in V_\alpha$. This means that $V_\beta \subseteq V_\alpha$. Furthermore, since $\beta < \alpha$, it follows that $\mathcal{P}(V_\beta) \subseteq V_\alpha$ by virtue of definition (C1.4), from which it follows that $V_\beta \in V_\alpha$. From this, by the Axiom of Regularity, we get that $V_\beta \neq V_\alpha$.[11]

Along the way, we have also proven a property of universes analogous to a property of ordinals: $V_\beta \in V_\alpha \leftrightarrow V_\beta \subsetneq V_\alpha$. Acting within the framework of universal algebra, we would say that the class of ordinals and the class of universes are order-isomorphic, where order in both cases is understood as the membership relation. And at the same time, we have obtained another variety of transitive sets—the universes V_α (although their elements are not always transitive, so the universes, except for V_0, V_1, are not ordinals).

We have already noted before, and we repeat now, that the universe V_ω is a model of the theory HF of hereditarily finite sets. It is precisely all its elements, and only them, that are represented by finite bracket notations. It is also worth noting that the universe $V_{\omega+\omega}$ is a model for Zermelo's theory Z, and Zermelo's theory is not capable of proving the existence of this universe, just as HF is not capable of proving the existence of the universe V_ω. This circumstance emphasizes that the Axiom of Replacement is a fundamentally stronger principle than Separation. Separation is by its nature "limited"—it

[11] The latter can be obtained without the Axiom of Regularity, using Cantor's theorem and the Cantor-Bernstein-Schroeder theorem: indeed, if $V_\beta \subseteq V_\alpha$, then under the assumption $V_\beta = V_\alpha$, we would have a bijection between $\mathcal{P}(V_\alpha)$ and V_α by the Cantor-Bernstein-Schroeder theorem, which is impossible by Cantor's theorem.

only creates subsets of already existing sets. Replacement, on the other hand, allows for the creation of potentially "new" sets, whose elements may not have a simple connection (through membership or rank) with the elements of the original set.

Theorem C1.4. *Every set is an element and, therefore, a subset of some universe in the von Neumann hierarchy.*

The proof of this theorem uses the concept of the transitive closure of a set and is more technical than educational in nature, so we will leave it for Part II (see Theorem D4.7).

For us, this theorem means that we can give the following definition: the **rank of a set** x is the least ordinal α such that $x \subseteq V_\alpha$.[12] Notation: $\operatorname{rank}(x)$.

The rank of a set has the following properties.

1° if $a \in b$, then $\operatorname{rank}(a) < \operatorname{rank}(b)$;

2° $\operatorname{rank}(\emptyset) = 0$, $\operatorname{rank}(n) = n$ for $n \in \omega$, $\operatorname{rank}(\omega) = \omega$, and in general $\operatorname{rank}(\alpha) = \alpha$;

3° $(\operatorname{rank}(A) < \alpha) \leftrightarrow (A \in V_\alpha)$;

4° $\operatorname{rank}(\mathcal{P}(A)) = S(\operatorname{rank}(A))$;

5° if $A \neq \emptyset$, then $\operatorname{rank}(\bigcup A) \leqslant \operatorname{rank}(A)$;

6° $\operatorname{rank}(\{a_1, \dots, a_n\}) = \max\{\operatorname{rank}(a_1), \dots, \operatorname{rank}(a_n)\} + 1$.

These properties make rank a powerful tool for analyzing the structure of sets and the entire von Neumann universe.

Returning to the representation of a set as a tree-like expansion of the membership relation, which we talked about in section B2.2, we can now specify the exact correspondence between the rank of a set and a numerical characteristic of such a tree, called the height of the tree. Indeed, the last property allows us to recursively calculate the rank of a hereditarily finite set, each time choosing the element with the highest rank and adding one. Thus, the rank is equal to the maximum length of a root path in the tree-expansion of the membership relation!

Moreover, if we represent the ordinal ω as such a tree, then for each element $n \in \omega$, the height will be equal to n, which means that to calculate the rank of ω, we need to take the supremum of $\{n + 1 \mid n \in \omega\}$, which is ω, as we expect from an infinite tree. Finally, if we move to the set $\{\omega\}$, then according to the formula for calculating the rank, it turns out that $\operatorname{rank}(\{\omega\}) = \omega + 1$,

---

[12] We do not use $x \in V_\alpha$ here, so that the rank of the empty set turns out to be zero, not one.

and this ordinal is strictly greater than ω. Thus, we have closed the problematic issue of how to jump over limit cases—when an additional tail in the form of a transition to $\{\omega\}$ grows on top of the tree of ω. It turns out that we need to act in stages, level by level: first calculate the ranks of the elements of the set, and then move on to the rank of the set itself. Finitary intuition, however, suggested a different scheme to us: to count the lengths of all root paths at once, regardless of the degree of branching. And this, as already noted, leads to the fact that it would be difficult to distinguish the ranks of infinite sets from ω.

The tree-like expansion also suggests to us the concept of a transitive set: in the tree-like expansion of a transitive set, there are direct edges from the root to all, without exception, vertices of this tree. Using transport semantics, one could say that the transitivity of the transport network of a certain city means that all suburbs can be reached by direct shuttles. This property makes the tree of a transitive set "flat" or "contracted"; in graph theory, we would say that the graph has a universal vertex and that a star graph can be taken as one of the spanning trees of our set-graph.

From this, it is easy to understand how the transitive closure of a set is constructed: all root paths must be duplicated with direct edges (introduce shuttles into the transport network). The result will be the graph of the transitive closure.

By the way, from this same concept, it is easy to see that in a transitive set, the $\in$-minimal element will always be $\emptyset$ (within the framework of the axiom of regularity, i.e., in the theory ZF). However, this is clear even without graphs, because if X is transitive, and $a \in X$ is such that $X \cap a = \emptyset$, then by virtue of transitivity, we have $\emptyset = X \cap a = a$.

From this point of view, one cannot help but look at ordinals in a new way. Recall that an ordinal is a transitive set, each element of which is transitive. This means that in our transport system, there are shuttles not only from the center to the suburbs, but also from all closer points to more distant ones, i.e., in the direction away from the center. In other words, if in the tree-expansion of an ordinal there is a path from vertex u to vertex v, then there must also be an edge uv. The graph of such a set is called transitively closed. This means that all such paths that start at the root, end at zero, and pass through any set of intermediate-rank vertices are present, which allows us to determine the number of leaves of such a tree. Thus, in the tree-expansion of the ordinal n, there are as many leaves as there are elements in the set $\mathcal{P}(\{0, \ldots, n-1\})$, i.e., 2^n. And here we again notice a deep connection between ordinals and the von Neumann universes.

The existence of the von Neumann hierarchy with the specified properties is key to understanding the structure of the universe of sets in the axiomatics of ZF. It shows how, with the help of ordinals, one can introduce a strict measure of the complexity of sets (rank) and how the entire universe can be represented as a hierarchy V_α, constructed using the operations of taking the power set and union at limit stages. The concepts of transfinite induction and recursion, based on ordinals, are the most important tools for working in set theory. The representation of the universe of sets as the limit of the von Neumann hierarchy is closely related to the Axiom of Regularity (Foundation), which guarantees that every non-empty set has an $\in$-minimal element and excludes infinite descending chains of membership, ensuring a "layered" structure of the universe.

The concept of rank, thus, is not just a technical tool, but a fundamental aspect of the modern understanding of set theory. It reflects the idea that the universe of sets is built iteratively and orderly, which allows avoiding many paradoxes and provides a solid foundation for mathematics.

C1.4 Constructions

C1.4.1 Structures

Now that we have introduced the concept of an arbitrary Cartesian product of sets and an arbitrary ordered tuple, we can generalize the concept of a relation. We will say that any subset $R \subseteq A_1 \times \cdots \times A_n$ is an *n-ary relation*. In this case, if $n = 1$, then R is simply a subset of A_1 (a *unary* relation), if $n = 2$, R is a *binary* relation, coinciding with what we defined earlier (a subset of $A_1 \times A_2$), and we exclude the case $n = 0$.

Similarly, if a function $f : A_1 \times \cdots \times A_n \to B$ is given, we call it an *n-ary function* and instead of $f(\langle a_1, \ldots, a_n \rangle)$ we write more simply: $f(a_1, \ldots, a_n)$. We also exclude the case $n = 0$ for uniformity with relations, although it is often customary to consider 0-ary functions as constants.

A function of the form $f : A^n \to A$ is called an n-ary *operation* on the set A.

Note that n-ary relations, functions, and operations exist independently of the Axiom of Choice, since a choice function with a finite domain exists regardless of the acceptance or rejection of the Axiom of Choice: a finite choice can always be made, simply by writing a formula with the corresponding number of existential quantifiers. The essence of the Axiom of Choice is precisely that it allows for a simultaneous infinite choice, which cannot be guaranteed by the finite means of predicate logic. One should not contrast the Axiom of Choice with a *definable* choice, i.e., one where the chosen elements can be explicitly calculated. No, even in the finite case, the choice provided by predicate logic does not have to be specific. The Axiom of Choice is something like a way to assert infinitely many existential quantifiers in a row, without resorting to formulas of infinite length.

Finally, we arrive at the central place of the conceptual apparatus of mathematics. An ordered tuple $\mathcal{M} = \langle M, \mathcal{R}, \mathcal{F}, C\rangle$, where M is an arbitrary *non-empty* set, $\mathcal{R}$ is a set of relations (of arbitrary arity) on M, $\mathcal{F}$ is a set of operations (of arbitrary arity) on M, and C is a subset of M, called the set of constants, is called a **structure**, where the set M is called the **universe** of this structure.

It is worth noting that there are discrepancies in the terminology of structures in the literature. In logic and model theory, the term **structure** is fundamental. Such a structure $\mathcal{M}$ becomes a **model** only in the context of a specific formal theory interpreted within it (i.e., it is a model *of a theory*). Often, logical structures do not contain operations at all and are called *relational structures*. However, in textbooks on algebra and other disciplines, the term "algebraic structure" is applied only to structures with a non-empty set of operations and a single relation—equality. Although often (for example, in universal algebra) algebraic structures are supplemented with an order relation, and in this case, they are called ordered algebraic structures. We will use the concise term *structure* in the general case.

Note that any operation $f : A^n \to A$ can be considered as an $n + 1$-ary relation, since there is a natural correspondence between the ordered tuples $\langle\langle a_1, \ldots, a_n\rangle, b\rangle$ and $\langle a_1, \ldots, a_n, b\rangle$, and in general, any function is by definition a relation. So one could say that in fact all structures are relational. However, explicitly distinguishing sets of relations, operations, and constants helps to better understand the composition of a structure, as well as to model the terms of a formal language, which is a critically important factor for conceptual clarity and convenience in constructing and interpreting formal theories. The irony of reducing operations to relations is that relations themselves are, by definition, sets of (choice) functions, which once again emphasizes the absence of a primary status for these concepts.

In the foundations of mathematics, structures are used to build a model of a formal mathematical theory, thereby reducing its consistency to the consistency of ZF, in which we have unconditional faith, because the intuitive clarity and constructiveness of both the theory HF and its bracket model give us confidence in the set-theoretic approach itself, which is generalized in ZF. In the theory ZF, we can easily construct models of HF and Zermelo's theory (without the axiom of replacement) using the universes V_ω and $V_{\omega+\omega}$, but such a justification for the consistency of theories looks more like a trick, as it refers to the consistency of a stronger theory that generalizes them.

Therefore, we will consider our reasoning with bracket notations for sets to be convincing enough to regard the theory of hereditarily finite sets as consistent. Moreover, both the bracket model of hereditarily finite sets and all the grammars of formal languages in a finite alphabet that we have considered represent regularly arranged *infinite* sets of *finite* strings, which are themselves witnesses to the possible satisfiability of the axiom of infinity. This gives us the right, from a purely philosophical standpoint, to consider the ZF axiomatics as a reliable way to formalize the general mathematical set theory that is implicitly assumed everywhere in mathematics as a given.

Thus, we construct the foundation of mathematics from two pillars: formal set theory and predicate logic. Each of these theories relies on the other, allowing the entire system of mathematical constructions to be held "in a vacuum," much like the Earth and the Moon are held together in space. Logic provides the language and rules of inference for formulating and proving theorems in set theory, and set theory provides the semantics (models) for interpreting logical formulas.

To avoid repetition, we will not reproduce here the examples of models for various theories mentioned earlier in Section C1.4.3, leaving this as an exercise for the reader. The only difference between the previously constructed models and "real" set-theoretic structures is that as the base set, we must now choose specific sets definable in ZF (and not boxes of pencils).

Instead, we will provide a general methodological approach to constructing various structures within ZF (ZFC) and build several specific examples of models of numerical structures.

C1.4.2 Principles of Structure Building

So, to begin with, let us note that any more or less significant object in mathematics has the characteristics of a structure, i.e., this object is built on some base set by defining relations and operations on it. Often this construction (like any construction) goes, so to speak, beyond the budget—it begins to use elements not only from the original universe set but also some additional constructions. Of course, there are methods that allow one to quite cleverly turn extra-structural constructions into structural ones, by expanding the base set in some way, and we will soon see such rather vivid examples. However, in general, the whole ideology revolves around a correctly chosen universe and the relations and operations defined on it.

In what ways can we construct sets? We do not have such a large arsenal of tools for this, which can be divided into two classes: those that increase the rank and those that do not. In Table C1.2, we have listed several elementary operations, breaking them down by these two classes.

Table C1.2. Operations and their relation to set ranks

Rank-increasing	Non-rank-increasing
$\{a\}, \{a, b\}, \ldots$	$\cup, \cap, \setminus, \triangle$
$\langle a, b\rangle, A \times B$	choice
$\mathcal{P}(A)$	$\bigcup A$
replacement	separation

Note that operations such as forming a pair, an ordered pair (Kuratowski's), and the Cartesian product of sets, although they increase the rank, do so insignificantly—by 1-2 steps. And only the axiom of replacement allows one to rapidly increase the rank, as it allows a transition from a set with a small rank to a set with any rank whatsoever. As has been noted, for example, the construction of the mapping $n \mapsto \omega + n$ already throws us from the realm of finite ranks into infinite ones. But definable mappings used as a generator for transfinite recursion allow for very high jumps up the ordinal scale. For example, the enumeration of alephs $\aleph_\alpha$ is an enumeration of infinite cardinalities (cardinals) using ordinals. Thus, $\aleph_0$ is the first infinite cardinal, i.e., ω, and $\aleph_1$ is the first uncountable cardinal. There is also the enumeration $\beth_\alpha$, which at each step passes from the cardinality $\beth_\alpha$ to the cardinality of the set $\mathcal{P}(\beth_\alpha)$. The sequence $\beth$ grows faster than $\aleph$ if we reject the generalized continuum hypothesis. There are other, much faster, transfinite sequences.

On the other hand, the decrease in our rank is usually not so rapid. Standard operations either do not change the rank at all (like the union of sets) or can lower it (like the intersection of sets), but only to the ranks of those sets that were already elements of the original ones. However, the axiom of regularity, for example, can drop the rank straight to zero by choosing the right element for the intersection. The axiom of separation obviously also does not increase the rank, acting by analogy with the intersection of sets. To be fair, it should be noted that the axiom of replacement can decrease the rank as much as one likes, so it can be considered a "mage beyond categories" or a queen in chess, which moves as it pleases.

By choice here, we mean the direct use of the axiom of choice, which from an initial set A creates a certain special subset in $\bigcup A$, the range of the choice function. Of course, one can say that we are talking about a function here, and therefore about an increase in rank, but one can also look at the axiom of choice as a way to choose a so-called *cross-section* of a family of sets.

If there is a family of non-empty sets $\{A_i \mid i \in I\}$, then a cross-section of this family is a set $\{s_i \mid i \in I\}$ such that $s_i \in A_i$ for all $i \in I$. Strictly speaking, we had one function $A(i)$ defined on I, and we got another function $s(i)$ defined on I, and if we compare the ranks of these functions, the rank of the second will not be higher, and may even be lower, than that of the first.

Knowing the elementary, i.e., given by axioms, tools for constructing sets, we can use their combinations to create already quite complex structures.

Suppose we have already somehow defined a base set M. To define some relation $R \subseteq M^n$ on it, we first go for a rank increase, creating the Cartesian product M^n (either as a combination of binary Cartesian products or through choice functions), and then we cut out a certain subset from the resulting set, without increasing the rank. A two-stroke cycle occurs: increase-decrease. And if the relation R is not total, but covers only part of the set M, then the rank of R can fall below the rank of M.

A similar action occurs when we define operations, which, as already noted, are essentially also relations on M. True, operations are always totally defined, so their rank will not fall below $\mathrm{rank}(M) + 2$ (the two is due to the ordered pair construction).

In addition, a structure may require unary relations, i.e., subsets of M. Here we can talk either about the direct application of the axiom of separation (we do not increase the rank), or we can first create the power set $\mathcal{P}(M)$, increasing the rank, and then cut out a certain part from the power set, without increasing its rank. This is how topology,[13] an algebra of sets, a filter,[14] a quotient set, and other systems of subsets of the set M are usually defined.

Why is it important for a logician to keep track of the described wandering along the rank scale when constructing various structures? The fact is that this allows us to track how powerful the von Neumann universes are required for the construction of these structures, and thus indirectly to assess the complexity of the set theory model in which we define these structures. The complexity of the model, in turn, indicates the axiomatic apparatus that may be required for the implementation of the corresponding structures.

Note that so far we fit well within the definition of a structure given above, as we do not stray far from the original set M.

But let us now see what the application of the axiom of replacement can give us. In fact, most mathematical structures are extremely difficult to create without it! And here is the first example—a metric space. As is known, this is a set with a metric defined on it. But a metric is a function $d : M^2 \to \mathbb{R}$, i.e., its range is outside any powers of the set M and, generally speaking, is not connected with it in any way. In other words, a metric is not an operation on M.

What is to be done? Firstly, the axiom of replacement allows the creation of such a function if there is a rule for associating a pair of elements of M with a real number. Secondly, one can use a "cheating" trick that allows one to cram the metric into the definition of the structure, namely: one must consider the set $M \cup \mathbb{R}$ as the universe of the structure, and then the metric will be a legitimate operation in such a structure (although it will have to be cleverly extended to this whole set, but that is a technical matter).

A second, even more sophisticated way to fit into the canonical definition of a structure is to add a huge number of separate symmetric relations R_α on M, consisting of all pairs of elements of M, the distance between which is equal to α, where $\alpha \in \mathbb{R}$. After all, no one forbade us in the definition of a structure to index relations and operations by some external sets. Thus, for each possible distance between points of M, we will have our own binary relation on M defined, and formally we will fit into the definition of a structure.

---

[13] The general concept of a topological space was introduced by Felix Hausdorff in his book "Grundzüge der Mengenlehre" (Basics of Set Theory) in 1914, which made it possible to unify the study of continuity and limit transitions in a wide variety of mathematical contexts.

[14] The concept of a filter was introduced by Henri Cartan in 1937 as a generalization of the concept of the limit of a sequence, which turned out to be very fruitful in general topology and other areas of analysis.

A similar trick can be used when we want to fit the concept of a "module over a ring" into the definition of a structure. Recall that a module over a ring is essentially two structures, connected in a certain close way. The first is an algebraic ring R (scalars), the second is a set M (vectors), which itself is an abelian group (i.e., a commutative vector addition operation, the taking of the opposite vector, and the constant 0 are defined). And between these structures, there is an operation of multiplying a vector by a scalar (or a scalar by a vector, if it is a left module) $P : M \times R \longrightarrow M$. And again we see that the operation falls out of both generating structures at once. Therefore, considering each scalar as a parameter, one can consider a set of operations $P_\lambda : M \longrightarrow M$ instead of a single operation P, thereby fitting the original operation into the definition of the structure.

Of course, such tricks do not do credit to set theory; they only emphasize that, if desired, the term "structure" can imply a rather simple thing—something that is generated by a base set with the involvement of some external parameterizing sets or variables. This allows for simplifying the formal-logical analysis of mathematics, avoiding the truly huge variety of mathematical objects that are present throughout mathematics.

Therefore, although Hilbert assured us that set theory is a paradise created by Cantor from which no one will expel us, this paradise has a bunch of dark cellars in the form of difficulties associated with the simplicity of the language and, as a consequence, the enormous complexity of formulas expressing rather simple mathematical facts, as well as the fact that many intuitively identical things turn out to be structured completely differently within ZF, sometimes even too inventively.

But on the other hand, the goal of formal set theory is rather to justify mathematics, to bring all its theories to a common logical basis, to "ground" them in first-order logic, and not to immerse the entire great variety of mathematical sciences into a single formal language. Here one can trace an analogy with programming languages: there is a huge number of them in the world, adapted to different user needs—from the internal languages of microcircuits to universal high-level languages like Python or Rust, but all of them can be grounded in Assembly or directly in machine code.

Therefore, mathematics is rather a multitude of formal theories than a multitude of structures within a single theory ZF. But the ability to build models for these theories within the framework of set theory is a categorically necessary skill for every mathematician.

So, as we see, the axiom of replacement allows us to move quite far from the universe of a structure M, involving various auxiliary sets like $\mathbb{R}$ or abstract rings. Therefore, the models of set theory with the axiom of replacement are huge models from the point of view of the von Neumann hierarchy; their rank is measured by special cardinal numbers that exceed all reasonable constructive ordinals.

On the other hand, if we know in advance how a particular auxiliary structure is arranged, then we have no need to use the axiom of replacement to its full power: it is sufficient to use its consequence—the axiom of separation, since a function of the form $P : M \times R \to M$ can be defined as part of the set $(M \times R) \times M$, and the rank increase here will be quite predictable and small. This means that when we build complex mathematical structures in stages, starting from, for example, the real numbers, we can in principle get by with Zermelo's theory Z, which is fully modeled in the universe $V_{\omega+\omega}$, which no longer seems to us to be something overwhelmingly complex and unimaginable. The only question is how to "ground" the basic mathematical structures, such as the field of real numbers, there.

C1.4.3 A Concrete Discussion about Models

Finally, to consolidate the skill described above of constructing various structures, let us build a few concrete examples.

To begin, we note that the structure $\mathbb{N} = \langle \omega, \emptyset, \{+, \cdot, S\}, \{\emptyset\} \rangle$, i.e., the first infinite ordinal ω as the universe, and the ordinal operations of addition and multiplication, as well as the operation S and the constant $\emptyset$, is a model of Peano arithmetic, which we considered in Section B2.1.

Here, in terms of constructing the structure itself, there is nothing extraordinary; the entire burden of construction complexity is transferred to the definition of ordinals and operations on them.

Next, it is worth mentioning the previously mentioned model V_ω of hereditarily finite sets. Essentially, this model only has the predicates of membership and equality, which play the role of themselves. It is not difficult to verify that all the axioms of HF are satisfied here, if the variables of the theory are restricted to the universe V_ω.

The universe $V_{\omega+\omega}$ is a model of the theory Z. As we remember, it must include the axiom of infinity, i.e., contain the ordinal ω, from which it follows that the model must also include all sets that are obtained from ω by the operations permitted by the axiomatics, and these are all the operations from Table C1.2, except for the one implemented by the axiom of replacement, as it is not in the system Z. But all such operations give only a finite rank increase, which means that all sets will have a rank of either n or $\omega + n$, where $n \in \omega$.

Quine's model is a special case of a model. It is a reflexive relation on a singleton. Let us take an arbitrary set s and set $M = \{s\}$, and then define the relation $\varepsilon = \{\langle s, s\rangle\}$ on it. Such a structure $\langle M, \varepsilon\rangle$ is a model of a very special piece of set theory without the axioms of regularity, infinity, empty set, and separation (see Table C1.1), where the loop-relation ε models a non-well-founded membership relation: $(s \,\varepsilon\, s)$. In such a model, equality is interpreted as equality, congruence and extensionality hold due to there being only one element, the axiom of pairing works, since one can construct the unique pair $\{s, s\}$, which is equal to $\{s\}$, the axiom of power set works ($\mathcal{P}(\{s\}) = \{\emptyset, \{s\}\}$), union ($\bigcup\{s\} = s$), the axiom of replacement works, since any definable mapping will define the correspondence $s \mapsto s$, and also the axiom of choice, since there is nothing to choose here but s.

Quine's model illustrates that the axiom of separation without the axiom of the empty set cannot be derived from the axiom of replacement.

Now let us consider the model $V_\omega \setminus \{\emptyset\}$, in which membership and equality model themselves. Such a model will satisfy only the following axioms: congruence, extensionality, pair, power set, union, regularity, replacement, choice, and also the negation of the axiom of infinity. Here we again see the absence of the empty set and, as a consequence, the axiom of separation. But with the presence of the axiom of regularity.

By constructing in this or a similar way sufficiently simple models of different sets of axioms of set theory, we can show their independence from each other or, conversely, emphasize some (partial) dependencies.

Let us now construct a model of the real numbers. More precisely, we will construct a model of the theory RCF of real closed (ordered) fields. RCF is given in the signature $\langle =, \leqslant, +, \cdot, 0, 1\rangle$ and has the following axiomatics:

RCF1 axioms of linear order for $\leqslant$;

RCF2 standard axioms of a field;

RCF3 consistency of order with operations:

$$(a \leqslant b) \to (a + c \leqslant b + c), \quad (0 \leqslant a) \wedge (0 \leqslant b) \to (0 \leqslant ab);$$

RCF4 $\forall y\ (0 \leqslant y) \to \exists x\ (x \cdot x = y)$;

RCF5 $\forall q_1, \ldots, q_n\ \exists x\ (x^n + q_1 \cdot x^{n-1} + \cdots + q_n = 0)$, where x^n is written as an $(n-1)$-fold composition of multiplication, and the number n is odd.

The last axiom is an axiom schema, since for each n we need to write our own separate formula with n variables. Axiom RCF4 states that any non-negative number is the square of some number.

It is known about the theory RCF that it is complete (any sentence of the language RCF is either provable or refutable within the listed axioms) and decidable (there exists an algorithm that determines, for any sentence, whether it is provable or refutable). This is a famous result by A. Tarski. The field of real numbers is the standard, but not the only, model of this theory.

In order to construct a model of RCF, we will first construct the standard model of the integers $\mathbb{Z}$. There are several approaches here; we will choose the most economical one. We will declare the universe of $\mathbb{Z}$ to be the set $Z = (\omega \times \{0\}) \cup (\{0\} \times \omega)$. Visually, these are two copies of the ordinal ω, glued together by a common zero (the pair $\langle 0, 0\rangle$). We will consider pairs of the form $\langle n, 0\rangle$ as negative numbers, and pairs of the form $\langle 0, n\rangle$ as positive numbers. For the integers, we need to define the order and operations.

Order: $\langle a_1, a_2\rangle \leqslant_{\mathbb{Z}} \langle b_1, b_2\rangle$ if $a_2 = b_2 = 0$ and $b_1 \subseteq a_1$, or $a_1 = b_1 = 0$ and $a_2 \subseteq b_2$, or $a_2 = b_1 = 0$.

That is, the comparison on the right copy of ω is standard, while on the left it is reversed, and any negative is less than or equal to any positive.

Addition:

$$\langle a_1, a_2\rangle +_{\mathbb{Z}} \langle b_1, b_2\rangle$$
$$= \begin{cases} \langle a_1 +_\omega b_1, a_2 +_\omega b_2\rangle, & (a_1 = b_1 = 0) \vee (a_2 = b_2 = 0), \\ \langle a_1 \dot{-} b_2, 0\rangle & (a_2 = b_1 = 0) \wedge (b_2 \subseteq a_1), \\ \langle 0, b_2 \dot{-} a_1\rangle & (a_2 = b_1 = 0) \wedge (a_1 \subseteq b_2), \\ \langle b_1 \dot{-} a_2, 0\rangle & (a_1 = b_2 = 0) \wedge (a_2 \subseteq b_1), \\ \langle 0, a_2 \dot{-} b_1\rangle & (a_1 = b_2 = 0) \wedge (b_1 \subseteq a_2). \end{cases}$$

Here, the point is to understand if the numbers have different signs (if not, we simply add them coordinate-wise) and what to do in the case of different signs, because when adding numbers with different signs, one needs to determine which one is larger in magnitude, find the difference, and set the correct sign for this difference. By the way, the subtraction operation on natural numbers (*monus*) is defined as follows: $a \dot{-} b$ is equal to their difference if $a \geqslant b$, and 0 otherwise (see Section D3.3).

Multiplication:

$$\langle a_1, a_2 \rangle \cdot_{\mathbb{Z}} \langle b_1, b_2 \rangle$$
$$= \begin{cases} \langle 0, (a_1 +_\omega a_2) \cdot_\omega (b_1 +_\omega b_2) \rangle, & (a_1 = b_1 = 0) \vee (a_2 = b_2 = 0), \\ \langle (a_1 +_\omega a_2) \cdot_\omega (b_1 +_\omega b_2), 0 \rangle, & \text{otherwise.} \end{cases}$$

Here it is simpler, as multiplying numbers with the same sign results in a non-negative number, and with different signs—a non-positive number.

Zero in Z will be $0_{\mathbb{Z}} = \langle 0, 0 \rangle$, and one will be $1_{\mathbb{Z}} = \langle 0, 1 \rangle$.

It can be verified that the relation $\leqslant_{\mathbb{Z}}$, the operations $+_{\mathbb{Z}}$, $\cdot_{\mathbb{Z}}$, and the constants $0_{\mathbb{Z}}$, $1_{\mathbb{Z}}$ defined in this way satisfy the axioms of an ordered discrete commutative ring with unity, of which the structure $\mathbb{Z} \overset{\text{def}}{\rightleftharpoons} \langle Z, =, \leqslant_{\mathbb{Z}}, +_{\mathbb{Z}}, \cdot_{\mathbb{Z}}, 0_{\mathbb{Z}}, 1_{\mathbb{Z}} \rangle$ is a model.

Next, one can proceed in different ways. For example, we can recall that a real number is the sum of an integer and a fractional part, which can be expressed as a countable binary sequence. The universe of such a model is easy to construct, but it is very difficult to define the operations of addition and multiplication in it. Therefore, we will take a different, more classical path, by first defining the set of rational numbers as equivalence classes of fractions.

Let $Q = \{\langle p, q \rangle \mid (p \in Z) \wedge (q \in Z) \wedge (q >_{\mathbb{Z}} 0_{\mathbb{Z}})\}$. On the elements of Q, we introduce an equivalence relation:

$$\langle p_1, q_1 \rangle \sim \langle p_2, q_2 \rangle \overset{\text{def}}{\leftrightarrow} p_1 \cdot_{\mathbb{Z}} q_2 = p_2 \cdot_{\mathbb{Z}} q_1.$$

Informally: $\frac{p_1}{q_1} \sim \frac{p_2}{q_2}$ if and only if $p_1 q_2 = p_2 q_1$. It is not difficult to check that $\sim$ is indeed an equivalence relation on Q. This means that Q can be partitioned into disjoint equivalence classes of the form

$$[\langle p, q \rangle]_\sim = \{\langle x, y \rangle \mid \langle p, q \rangle \sim \langle x, y \rangle\}.$$

Now we form the quotient of the set Q by the relation $\sim$, i.e., we consider the quotient set $Q' = Q/_\sim = \{[x]_\sim \mid x \in Q\}$. Factorization is another example of increasing the rank of sets, which combines two elementary steps: taking the power set and separating a subset from it.[15]

Each equivalence class now represents what we called a fraction or a rational number in school. We could have dispensed with equivalence classes and worked only with irreducible pairs of numbers $\langle p, q \rangle$, i.e., immediately include a restrictive predicate in the definition of Q expressing that the numbers p and

[15] The construction of quotient structures (quotient groups, quotient rings) via equivalence classes is a powerful method of abstract algebra, largely due to the work of Emmy Noether and her school in the 1920s, who systematized and generalized these constructions.

q are coprime. But this path leads to a complication in the definition of the operations.

Henceforth, we will omit the tilde in the notation of equivalence classes. Having dealt with the universe, we now define the order relation, constants, and operations on the elements of Q'.

Order: $[\langle p, q\rangle] \leqslant_{\mathbb{Q}} [\langle p', q'\rangle]$ if $p \cdot_{\mathbb{Z}} q' \leqslant_{\mathbb{Z}} p' \cdot_{\mathbb{Z}} q$ (the usual rule for comparing fractions). True, there is a subtle point here: we define the predicate on equivalence classes through their representatives, but will the same be true for other representatives of these classes (well-definedness)? This must always be checked, and in this case, it is not difficult to do so. Let $\langle p, q\rangle \sim \langle a, b\rangle$ and $\langle p', q'\rangle \sim \langle a', b'\rangle$, will the condition $a \cdot_{\mathbb{Z}} b' \leqslant_{\mathbb{Z}} a' \cdot_{\mathbb{Z}} b$ then be true?

Let us multiply the inequality $p \cdot_{\mathbb{Z}} q' \leqslant_{\mathbb{Z}} p' \cdot_{\mathbb{Z}} q$ by $b \cdot_{\mathbb{Z}} b'$ (the sign will not change, as these are positive denominators): $p \cdot_{\mathbb{Z}} q' \cdot_{\mathbb{Z}} b \cdot_{\mathbb{Z}} b' \leqslant_{\mathbb{Z}} p' \cdot_{\mathbb{Z}} q \cdot_{\mathbb{Z}} b \cdot_{\mathbb{Z}} b'$. From the definition of the equivalence of pairs, we extract that $p \cdot_{\mathbb{Z}} b = a \cdot_{\mathbb{Z}} q$ and $p' \cdot_{\mathbb{Z}} b' = a' \cdot_{\mathbb{Z}} q'$, from which, using the congruence of equality, we arrive at $a \cdot_{\mathbb{Z}} q \cdot_{\mathbb{Z}} q' \cdot_{\mathbb{Z}} b' \leqslant_{\mathbb{Z}} a' \cdot_{\mathbb{Z}} q' \cdot_{\mathbb{Z}} q \cdot_{\mathbb{Z}} b$, after which, by canceling the positive $q \cdot_{\mathbb{Z}} q'$, we obtain the required inequality. Undoubtedly, the entire algebraic arsenal of the commutative ring $\mathbb{Z}$ helps us here.

Addition: $[\langle p, q\rangle] +_{\mathbb{Q}} [\langle p', q'\rangle] = [\langle p \cdot_{\mathbb{Z}} q' +_{\mathbb{Z}} p' \cdot_{\mathbb{Z}} q, q \cdot_{\mathbb{Z}} q'\rangle]$. Here, the use of equivalence classes simplifies the definition, as we do not need to worry about the irreducibility of the fraction. But the well-definedness of the definition, of course, should also be checked.

Multiplication: $[\langle p, q\rangle] \cdot_{\mathbb{Q}} [\langle p', q'\rangle] = [\langle p \cdot_{\mathbb{Z}} p', q \cdot_{\mathbb{Z}} q'\rangle]$.

Constants: $0_{\mathbb{Q}} = [\langle 0_{\mathbb{Z}}, 1_{\mathbb{Z}}\rangle]$, $1_{\mathbb{Q}} = [\langle 1_{\mathbb{Z}}, 1_{\mathbb{Z}}\rangle]$.

Thus, we have defined the structure $\mathbb{Q} \overset{\text{def}}{\rightleftharpoons} \langle Q', =, \leqslant_{\mathbb{Q}}, +_{\mathbb{Q}}, \cdot_{\mathbb{Q}}, 0_{\mathbb{Q}}, 1_{\mathbb{Q}}\rangle$, which is the standard model of the field of rational numbers or, as an algebraist would say, the ordered field of fractions of the ring $\mathbb{Z}$.

Finally, we can move on to constructing the structure $\mathbb{R}$, which will be a model of the field of real numbers. Here too, there are several equivalent approaches, but we will choose the simplest from the point of view of defining operations. Let us consider the set $(Q')^{\omega}$ of all functions from ω to Q'. These are sequences of rational numbers, which we will traditionally write as $\langle x_n\rangle$. We say that a sequence $\langle x_n\rangle$ is **fundamental**[16] if

$$\forall \varepsilon >_{\mathbb{Q}} 0_{\mathbb{Q}} : \exists N \in \omega : \forall n, m > N : |x_n - x_m| <_{\mathbb{Q}} \varepsilon.$$

[16] Also known as a Cauchy sequence, after the French mathematician Augustin-Louis Cauchy, who laid the rigorous foundations of analysis.

Here we arrive at the familiar language of calculus from our school days. The given formula means that for sufficiently large indices, the terms of the sequence become arbitrarily close (the number ε is taken from Q', as we have nothing else). We do not provide the definition of the difference and absolute value functions for rational numbers, considering this exercise to be trivial.

Next, we define the set

$$F \overset{\text{def}}{\rightleftharpoons} \{x \mid x \in (Q')^{\omega} \wedge x \text{ is fundamental}\}$$

of all fundamental sequences. Finally, on this set, we introduce an equivalence relation:

$$\langle x_n \rangle \approx \langle y_n \rangle \overset{\text{def}}{\leftrightarrow} \forall \varepsilon >_{\mathbb{Q}} 0_{\mathbb{Q}} : \exists N \in \omega : \forall n > N : |x_n - y_n| <_{\mathbb{Q}} \varepsilon,$$

i.e., we identify sequences whose difference tends to zero. The equivalence class $[\langle x_n \rangle]_{\approx}$ under this relation will be the model of a real number. Let us set

$$R \overset{\text{def}}{\rightleftharpoons} F/_{\approx}.$$

It remains to define the constants, order, and operations.

$$0_{\mathbb{R}} \overset{\text{def}}{\rightleftharpoons} [\langle x_n = 0_{\mathbb{Q}} \rangle], \quad 1_{\mathbb{R}} \overset{\text{def}}{\rightleftharpoons} [\langle x_n = 1_{\mathbb{Q}} \rangle],$$

where $\langle x_n = r \rangle$ denotes the constant function with the value $r \in Q'$. In fact, all rational numbers are mapped into R in this way. For irrational numbers, more clever sequences are required. For example, if we want to define $\sqrt{2}$, we take a recursively defined sequence $\langle x_n \rangle$ such that

$$x_0 = 1, \quad x_{n+1} = \frac{1}{2}\left(x_n +_{\mathbb{Q}} \frac{2}{x_n}\right),$$

where the division is from the constructed field $\mathbb{Q}$. This is the so-called "Babylonian" sequence, which has the limit $\sqrt{2}$.

Further, as in the case of rational numbers, the order and operations on the equivalence classes are defined through their representatives (with proof of well-definedness).

Order:

$$[\langle x_n \rangle] <_{\mathbb{R}} [\langle y_n \rangle] \overset{\text{def}}{\leftrightarrow} \exists \delta >_{\mathbb{Q}} 0_{\mathbb{Q}} : \exists N \in \omega : \forall n > N : x_n +_{\mathbb{Q}} \delta \leqslant_{\mathbb{Q}} y_n,$$

i.e., the first sequence is strictly less than the second if, after a finite number of steps, they diverge by a distance of at least δ with the same inequality sign.

Addition and multiplication:

$$[\langle x_n \rangle] +_{\mathbb{R}} [\langle y_n \rangle] \overset{\text{def}}{\rightleftharpoons} [\langle x_n +_{\mathbb{Q}} y_n \rangle], \quad [\langle x_n \rangle] \cdot_{\mathbb{R}} [\langle y_n \rangle] \overset{\text{def}}{\rightleftharpoons} [\langle x_n \cdot_{\mathbb{Q}} y_n \rangle].$$

As we can see, the definitions are not complicated, and proving their well-definedness is also not difficult. Thus, we have completely defined the structure $\mathbb{R} \overset{\text{def}}{\rightleftharpoons} \langle R, =, <_{\mathbb{R}}, +_{\mathbb{R}}, \cdot_{\mathbb{R}}, 0_{\mathbb{R}}, 1_{\mathbb{R}} \rangle$. If we look at our constructions from the point of view of the ranks of sets, we have not gone far from the rank ω (to be more precise, $\text{rank}(Z) = \omega$, $\text{rank}(Q/_{\sim}) = \omega + 2$, $\text{rank}(R) = \omega + 5$). In the case of using Dedekind cuts, the rank would be slightly lower.[17]

We will not provide here (and not even in Part II) a proof of the truth of the RCF axioms in this model of the real numbers, as this is a standard procedure shown in many textbooks on real analysis (see, e.g., [63]). What is important for us is something else: having a very simple language of ZF, we have managed to construct specific set-theoretic structures that are models of the natural, integer, rational, and real numbers. If desired, we can continue the construction, moving to the structure $\mathbb{R}^n$ with the necessary signature, thereby obtaining complex numbers and Euclidean space, and then add the infinite-dimensional space $\mathbb{R}^\omega$, etc. In terms of the rank scale, we will not move far from ω (by a finite number of steps).

The existence of such models primarily indicates that all these number systems are consistent insofar as Zermelo-Fraenkel set theory is consistent.

Secondly, this construction is a classic educational technique for building models of any formal mathematical languages and theories.

Thirdly, we can notice that all our models are very strongly tied to specific sets, primarily the ordinal ω. But this does not at all mean that they are the only possible models. One can change both the very methodology of constructing the universes of these structures (say, choosing Dedekind cuts to construct $\mathbb{R}$ or a special kind of matrix to construct $\mathbb{Z}$), and the starting sets. The fact is that ultimately we are interested only in the algebraic properties of the newly created models, expressible in the language of the signature, and not at all in their internal structure and place in the von Neumann hierarchy. Moreover, knowledge about the structure of a specific model cannot be used to obtain theorems of the theory that we are modeling in it, because a theory is first and foremost a subset of a language. And if the language of the theory (as, for

[17] For the construction of $\mathbb{R}$ using Dedekind cuts, see [51].

example, in the language of RCF) does not have a membership predicate, then one cannot use facts from the model that contain such a predicate to work with the theory.

Of course, there are situations where a set-theoretic fact from a model can be translated into the language of the structure, i.e., expressed in the language obtained from the signature of the given structure, in which case we get a model that satisfies some sentence from the language of the theory. And if neither this sentence nor its negation is derivable from the axioms of the theory, then the model confirms its independence from these axioms.

In the case of the theory RCF, we know that this theory is complete, i.e., any sentence in its language is either provable or refutable, which means that in all its models, those and only those sentences that are derivable from the axioms are true. Everything else is from the evil one. More precisely, any other facts from the model cannot be written (expressed) in the language of the theory RCF.

A good example of such a "deceptive" fact is the Archimedean property of the standard model $\mathbb{R}$. The field that we constructed above has the Archimedean property, i.e., for any real numbers $a, b > 0$, there exists a natural number n such that $an > b$. This property cannot be expressed in the language of the signature $\langle 0, 1, \leqslant, +, \cdot \rangle$, because we cannot quantify over arbitrary natural numbers in it. Yes, we can express each specific natural number by a numeral of the form $1 + \cdots + 1$ (in a similar way we introduced numerals in PA), but we cannot write $\forall n$ or $\exists n$ and we cannot write in this language a formula whose truth domain would be only the natural numbers encapsulated in the model $\mathbb{R}$.

Since the Archimedean property is inexpressible in the language of RCF, it certainly cannot be either proven or refuted within the axioms of RCF, so there exist models of real closed fields both with and without this property.

Another example of such an inexpressible fact is the completeness of the set of real numbers. The completeness property states that any non-empty set of real numbers that is bounded above has a least upper bound. The model that we constructed has this property, but in the language of RCF, this statement is also inexpressible, as it requires the use of second-order logic (with quantifiers over sets of real numbers).

However, it is worth noting that it is precisely this second-order completeness property (often referred to as Dedekind completeness) that uniquely defines the real numbers up to isomorphism. In abstract algebra, it is a fundamental theorem that any two complete ordered fields are isomorphic. Therefore, while the first-order theory RCF admits a wide variety of non-isomorphic models (including non-Archimedean ones), the transition to second-order logic allows us to pinpoint the field of real numbers as a truly unique algebraic structure within the entire framework of set theory.

The given examples emphasize the fact that for the same theory in a given language, there can be a great variety of models, and these models may contain facts that are inexpressible and/or unprovable in the given theory.

Even with set theory itself, not everything is as smooth as Hilbert once wished. Yes, this language has colossal expressive power, but even for it, we can invent very different models, if, of course, we slightly weaken the axiomatics (recall the models of Quine and $V_\omega \setminus \{\emptyset\}$). However, with the same ZF axiomatics, we also have different models, constructed by Gödel and Cohen, confirming the independence of the axiom of choice and the generalized continuum hypothesis.

Nevertheless, in order not to be tied each time to some specific model, it is proposed to consider a class of isomorphic models that have the properties we need. An isomorphism of models is a bijection between their universes that preserves relations, operations, and constants.[18] Isomorphism is an equivalence relation on the class of all structures with a given signature, and from the point of view of abstract algebra or general topology, isomorphic structures are considered identical. For example, a group consisting of a single element is considered unique in the entire universe, despite the fact that in set theory, it has a whole proper class of isomorphic models.

At the same time, noting that structures that are models of a particular theory may have certain useful properties (for example, being Archimedean and complete) that are inexpressible in this theory, we can single out a special subclass of isomorphic models that have precisely these properties, and declare this subclass, say, the standard model of the theory. Moreover, specifically for this subclass of models, we can make the language of the theory more expressive (introduce variables for sets for it, for example), so that these "good" properties of the models from our point of view can be expressed in this language. After that, we can study the so-called *elementary theory* of this model (more precisely, of the subclass of isomorphic models), which by definition will be complete, as it includes all sentences true in the model.

Such theories usually include $\mathrm{Th}(\mathbb{N})$—the complete theory of the standard model of arithmetic, $\mathrm{Th}(\mathbb{Z})$—the complete theory of the standard model of the ring of integers, $\mathrm{Th}(\mathbb{Q})$—the complete theory of the standard model of the field of rational numbers. All three theories are given in the first-order

[18] The concept of isomorphism crystallized in the 19th century in works on abstract algebra (for example, by Arthur Cayley, who introduced the concept of an abstract group).

language of the signature $\langle =, <, +, \cdot, 0, 1 \rangle$,[19] they are complete by definition, but the set of their theorems is non-enumerable. It is all the more interesting that the complete first-order theory of the real numbers $\mathrm{Th}(\mathbb{R})$ *in the same signature* coincides with the theory RCF, i.e., is enumerable.

It might seem at first glance that there is a certain contradiction here: it seems that each time the numerical structure becomes stronger in an algebraic sense, and one would expect the same from its theory, but the theories, on the contrary, become simpler in some sense, at least this can be said about the theory $\mathrm{Th}(\mathbb{R})$. Why is that?

The fact is that all four theories, being complete, pick out different subsets of closed formulas from their common language. In other words, each of them contains sentences that are not provable in the other theories. Their sets of true sentences have no continuity; they do not follow the inclusion of sets $\mathbb{N} \subsetneq \mathbb{Z} \subsetneq \mathbb{Q} \subsetneq \mathbb{R}$. Here are examples of sentences unique to each theory:

- $\forall y \geqslant 0 : \exists x\ (y = x^2)$ (a square root can be extracted from any non-negative number)—true in $\mathrm{Th}(\mathbb{R})$ (axiom RCF), but false in $\mathrm{Th}(\mathbb{N})$, $\mathrm{Th}(\mathbb{Z})$, $\mathrm{Th}(\mathbb{Q})$;
- the statement about the existence of a number x such that $0 < x < 1$ and x is not the square of some y, is true in $\mathrm{Th}(\mathbb{Q})$, but false in $\mathrm{Th}(\mathbb{N})$, $\mathrm{Th}(\mathbb{Z})$, $\mathrm{Th}(\mathbb{R})$;
- the statement about the existence of a negative prime number is true in $\mathrm{Th}(\mathbb{Z})$, but false in $\mathrm{Th}(\mathbb{N})$, $\mathrm{Th}(\mathbb{Q})$, and $\mathrm{Th}(\mathbb{R})$;
- $\forall x\ (x \geqslant 0)$ is true in $\mathrm{Th}(\mathbb{N})$, but false in $\mathrm{Th}(\mathbb{Z})$, $\mathrm{Th}(\mathbb{Q})$, and $\mathrm{Th}(\mathbb{R})$.

We deliberately do not go into stronger logical formalisms here (for example, languages with quantification over sets of reals), so as not to overload the exposition. For our purposes, it is enough to keep in mind that RCF captures exactly the first-order part of the theory of real numbers in the signature $\langle =, <, +, \cdot, 0, 1 \rangle$, while properties such as order-completeness require a stronger language.

The theory RCF, being a complete theory,[20] is at the same time the theory of *all* its models, the class of which is significantly wider than the standard Archimedean complete field of real numbers. For detailed constructions and classical analysis background, see [51, 63].

[19] In the case of $\mathbb{N}$, the signature is slightly different (there is no 1, but there is a successor), but it can easily be rewritten in this same signature by replacing $S(x)$ with $x + 1$.

[20] It is not Gödelian, if only for the reason that it cannot interpret arithmetic.

C1.5 Summary and Self-check Questions

With this, we conclude our journey into the world of formal theories and their models. We have seen how, from a rather modest set of axioms of set theory, like building blocks, one can erect complex and diverse mathematical structures that serve as interpretations for other, more specific theories—from arithmetic to analysis. The construction of models not only demonstrates the relative consistency of these theories (assuming the consistency of set theory itself) but also allows for a deeper understanding of their nature, revealing properties that may be true in some models and false in others, or even inexpressible in the language of the original theory.

The idea of matching a formal theory with a concrete mathematical structure in which this theory "lives" has a rich history. Although the intuitive concept of a model was used in mathematics long before formalization (for example, in Euclid's geometry or in the construction of models of non-Euclidean geometries by Beltrami, Klein, and Poincaré in the 19th century), the explicit formulation of the connection between syntax (axioms and rules of inference) and semantics (interpretations, models) became possible only at the turn of the 19th-20th centuries with the development of mathematical logic.

The works of Leopold Löwenheim (1915) and Thoralf Skolem (1920s), culminating in the Löwenheim-Skolem theorem, showed that if a first-order theory has an infinite model, then it has models of any infinite cardinality. This led to the understanding that first-order axioms cannot uniquely characterize structures such as the natural or real numbers, giving rise to the concept of "non-standard models."

A key moment was Kurt Gödel's proof of the completeness theorem for first-order logic (1929), which states that any consistent first-order theory has a model. This fundamentally linked provability (a syntactic concept) with truth in all models (a semantic concept). His incompleteness theorems (1931) for sufficiently rich theories, such as Peano arithmetic, demonstrated the existence of true but unprovable statements, which also stimulated the study of non-standard models of arithmetic.

Alfred Tarski in the 1930s-50s laid the foundations of modern model theory as an independent discipline, introducing a rigorous definition of truth in a model and developing methods for investigating elementary theories. The Tarski-Seidenberg result on the completeness and decidability of the theory of real closed fields, which we mentioned, is one of the classic achievements of this era.

Finally, a revolutionary breakthrough in set theory was the development of the forcing method by Paul Cohen in the 1960s, which made it possible to construct models of set theory in which statements such as the axiom of choice or the continuum hypothesis are independent. This demonstrated the incredible flexibility and richness of the von Neumann universe and its possible "alternative versions."

Alfred Tarski

Thus, the concept of a model has come a long way from an intuitive tool to a central object of study in mathematical logic. It allows not only to build and classify mathematical structures but also to gain a deeper understanding of the limits and possibilities of formal systems, as well as the nature of mathematical truth. We hope that the introduction to this approach, presented in this book, will serve as an incentive for the reader to further study these fascinating and fundamental questions of mathematics and logic.

As usual, at the end of another "linguistic" level, we offer a few questions for self-control.

Q1 What is the difference between the various axiom systems of set theory (ZF, ZFC, Z)? [p. 172]
Q2 Explain the conceptual significance of the von Neumann universe hierarchy (C1.4) and the concept of the rank of a set. How does this hierarchy help to structure the universe of sets and understand the nature of infinite sets? [p. 198]
Q3 Discuss the conceptual impact of the Axiom of Choice on mathematics. What are its positive and negative aspects? [p. 179]
Q4 Provide examples of statements equivalent to AC.
Q5 Provide examples of statements that are weaker than AC.
Q6 Provide examples of statements that are stronger than AC or are alternatives to AC.

Exercises

C1.1 Translate the following statements into formulas of the language of ZF using the membership predicate $\in$ and logical connectives:

1. Set A is a subset of set B.
2. Sets A and B have no common elements (are disjoint).

3. Set A contains exactly two elements.
4. The set P is the power set of A (i.e., $P = \mathcal{P}(A)$).

C1.2 Formulate the following concepts and theorems as single formulas in the language of Set Theory (using auxiliary predicates if necessary):

- Cantor's theorem: there is no surjection from A to $\mathcal{P}(A)$.
- The Axiom of Countable Choice (AC_ω).
- Definition of a Filter: A family $\mathcal{F} \subseteq \mathcal{P}(S)$ is a filter on S if:

  1. $S \in \mathcal{F}$ and $\emptyset \notin \mathcal{F}$;
  2. $\mathcal{F}$ is closed under intersection (if $A \in \mathcal{F}$ and $B \in \mathcal{F}$, then $A \cap B \in \mathcal{F}$);
  3. $\mathcal{F}$ is closed under supersets (if $A \in \mathcal{F}$ and $A \subseteq B \subseteq S$, then $B \in \mathcal{F}$).

- Hausdorff's maximal principle: in any partially ordered set, every chain is contained in a maximal chain.

C1.3 Let S be an infinite set. Consider the collection $\mathcal{F}$ of all subsets $X \subseteq S$ such that $S \setminus X$ is finite. Prove that $\mathcal{F}$ is a filter.

C1.4 Let $A = \{a, b, c\}$. Determine which of the following relations on A are reflexive, symmetric, antisymmetric, or transitive:

1. $R_1 = \{\langle a, a\rangle, \langle b, b\rangle, \langle c, c\rangle\}$.
2. $R_2 = \{\langle a, b\rangle, \langle b, a\rangle, \langle a, a\rangle\}$.
3. $R_3 = \{\langle a, b\rangle, \langle b, c\rangle, \langle a, c\rangle\}$.

C1.5 Let $J : \omega \times \omega \to \omega$ be defined by the formula:

$$J(a, b) = 2^a(2b + 1) - 1.$$

Prove that J is a bijection.

C1.6 Using the Axiom of Choice, prove that for any surjective function $f : A \to B$, there exists a function $g : B \to A$ (called a right inverse or a section) such that $f(g(y)) = y$ for all $y \in B$.

C1.7 Let A be a finite non-empty set with a partial order $\leqslant$. Let $n = |A|$. Describe a recursive algorithm that constructs a bijection $f : A \to \{0, 1, \ldots, n - 1\}$ such that:

$$x < y \implies f(x) < f(y).$$

Show that the order defined by $x \preccurlyeq y \iff f(x) \leqslant f(y)$ is a linear order on A that extends $\leqslant$.

C1.8 Determine whether the following sets are ordinals (according to the definition: a transitive set well-ordered by $\in$):

1. $A = \{\emptyset, \{\emptyset\}, \{\emptyset, \{\emptyset\}\}\}$.
2. $B = \{\emptyset, \{\{\emptyset\}\}\}$.
3. $C = \omega \cup \{\omega\}$.

C1.9 Calculate the rank (rank) of the following sets:

1. $\emptyset$.
2. $\{\emptyset\}$.
3. $\{\{\emptyset\}\}$.
4. $\{\emptyset, \{\emptyset\}\}$.

C1.10 1. Using the definition of the Kuratowski pair $\langle a, b\rangle = \{\{a\}, \{a, b\}\}$, express $\mathrm{rank}(\langle a, b\rangle)$ in terms of $\alpha = \mathrm{rank}(a)$ and $\beta = \mathrm{rank}(b)$.

2. Determine the rank of the set of integers $\mathbb{Z}$, if integers are defined as equivalence classes of pairs of natural numbers.

C1.11 Determine to which universe V_α the following sets belong for the first time (find the least α such that $x \in V_\alpha$):

1. $x = \emptyset$.
2. $x = \omega$.
3. $x = \mathcal{P}(\omega)$.

C1.12 Let a function $F : \mathrm{Ord} \to \mathrm{Ord}$ be defined by the condition $F(\alpha) = \sup(\{F(\beta) + 1 \mid \beta < \alpha\})$. What is $F(\alpha)$ equal to?

C1.13 We want to define the Fibonacci sequence F on natural numbers ω (viewed as finite ordinals) such that $F(0) = 0, F(1) = 1, F(n) = F(n-1) + F(n-2)$. Explicitly describe the generator function $\Phi(g)$, which takes a history function g (defined on some ordinal $\alpha < \omega$) and returns the next value.

C1.14 Construct an explicit bijection between the set of real numbers $\mathbb{R}$ and the closed unit interval $[0; 1]$.

C1.15 Show that the set of points in the open interval $(0, 1)$ is equinumerous to the set of points in the union of intervals $(0, 1) \cup (2, 3)$.

C1.16 Let $A = B = \omega$ (the set of natural numbers) with the standard order $\leqslant$. Consider the product $P = A \times B$. Compare the pairs $x = \langle 2, 100 \rangle$ and $y = \langle 3, 1 \rangle$ using the following relations. For each relation, determine if it is a linear order and if it is a well-ordering on P:

1. Component-wise order $\leqslant_{prod}$: $\langle a, b \rangle \leqslant_{prod} \langle c, d \rangle \overset{\text{def}}{\leftrightarrow} a \leqslant c \wedge b \leqslant d$.
2. Lexicographical order $\leqslant_{lex}$: $\langle a, b \rangle \leqslant_{lex} \langle c, d \rangle \overset{\text{def}}{\leftrightarrow} a < c \vee (a = c \wedge b \leqslant d)$.

C1.17 Show that the axiom of regularity prohibits the existence of a set x such that $x = \{x\}$.

C1.18 Calculate the following sums and products of ordinals (recall that operations are non-commutative):

1. $1 + \omega$.
2. $\omega + 1$.
3. $2 \cdot \omega$.
4. $\omega \cdot 2$.

C1.19 Check the right distributivity law $(\alpha + \beta) \cdot \gamma = \alpha \cdot \gamma + \beta \cdot \gamma$ for $\alpha = 1, \beta = 1, \gamma = \omega$.

C1.20 Find all ordinals x satisfying the equation:

$$\omega + x = \omega \cdot 2.$$

C1.21 Compare the ordinals $\alpha = \omega^2 + \omega \cdot 3 + 1$ and $\beta = \omega^2 + \omega \cdot 2 + 5$. Which one is smaller?

C1.22 Consider the set $P = \omega \times \omega$ (pairs of natural numbers) equipped with the lexicographical order $\leqslant_{lex}$ from the exercise 16. Determine the order type of the structure $\langle P, \leqslant_{lex} \rangle$. Express the result using ordinal multiplication. *Hint: Visualize the order as a sequence of "rows" $R_n = \{\langle n, m \rangle \mid m \in \omega\}$, ordered by the index n.*

C1.23 Why can we not talk about the "set of all sets" in ZFC? Which axiom would be violated if such a set existed?

C1.24 Assume the field of real numbers $\mathbb{R}$ is constructed. Define the set of Complex Numbers $\mathbb{C}$ as the Cartesian product $\mathbb{R} \times \mathbb{R}$. Define the operations $+_{\mathbb{C}}$

and $\cdot_{\mathbb{C}}$ formally as subsets of $(\mathbb{C}\times\mathbb{C})\times\mathbb{C}$. Verify that for the element $i = \langle 0, 1\rangle$, the condition $i \cdot_{\mathbb{C}} i = \langle -1, 0\rangle$ holds.

C1.25 Let $\langle A, \leqslant\rangle$ be a partially ordered set. We say that a relation $\preccurlyeq$ is an extension of $\leqslant$ if $\leqslant\subseteq\preccurlyeq$. Use Zorn's Lemma to prove that there exists a linear order on A that extends the partial order $\leqslant$ (Szpilrajn's extension theorem). *Hint: Consider the set of all partial orders on A extending $\leqslant$, ordered by set inclusion $\subseteq$.*

Part II
Proof-Theoretical Basis

Introduction to Part II

As we transition from Part I to Part II, the reader will notice a distinct shift in tone and complexity. While Part I was designed as a highly accessible narrative aimed at building intuition, Part II serves as a rigorous formal reference. Here, we delve into the logical foundations that formally justify the concepts previously discussed. Please approach this section not as a continuous story, but as a structural blueprint—a formal approval of the mathematical architecture we have been exploring.

Chapter D1
Language

You must perceive not what I say or write, but what I think.
— Nikolai A. Vavilov

D1.1 Elementary Particles of Language

Since our language $\mathfrak{Math}$ is purely written, we view its possible implementations as texts or, more generally, texts with additional graphical materials, independent of their verbalization in any natural language. Both the texts and the additional graphics are composed of **elementary graphemes**, i.e., certain indivisible graphical images whose scale and minor variations in style are not significant for the language $\mathfrak{Math}$ (with rare exceptions). We will consider graphemes from different fonts (typefaces), as well as graphemes of different cases (lowercase and uppercase), to be distinct, so as to avoid introducing the concept of an *allograph*.

The concept of an indivisible graphical image is rather conventional, but in general, it can be defined as a composition of graphical elements that is not separated or overlapped by other elements in any implementation of the language $\mathfrak{Math}$ (in other words, an indivisible grapheme cannot be split into two parts such that both parts have some meaning in mathematics, except for the special case of rigid graphemes, which are forcibly classified as elementary).

On the other hand, an elementary grapheme can be quite complex and contain a simpler elementary grapheme as a part of it. Therefore, elementary

N. I. Kazimirov, *Mathematics as a Foreign Language*, Mathematics in Mind,
https://doi.org/10.1007/978-3-032-30721-7_6

graphemes must be structured by classifying them into classes and types to obtain a certain "algebra" for constructing elementary graphemes.

First of all, let us define the **alphabet**, or the list of **basic graphemes**.

First and foremost, the list of basic graphemes includes the *Latin letters* from A, a to Z, z. Some lowercase and uppercase letters of the Latin alphabet are difficult to distinguish without knowing their size; however, within a sufficiently long text, one can always determine the average size of lowercase letters and thereby identify which variant of a Latin letter is being used.

Next, Greek letters from α to ω are undoubtedly basic graphemes, as well as their uppercase variants that differ in appearance from their Latin counterparts, for example, Λ, Σ, Ω.

Furthermore, the language $\mathfrak{Math}$ less frequently uses letters from other alphabets, primarily Hebrew, such as $\aleph$ or $\beth$, as well as various modifications of letters through significant font changes: for example, we consider the letters F, $\mathbb{F}$, F, and $\mathcal{F}$ to be different basic graphemes.

The next type of basic graphemes in our alphabet is the digits $0, 1, 2, 3, 4, 5, 6, 7, 8, 9$, which can also undergo font modifications, generating new elementary graphemes (though this is an extremely rare phenomenon).

We will group the alphanumeric basic graphemes into a single class—the *independent basic graphemes*. These are typically symbols used to form proper names and notations for objects and variables in the language $\mathfrak{Math}$. The class of independent graphemes can also be extended with a small set of special symbols like ∞ or %, if it is essential to use them in object names.

We also include in the alphabet of the language $\mathfrak{Math}$ various non-alphabetic indivisible graphemes, which can be conventionally divided into three types: a) grouping; b) separating; c) operator. We will combine these three types of elementary graphemes into the class of *dependent* graphemes. They generally serve to denote functions and relations and are therefore used in combination with independent graphemes.

Grouping graphemes (or *groupers*) are most often paired symbols intended to set apart a group of graphemes. This is a soft way to assemble a new lexical unit from simpler ones. I call it soft because the created group does not represent a new, single object of the language and can be broken down into its constituent parts during contextual analysis.

Examples of groupers include: parentheses ( ), square brackets [ ], curly braces { }, angle brackets $\langle\ \rangle$, quotes (single, double), as well as vertical bars | | and double vertical bars $\|\ \|$. Sometimes, the symbols of a grouper can be asymmetrical, for example, the notations for bra- and ket-vectors: $\langle\, |$ and $|\, \rangle$.

Separating graphemes (or *separators*) are essentially the punctuation marks of the language $\mathfrak{Math}$, designed to separate one part of a complex lexeme from another. These include: comma (,), colon (:), semicolon (;), vertical bar (|), etc. (here, the parentheses are used to set off the symbol being defined and are not part of it).

The richest arsenal of the language $\mathfrak{Math}$ is represented by operator graphemes, which are used to denote all sorts of actions (operations) and connections (relations). In essence, operator graphemes are the verbs of the language $\mathfrak{Math}$, or more precisely, only the basic part of the verbs, since we can form more complex operator lexemes from them, which can be considered analogous to more complex verbs (here, the word formation of the German language is an appropriate analogy).

Operator graphemes, or *operators*, constitute the "hieroglyphic" component of the language $\mathfrak{Math}$. As widely known examples of operator graphemes, we can cite the following: $+$, $\int$, $<$, $\Rightarrow$, $\cup$, $\square$, $\vdash$. The number of operator graphemes used in mathematics is orders of magnitude greater than the number of letters used. Moreover, there are operator graphemes that are obtained from letters by inverting them, for example, $\forall$, $\exists$, ∇ (this sometimes helps to explain the meaning of the chosen notation).

Furthermore, some operator graphemes may visually resemble other symbols of the alphabet, for example, the set union symbol $\cup$ is similar to the Latin letter U (as it originates from the word *unite*), the summation symbol $\sum$ is directly derived from the Greek letter Σ, and the divisibility relation $|$ is indistinguishable from the separating symbol 'vertical bar' and the grouping 'vertical bar'. The subtle differences between these symbols are not so crucial for us (although one could say, for example, that the operator grapheme $\sum$ never appears without operands, and the divisibility sign has half-spaces around it, unlike the vertical bar), since, if desired, we can consider that the same basic symbol participates in the formation of more complex lexemes in different roles, and its meaning is determined by context.

Many operator graphemes are obtained from other operator graphemes by reflection or rotation. For example, we can take one arrow $\longrightarrow$ and create several new symbols from it: $\longleftarrow$, $\uparrow$, $\nearrow$. Another example: mutually inverse relations $\in$ and $\ni$, $\subset$ and $\supset$, $<$ and $>$. Syntactically, these are different graphemes (they are read differently and are sometimes used for different purposes[1]), although they are produced from a single basic graphical element.

[1] For instance, the symbol $\supset$ is sometimes used to denote implication instead of $\longrightarrow$.

This concludes the description of the set of basic graphemes (the alphabet) of the language $\mathfrak{Math}$.

The next block of elementary graphemes is the **auxiliary graphemes**. These are symbols that are generally not used as basic graphemes but serve to form a potentially infinite series of elementary graphemes.

Auxiliary graphemes include, for example, the following: tilde ($\tilde{o}$), apostrophe (o'), acute accent ($\acute{o}$), grave accent ($\grave{o}$), circumflex ($\hat{o}$), umlaut (ö), macron ($\overline{o}$), where the letter o is included for illustrative purposes and is not part of the listed auxiliary graphemes. Typically, these are diacritical marks from spoken languages, but other types of auxiliary graphemes are also found, such as the asterisk (*), which can act as an accent to denote a variable, and as an operator symbol to denote an adjoint operator or multiplication. Auxiliary graphemes themselves are not names of objects or variables (except for those that graphically coincide with operators).

Table D1.1. Elementary Graphemes

Class of Graphemes	Type of Graphemes	Examples
	Basic Graphemes (Alphabet)	
Independent	Letters	$A \ldots z, \alpha \ldots \omega, \aleph, \beth$
	Digits	$0, 1, 2, 3, 4, 5, 6, 7, 8, 9$
	Special Symbols	$\infty,\ \emptyset,\ \top,\ \bot,\ \hbar$
Dependent	Groupers	$(),\ [\,],\ \lvert\,\rvert,\ \langle\rangle,\ \{\},\ \langle\,\rvert,\ \lvert\,\rangle$
	Separators	$, : ; . ! \vert$
	Operators	$+,\ \int,\ <,\ \Rightarrow,\ \cup,\ \square,\ \vdash,\ \uparrow$
Auxiliary Graphemes		˜ ′ ´ ` ^ ‚ ¯ * →
	Derived Graphemes	
Independent	Letters	$A' \ldots \hat{\hat{z}}, \tilde{\alpha} \ldots \omega^*, \bar{\bar{\aleph}}, \bar{\beth}$
	Digits	$\overline{0}, \overline{1}', 2'', 3''', 4'''', {}_{\prime}5, {}_{\prime}6^*, 7^{**}, \vec{8}, \tilde{9}$
	Special Symbols	$\overline{\infty}$
Dependent	Operators	$+',\ <'',\ \square'$

Finally, the last part of elementary graphemes is the **derived graphemes**. These are graphemes obtained by combining one basic grapheme with one or more auxiliary graphemes. Derived elementary graphemes inherit the division into classes and types from the basic ones, and their names: independent (alphabetic, numeric, special) and dependent (mostly operators).

Thus, **elementary graphemes** are basic (alphabetic) graphemes either by themselves or combined with auxiliary graphemes, i.e., derived graphemes.

Table D1.1 summarizes the classes and types of elementary graphemes of the language $\mathfrak{Math}$ with examples.

D1.2 Lexemes

Having dealt with the initial set of symbols, we now turn to the rules of "text construction". In other words, we will specify a number of simple principles by which anything can be assembled into some text in our language $\mathfrak{Math}$. This does not mean that all lexemes obtained in this way are used in practice, i.e., are in any way common, but we can say with certainty that all "correct" text in the language $\mathfrak{Math}$ can only be constructed in this way and no other.

Recall that spoken language also has certain rules of word formation (this very word is built according to one of them), but this does not mean that we actually use all the words that can be constructed this way. There are regular words, rare ones, vulgar ones, and non-existent ones—among those that can be built according to the rules.

We will distinguish two main methods of "text construction". All text fragments of the language $\mathfrak{Math}$ obtained by these methods will be called *lexemes*. Any text composed of elementary graphemes that is not a correctly formed lexeme is considered incorrect. A precise definition of a lexeme will be given below.

First, we can concatenate any independent elementary graphemes, as well as their derived graphemes, in any quantity and in any order. In other words, a sequence of letters and digits (and perhaps special symbols), some of which may be endowed with auxiliary symbols (in any quantity), is a well-formed grapheme of $\mathfrak{Math}$.

In this way, for example, multi-letter names of functions like sin and cos appear, as well as representations of numbers like 123 (decimal notation), 1EF0 (hexadecimal notation), MMXXII (Roman numerals).

Graphemes obtained in this way we will call *rigid graphemes* in the sense that they represent syntactically indivisible names of objects and variables (for example, no one in their right mind would read the combination sin as the product of three variables). Moreover, while there are infinitely many purely numeric (decimal or hexadecimal) graphemes (for each natural number there is its own grapheme), there is only a finite set of alphanumeric ones, since in practice we use a limited number of names for functions and relations.

We also include as rigid graphemes those that are obtained from already constructed rigid graphemes by adding auxiliary symbols. For example, let's say we first constructed the rigid grapheme lim, and then decided to add an accent in the form of a macron, resulting in the grapheme $\overline{\lim}$. Such a grapheme is also considered rigid.

Sometimes one can find the use of auxiliary symbols as an operator notation, for example, complex conjugation or a derivative, and in this case, this auxiliary symbol can apply to an arbitrarily long text fragment (lexeme). But, as noted above, this only means that the auxiliary symbol is graphically very similar to an operator symbol, although in essence, they are two symbols with different purposes (therefore, $\overline{\lim}$ is not the complex conjugate of a limit, but a whole derived operator grapheme).

As one can easily see, all independent elementary graphemes are rigid graphemes. Thus, we have built an extension of the class of independent elementary graphemes to rigid ones. Therefore, it is also appropriate to call rigid graphemes independent. Note that we do not classify operator elementary graphemes and special symbols (both basic and derived) as rigid (independent) graphemes.

If we draw parallels with spoken language, rigid graphemes can probably be compared to words, with multi-letter and numeric rigid graphemes being proper nouns, and single-letter ones being common nouns (usually variables). Operator graphemes are more like verbs, as they symbolize action.

The second way to construct lexemes in the language $\mathfrak{Math}$ is through *substitution templates* or simply *templates*. Essentially, this is a complex graphical symbol composed of rigid graphemes, dependent graphemes, and their derivatives (we are talking about accented operator symbols), as well as the special space symbol, which in this case is called a *placeholder*.

To be more precise, let's first define a **simple template**, which is formed by at most one pair of groupers (parentheses) and some (possibly zero) number of separators and spaces located between the groupers (if any). It is important to note that a template is, generally speaking, a two-dimensional figure,[2] and its placeholders can be arranged in any way relative to each other, with the only prohibition being the overlapping of placeholders. The template is needed

[2] In fact, there are even three-dimensional templates, an example of which is the Levi-Civita symbol in tensor calculus.

precisely to group and arrange rigid graphemes and operators in a specific order and manner.

A typical simple substitution template has a linear form:

$$\sqcup(\sqcup, \sqcup, \ldots, \sqcup),$$

that is, one placeholder, followed by some (non-zero) number of placeholders inside parentheses, separated by commas. This prefix way of writing an operator and its arguments is called *Polish notation* by Jan Łukasiewicz and, by and large, does not even require parentheses to list the arguments, since the order of operations is uniquely determined.[3]

Note that in this case, the ellipsis is not a symbol of the language $\mathfrak{Math}$, but merely an indication that there can be any number of placeholders and commas inside the parentheses, i.e., the notation above actually covers an infinite series of simple templates. This way of presenting a series of similar objects is called a *schema*. The concept of a schema is often used to write down the list of axioms of a formal theory.

A more complex example is a matrix, formed by a pair of parentheses and a number of placeholders arranged in columns and rows:

$$\begin{pmatrix} \sqcup & \sqcup & \sqcup \\ \sqcup & \sqcup & \sqcup \end{pmatrix}.$$

To move forward, we will need the concept of a **recursive definition**. Such definitions are extremely useful and therefore common in mathematics. They usually consist of two steps: a base case and a recursive (or productive) step. The base case of a recursive definition specifies the initial set of objects belonging to this definition, and the recursive step provides an algorithm for generating new objects from already constructed or basic ones. Thus, by repeatedly applying the recursive step of a recursive definition, we can construct more and more complex objects, starting from the base set. The concept of a "Recursive definition" or, more simply, *recursion*, gets a more precise definition in specific areas of mathematics. For example, in arithmetic, we deal with *recursive functions*, and in set theory, with *transfinite recursion*.

Let's give a simple example of a recursive definition. Suppose we have a set of letters a, b, c. We say that each of the letters a, b, c is a word, and also, any word is a concatenation of two words. Here, the base case of the

[3] There is also *reverse Polish notation*, where the operator sign is placed after the arguments. This method of writing operations was formerly used on programmable calculators.

recursive definition of a "word" is classifying the letters a, b, c as words, and the recursive step is obtaining new words by concatenating existing ones. Thus, this definition tells us that words are any finite sequences of letters from the list a, b, c.

So, let's give a recursive definition of a **substitution template** (or, for short, a **template**). First, we say that every simple template is a (substitution) template. Second, if templates are substituted into the placeholders of a template, the result is also a template. Thus, a template is anything that can be obtained from simple templates by substituting templates into templates. What we get is a kind of "Matryoshka doll" of simple templates.

An example of a template corresponding to the given definition:

$$\left[\begin{pmatrix} (\sqcup,\ \sqcup) & \sqcup & \sqcup \\ \sqcup & \sqcup & \langle \sqcup; \sqcup \rangle \end{pmatrix} \mid \{\sqcup \mid \sqcup\}\right].$$

But this is actually not enough for us, because we would like to have more meaningful templates, in which some operator graphemes are predefined, allowing us to understand how this template is used in a specific implementation of the language $\mathfrak{Math}$. Therefore, we will supplement the definition of templates given above with the concept of a **derived template**. Derived templates include all constructions that can be obtained from templates by substituting arbitrary rigid graphemes, placeholders, and operator symbols into their placeholders.

Figuratively speaking, if we take a Matryoshka doll and put a candy in its smallest part, we get a derived template. However, as is known, every analogy is flawed, and in our example with the Matryoshka doll, we have only a linear construction with a single placeholder where we put the candy. In reality, templates usually have a branched structure, so several candies can be hidden in them. Unfortunately, I have not come across Matryoshka dolls that have several chambers for placing smaller Matryoshka dolls with several chambers themselves.

But let's return to the language $\mathfrak{Math}$. For example, by taking the template $\sqcup(\sqcup)$ and substituting the grapheme sin for the first placeholder, we get a derived template with one placeholder $\sin(\sqcup)$, which is used to write the sine function. If we take the initial template of the form

$$\underset{\sqcup}{\overset{\sqcup}{\sqcup}}\ \sqcup\ \sqcup\ \sqcup,$$

then by substituting the operator symbol $\int$ and the letter d into it, we get a derived template for denoting a definite integral:

$$\int\limits_{\sqcup}^{\sqcup} \sqcup \, d \sqcup .$$

Derived templates are the basis for the syntactic definition of all mathematical operations, relations, and functions. Here we can give such examples:

$$\sqcup + \sqcup, \quad -\sqcup, \quad \frac{\sqcup}{\sqcup}, \quad \sqrt{\sqcup}, \quad \sqcup^{\sqcup}, \quad \sqcup = \sqcup, \quad \cos(\sqcup), \quad \ldots$$

As another example, one can cite the template $\sqcup.\sqcup$, which allows writing numbers with a decimal point separating the integer part. This template can be extended to create a template for denoting floating-point numbers in exponential form $\sqcup . \sqcup \mathrm{E}\sqcup$.

Using the example of defining derived templates, we can demonstrate a very important feature of mathematical definitions, namely: the maximum coverage of edge cases by a single definition. Thus, in the definition of a derived template, we allowed placeholders to be placed in the placeholders of templates. In essence, this means that we can do nothing, and therefore *every substitution template is a derived template* by definition. In other words, the concept of a derived template extends the concept of a substitution template.

Further, since in the definition of a derived template we did not specify whether all placeholders must be filled during its construction or if some must remain empty, we can consider another edge case—when the placeholders are completely eliminated, i.e., there are no placeholders left in the resulting derived template. In this regard, we give the following definition. A **lexeme** of the language $\mathfrak{Math}$ is a derived template with no placeholders.

Thus, derived templates include lexemes as a special case. That is, our Matryoshka doll, in which all voids are filled with candies, is a lexeme. Accordingly, lexemes will be such graphical constructions as, for example,

$$a + \hat{b}, \quad -x^*, \quad \frac{32.33}{127.87}, \quad \sqrt{5}, \quad e^{\pi}, \quad A = \tilde{B}, \quad \cos(x), \quad \int\limits_{0}^{\infty} u dt.$$

Here one can draw a direct analogy with formulas and sentences. In formulas, as in derived templates, there are undefined placeholders—free variables, while sentences are closed formulas in which all free variables are somehow eliminated (replaced by closed terms or bound by quantifiers). So lexemes are, in a sense, closed derived templates.

By the way, if we use the template $\sqcup$ (consisting only of a single placeholder, which is not forbidden by the definition), we thereby classify all rigid graphemes (and therefore all independent graphemes with and without accents) and operator symbols as lexemes. For example, the letter x is a lexeme, the letter combination cos is also a lexeme, the integral sign is also a lexeme, and so on. But in the future, we will agree, as an exception to the rule (although mathematicians do not like exceptions!), purely for aesthetic reasons, not to consider dependent elementary graphemes as lexemes.

Finally, let us note one more remarkable fact: if we substitute derived templates into a derived template instead of placeholders, we will get a derived template. Let us formalize this deductive consideration as a theorem.

Theorem D1.1. *Derived templates form a set of graphemes that is closed under the operation of substitution.*

Verifying this property is quite simple, by reducing the closure of the substitution operation on derived templates to the analogous operation on substitution templates. And for the latter, closure is guaranteed by definition.

Let's demonstrate this theorem with an example. Suppose we had the template $\sqcup(\sqcup)$, then we obtained the derived template $\sin(\sqcup)$ from it, and then we substituted this derived template into itself, resulting in the construction: $\sin(\sin(\sqcup))$. Why is such a construction a derived template?

To answer this question, we move from substituting a derived template into a derived one to substituting a template into a template. Specifically, we substitute $\sqcup(\sqcup)$ into itself at the second placeholder, resulting in a new template $\sqcup(\sqcup(\sqcup))$. From this template, by substituting the rigid grapheme sin into the first two places, we obtain the required derived template.

In other words, we can substitute templates and derived templates into each other as we please, without going beyond the scope of permitted constructions!

From this, in particular, it follows that we can substitute any lexemes into a derived template, and as a result, we will get a derived template, and if there are no free placeholders left, we will get a new lexeme. In particular, using the derived template $\sin(\sqcup)$ and the lexeme $i\sqrt{\pi}$, which is obtained from the multiplication operation template $\sqcup\ \sqcup$ and the square root template $\sqrt{\sqcup}$, we can obtain a new lexeme $\sin(i\sqrt{\pi})$.

In this clever way, using a recursive definition, we have arrived at the main descriptive definition of the language $\mathfrak{Math}$—the concept of a *lexeme*. All text written in the language $\mathfrak{Math}$ is a sequence of correctly constructed lexemes!

Fig. D1.1 shows a brief diagram of the formation of lexemes from the alphabet using the methods we have described.

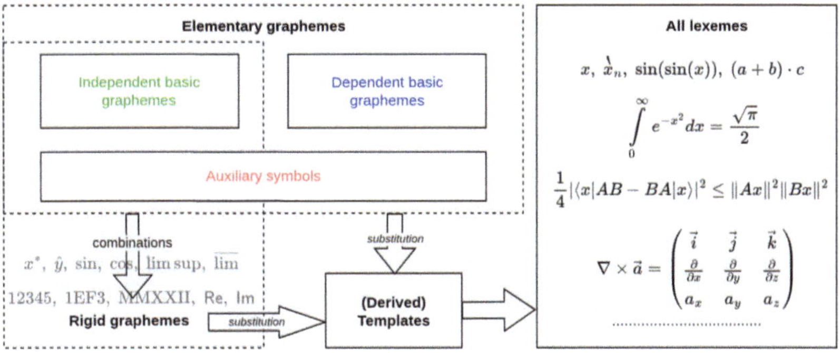

Fig. D1.1. Scheme of lexeme construction.

D1.3 Syntactic Analysis

Substitution templates are perhaps the main and most convenient tool for "text construction" in the language $\mathfrak{Math}$. Firstly, it is excellent because it is very simple to describe and use—just take and combine templates according to the Matryoshka principle. Secondly, any lexeme constructed with substitution templates can be represented as a so-called **parse tree**, which shows exactly how the given lexeme was constructed. True, a single lexeme may correspond to several parse tree variants, and to exclude this, in formal languages the rules for constructing texts become more rigid than in our general version of the language $\mathfrak{Math}$.

It is this rigidity that makes formal languages accessible to machine processing. In linguistics and computer science, there is the so-called Chomsky hierarchy, which classifies formal languages by the complexity of their grammar. The simplest of them, regular languages, describe simple templates, while more complex, context-free languages, are capable of describing nested structures like those we see in arithmetic expressions with parentheses or in programming languages. The parse tree is a key tool precisely for the analysis of context-free grammars, on which most languages "understood" by computers are based.

An example of a parse tree for a well-known formula is shown in Fig. D1.2.

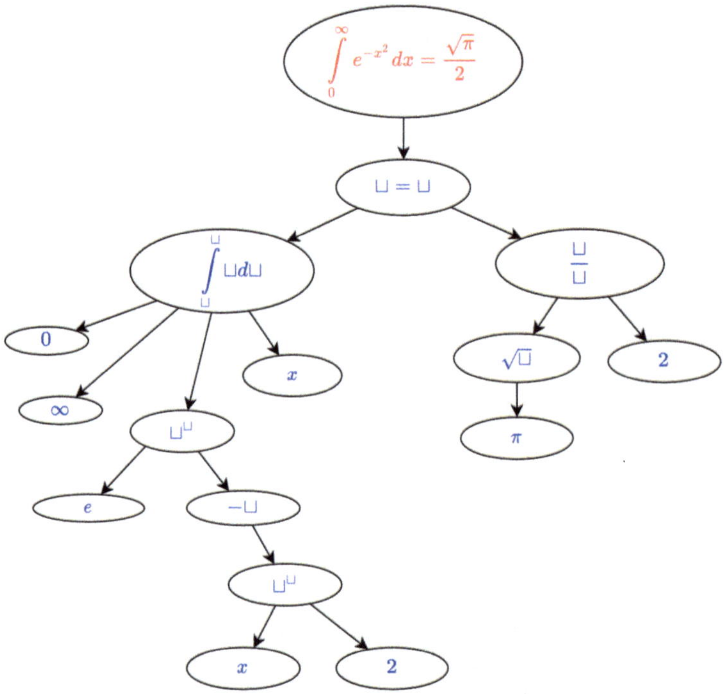

Fig. D1.2. Parse tree for the Poisson integral formula.

For convenience, the final result, i.e., the lexeme being parsed, is placed at the root of the tree. As we can see, we mainly use derived templates with two or one placeholders here (the equality operator, fraction, power, root, negation operator), but there is also one more complex derived template for creating an integral expression with four placeholders. The parse tree, as it should, ends with leaves (nodes without descendants), which contain rigid graphemes represented by basic graphemes (the digits 0 and 2, the letters e, x, π, and the special symbol ∞).

Note that this tree can be further detailed if we do not use any templates in it other than simple substitution templates. In this case, the tree will have leaves containing operator symbols (in our case, these are the integral, equality, fraction, and square root). However, such a tree would be very difficult to read, as templates with operator symbols (e.g., the integral and the root) allow one to grasp their semantics at a glance. Furthermore, derived templates with operators correspond to specific mathematical concepts, which also makes them more attractive for use in the syntactic analysis of a lexeme.

Understanding the internal structure of a formula through a parse tree is not merely an exercise in visualization. It is the fundamental principle used by computer systems to process mathematical and programming languages. Compilers build such trees (Abstract Syntax Trees, AST) to translate human-readable code into machine instructions, Computer Algebra Systems (CAS) rely on them to perform symbolic manipulations like integration or simplification, and modern Artificial Intelligence models use similar structural representations to "understand" and generate valid mathematical texts.

In the example given, we did without grouper symbols, i.e., without parentheses, so the reader is invited to construct a parse tree for the Robertson–Schrödinger uncertainty relation as an independent exercise:

$$\frac{1}{4}|\langle x|AB - BA|x\rangle|^2 \leqslant ||Ax||^2||Bx||^2,$$

using derived templates with operator symbols.

D1.4 Simplifications

Above, we described the construction of lexemes in the language $\mathfrak{Math}$ as it is accepted in mathematics. However, we can notice that this language can be significantly simplified if we consider only linear templates of the form

$$\sqcup(\sqcup, \ldots, \sqcup).$$

as simple templates.

Indeed, no matter how complex and multidimensional a template may be, it represents a certain number of placeholders connected by a specific "geometry". No matter how many such templates there are, they can all be numbered with specially invented rigid graphemes like Sch1, Sch2, Sch3, etc. Moreover, in each of them, a certain order of filling the placeholders can be defined, for example, in the template for an integral, we would say that first the integral sign is written, then the lower limit, then the upper limit, then the integrand, then the letter d, and finally the integration variable.

This means that instead of the huge zoo of templates we outlined earlier, we can leave only linear templates of the form:

$$\text{Sch1}(\sqcup, \ldots, \sqcup), \quad \text{Sch2}(\sqcup, \ldots, \sqcup), \quad \text{Sch3}(\sqcup, \ldots, \sqcup), \quad \ldots$$

Moreover, we can perform such an operation with most derived templates (except for lexemes), so that our entire language will consist exclusively of lexemes obtained by substituting rigid graphemes into templates of the specified form. In essence, we will get a language where every lexeme is a Polish notation of a function with arguments, for example,

$$\text{Sch1}(a, b, c),$$
$$\text{Sch2}(\int, 0, \infty, \text{Sch4}(\sin, x), d, x),$$
$$\text{Sch3}(\sqrt{}, \text{Sch5}(+, i, \lambda)).$$

Finally, we can simplify these templates as well, by agreeing that in our alphabet we have only one letter x, one digit 0, one special symbol λ, one pair of groupers (parentheses), one separator (comma), one auxiliary symbol (prime), and an infinite set of template names Sch1, Sch11, Sch111, etc. At the same time, we will obtain independent derived graphemes by unboundedly adding a prime to the letter x and the digit 0, all derived operators—by unboundedly adding a prime to the symbol λ, and we will also agree that in a template, we always substitute letter graphemes first, then numeric graphemes, then operator graphemes. Then the lexemes of our terribly simplified language will take the following form

$$\text{Sch1...1}(x, x'', x''''', \ldots, 0'', 0''', \ldots, \lambda''', \lambda'''', \ldots).$$

We will get a very simple language, equivalent to our language $\mathfrak{Math}$, but terribly unreadable, which is convenient for analyzing the language as a subject of research, but inconvenient for practical application.

Such a method of language simplification (reduction) is often used precisely for analyzing the expressive capabilities of a language, for proving some general syntactic properties. For example, dealing with such a simplified version of the language $\mathfrak{Math}$, it would be much easier for us to prove Theorem D1.1 on the closure of derived templates under the substitution operation. To be honest, we did something similar when we "proved" this theorem using the example of the simplest template $\sin(\sqcup)$.

D1.5 Graphics

At the beginning of our discussion about the language $\mathfrak{Math}$, we noted that in addition to texts, the language allows for additional graphical images. Let us immediately note that any graphical materials in mathematics are merely illustrations of concepts, relations, ideas, etc. No reasoning in mathematics will be accepted as a proof if it relies solely on "graphics".

Of course, there are exceptional cases when explaining a proof "on one's fingers" is much easier than writing down its formal text, but in such cases, it is assumed that any recipient of such a proof can uniquely and completely restore the formal reasoning hidden under the illustration. We often encounter such a situation in simple geometric problems, where some additional construction or rotation of figures suggests the idea of the proof and leads us to previously proven theorems of geometry.

Nevertheless, any illustrative material in mathematics should be considered no more than auxiliary.

What kind of graphical material can we note? Let's give such a classification:

- geometric constructions in planimetry and stereometry;
- finite or countable graphs;
- Venn–Euler diagrams;
- commutative diagrams;
- plots;
- tables and charts.

The types of auxiliary graphical materials are listed in descending order of their importance for the evidentiary basis of reasoning in mathematics (in the author's opinion).

The question arises: if we still consider graphical elements as part of the language $\mathfrak{Math}$, then how do they fit into all the previous constructions? After all, it is obvious that they cannot be defined by any lexeme.

Yes, on the one hand, this is true, and it is hardly possible to convey the structure of, for example, a function graph with asymptotes using derived templates and rigid graphemes. However, if we approach the matter purely syntactically, at least two approaches come to mind.

The first approach is that any graph, being drawn by computer means, is a finite matrix whose cells contain zeros or ones (zero—the cell is white, one—black). In other words, any drawing can be considered a pixel matrix, and then we can define it with the simplest matrix template, filling its placeholders only where the black dots of our drawing should be. The problem is that such an approach is too general, because in this way one can depict not only a function graph with asymptotes, but also the portrait of Mona Lisa and any formula of the language $\mathfrak{Math}$. At the same time, the syntactic structure of the text would be completely lost, because the parse tree would contain only one template (a matrix) and leaf-ones, which are placed in some cells of this matrix. It is clear that we would not want to lose the structure at all. Therefore, we will consider this path a dead end.

The second approach is that we can allow adding any graphical backgrounds (background images) to any templates. Then we will get a kind of hybrid of templates and graphics, and the template will be responsible for where on the picture one can insert lexemes to fill the graph with necessary text insertions. True, in this case, one should adopt the rule that templates with different backgrounds should be considered different, even if the number and relative arrangement of placeholders, parentheses, and separators in them are the same.

However, there are a number of inconvenient problems in the second approach as well. For example, no one prevents us from taking a graph of a parabola as a background image, and in the captions, using a template, write the same sine function, as a result of which such a graphical lexeme will be contradictory. It will no longer be possible to link purely textual lexemes with a graphical background without special software tools. And this, in turn, means that within the language $\mathfrak{Math}$ we will need to build some special subset of lexemes that will be a kind of programming language allowing us to build graphs of functions, and only after that will it be possible to define a special type of templates (computing, drawing templates) that will "understand" the text written in a specific cell as program code and use it to build a background image in the form of a function graph or anything else that can be defined analytically on the Euclidean coordinate plane.

In short, this path is not easy, but it is productive. And yet, within the framework of the book, we will adhere to the paradigm that we have outlined above, namely, that *we consider all illustrative material to be purely auxiliary and do not consider it part of the language* $\mathfrak{Math}$. Although, we repeat, our language has certain possibilities for implementing graphics.

By the way, tables, which ended up in last place in terms of importance for mathematics, are much easier to implement with templates than all other graphical materials, since a table is essentially a matrix template, and quite meaningful in contrast to the pixel matrix we mentioned above. But if tables in mathematics help with anything, it is, perhaps, to beautifully present or summarize some already proven classification of some objects and concepts. Tables do not allow one to arrive at this classification or to prove some facts about it.

D1.6 Variables

We have previously mentioned some analogies between the language $\mathfrak{Math}$ and spoken languages. In particular, operator graphemes can be considered analogues of verbs, since they usually denote actions, while alphanumeric rigid graphemes can be considered analogues of nouns, with single-letter ones being common nouns and multi-letter ones being proper nouns.

In reality, of course, things are not so straightforward, since operators can be denoted by much more than just operator graphemes; more precisely, they are always produced using derived templates containing operator graphemes. For example, our familiar template for a definite integral is assembled from the operator symbol for the integral and reserves free places for the limits of integration, the integrand, and the variable of integration. In addition, a number of templates that use rigid graphemes to denote an operator can also be classified as operator templates. For example, sin is a rigid grapheme (an analogue of a noun), while $\sin(\sqcup)$ is an operator template for writing the sine function.

The situation is no simpler with the correct correspondence between proper nouns and common nouns with the lexemes of the language $\mathfrak{Math}$. Firstly, not only rigid graphemes, but also some lexemes can be proper names. First of all, such lexeme-names are assembled from rigid graphemes and numerical indices. For example, the standard notation for the first infinite cardinal $\aleph_0$ is a lexeme obtained from a letter and a numerical index using the template $\sqcup_\sqcup$. From the point of view of ordinary language, $\aleph_0$ is a proper name, since this notation is assigned to a single object throughout all of mathematics. One could say that proper names are notations for constants, but here too the analogy would be incomplete, since in mathematics we very often use temporary notations for constants.

Finally, our common names can also be not only rigid graphemes, but also lexemes, since we very often use numbered variable names.

Let's consider in more detail the case of lexemes that serve to denote **variables**, i.e., analogues of common nouns in a spoken language. In most cases (in mathematics—in almost 100% of cases), variable names are specified in the following ways:

V1 a single-letter basic grapheme, i.e., simply a letter, most often from the Latin or Greek alphabet;

V2 a single-letter derived grapheme, i.e., a letter with auxiliary graphemes;

V3 a lexeme obtained by substituting a single-letter grapheme into the first placeholder of the template $\sqcup_\sqcup$, and a numeric grapheme into the second.

Examples of variables: $x, y, z, \hat{x}, y_1, z''_{100}$. Of course, not all such lexemes are used as variables; among them there are also proper names, but they are usually distinguished by a special typeface. These include the notations for number sets $\mathbb{N}, \mathbb{Z}, \mathbb{Q}, \mathbb{R}$, the probability symbol P, the ordinal ω, and some others.

In formal languages (in particular, programming languages), it is very often necessary to distinguish variables by type. First, as we have already noted in Part I, variables can be divided into bound and free. Bound variables serve to define bindings, i.e., lexemes that carry some iterative description of an object, for example, the set $\{x \mid \mathcal{A}(x)\}$ or the formula $\forall x\ \mathcal{A}(x)$. Free variables are parameters; they serve to describe a whole class of objects with a single lexeme, each of which is obtained by replacing all parameters with some constants.

Second, all variables (both free and bound) can be divided into several types, thereby prohibiting the substitution of anything anywhere when constructing lexemes. For example, we might have string and numeric variables (in a programming language), or variables for natural numbers and sets of natural numbers (second-order arithmetic). Such typification restricts the freedom of lexeme construction in the language $\mathfrak{Math}$ to the needs of a specific formalism.

Free variables can be thought of as named placeholders in templates. In this sense, a template can be perceived as a special case of a lexeme. At the same time, variables have an advantage over empty placeholders because they allow some of these places to be identified. Indeed, looking at the template $\langle \sqcup, \sqcup, \sqcup \rangle$, we only see the possibility of assembling a lexeme from three arbitrary notations, without focusing on what will result and how well it corresponds to our goals. If, however, we write the lexeme $\langle x, x, y \rangle$, we already understand that only one thing can be substituted in the first two places, since identical lexemes must be substituted for identical variables. Moreover, if a type is assigned to the variables, we immediately understand what type of objects can be substituted for them in the given lexeme.

Usually, the procedure of substituting a lexeme for a variable is written as follows: $[x/a]$ (replace x with a). This is nothing other than the notation for the *syntactic substitution operator*, which tells us in the text that in the lexeme preceding this operator, the variable x should be replaced everywhere with the lexeme a. In logic and mathematical theories (see below), we will use it in a more specific understanding of lexemes and variables.[4]

The presence of types in variables simultaneously requires that all lexemes be distributed among these same types, otherwise we simply will not be able to substitute lexemes for variables. Therefore, when describing any formalism, along with specific rules for constructing lexemes through templates, one must also indicate the type of the resulting lexemes. Usually in mathematics, all lexemes are divided into formulas and terms. **Formulas** are lexemes of a logical type (the value is either true or false), and **terms** are lexemes of an object type, intended to denote the objects under study. Terms, in turn, can be divided into types, such as numbers and strings, numbers and sets, etc.

Given all of the above, we note that although templates with placeholders are the basic units for constructing lexemes, in the definition of each specific formalism, a more advanced technology is used, namely: *typed lexemes*. These are lexemes with free variables, where a specific type is assigned to both these variables and the lexeme itself. The set of these types is called the *signature* of the typed lexeme. For example, the lexeme $a + b$ in arithmetic is accompanied by the following signature: a, b are natural numbers, the result $a + b$ is also a natural number. In the Python language, the explicit specification of a signature when defining a function is introduced as follows:

```
def Func(x: int, y: str) -> bool:
```

Here **int**, **str**, **bool** constitute the signature of the function Func.

In conclusion, we provide a comparative table of variables and constants (proper names of objects)—see Table D1.2.

Exercises

D1.1 Classify the following elementary graphemes by type (independent/dependent, basic/derived): a) π; b) $\subseteq$; c) $\hat{y}$; d) ∇; e) { }.

D1.2 Identify the type of the following elementary graphemes (Basic/Derived, Independent/Dependent, Operator/Separator/etc.): a) Σ (capital Greek sigma); b) $\sum$ (summation symbol); c) $\overline{z}$ (variable with macron); d) ; (semicolon); e) $\rightarrow$ (arrow).

[4] Some sources use the symmetric notation for the substitution operator: $[a/x]$ (a for x).

Table D1.2. Variables and Constants

	Variables	Proper Names
Definition	general notation for elements of a class	notation for a specific object
Structure	single-letter rigid graphemes [+ numeric indices]	multi-letter (rarely single-letter) rigid graphemes [+ numeric indices]
Purpose	bindings and parameters	notations for functions, operators, sets, classes
Spoken Language	common nouns, pronouns	proper nouns
Examples	x, y, $\hat{z}$, t_1, w_2'	sin, P, $\aleph_0$, $\mathbb{Z}$, Ord

D1.3 Which of the following are considered rigid graphemes (syntactically indivisible names)? a) arctan; b) x_1; c) $a + b$; d) const.

D1.4 Identify the placeholders in the following templates. How many arguments does each template accept?

1. $\lim_{\sqcup \to \sqcup} \sqcup$;
2. $\log_{\sqcup} \sqcup$;
3. $|\sqcup|$;
4. $\frac{\partial \sqcup}{\partial \sqcup}$.

D1.5 Construct a parse tree for the logistic function (sigmoid) formula:

$$\sigma(x) = \frac{1}{1 + e^{-x}}$$

using derived templates for division, addition, exponentiation, and negation.

D1.6 Convert the following 2D-templates into a linear format (like Polish notation or function calls 'Template(arg1, arg2...)'):

1. $\frac{a}{b}$;
2. a^b;
3. $\sqrt[n]{x}$.

D1.7 Translate the standard arithmetic expression $(a + b) \cdot (c - d)$ into:

1. Prefix notation (Polish notation);
2. Postfix notation (Reverse Polish notation).

D1.8 Perform the syntactic substitution operation $[x/T]$ for the following lexemes:

1. $(a + x) \cdot y$, where $T = \sin(\omega t)$;
2. $\int_0^x f(t)dt$, where $T = x^2$ (Note: distinguish between free and bound variables in the context of typical mathematical notation, although formally in the template they are just placeholders);
3. $P(x, y) \to Q(x)$, where $T = g(z)$.

D1.9 Let $T = f(u)$. Perform the syntactic substitution $[x/T]$ in the following lexemes. Ensure parentheses are added where necessary to preserve precedence.

1. x^2;
2. $\sin(2 \cdot x)$;
3. D_x (derivative operator acting on x).

D1.10 We stated that "substituting a lexeme into a derived template yields a lexeme". Demonstate this by substituting the lexeme $2k + 1$ into the derived template $\sum_{k=0}^{\infty} \sqcup$.

D1.11 Identify the signature (types of variables and result) for the following typed lexemes:

1. $\vec{a} \cdot \vec{b}$ (scalar product);
2. $||\vec{x}||$ (norm);
3. $A \subset B$.

D1.12 Determine the signature (input types $\to$ output type) for the following operators in the context of standard Vector Algebra (where S is a scalar, V is a vector):

1. Operator $\cdot$ (dot product);
2. Operator $\times$ (cross product);
3. Operator $\cdot$ (scalar multiplication, e.g., $\alpha\vec{v}$);
4. Operator $| \sqcup |$ (magnitude).

D1.13 Consider the lexeme $\sin x + y$.

1. Draw a parse tree corresponding to the interpretation $(\sin x) + y$.
2. Draw a parse tree corresponding to the interpretation $\sin(x + y)$.
3. Which one is the standard convention?

D1.14 Let the definition of a "List" be:

1. () is a List.
2. If L is a List and a is an atom, then (a, L) is a List.

Construct the list containing atoms $1, 2, 3$ using this definition.

D1.15 Construct three different derived graphemes based on the letter x using different auxiliary symbols (accents, indices).

D1.16 In the logic formula $\forall x(P(x) \to Q)$, identify the types of the lexemes:

1. x;
2. $P(x)$;
3. $\to$;
4. $\forall x(\ldots)$.

(Options: Variable, Formula, Connective, Quantifier/Binding).

D1.17 Given the logical lexeme $\neg\mathcal{A} \wedge \mathcal{B} \to C$. Place parentheses to reflect the standard order of operations in logic (Negation > Conjunction > Implication).

D1.18 In the lexeme $\int_a^b x^2 dx$, replace the bound variable x with t. Does the meaning of the lexeme change? What if we replaced x with a?

D1.19 Deconstruct the lexeme $\sum_{i=1}^{n} x_i^2$ into its constituent elementary graphemes and simple templates.

Recommended Reading

Since the central theme of our book is the view of mathematics as a language, the literature review should be built around its two hypostases: mathematical language as a rigorous formal calculus and as a phenomenon closely related to natural languages and human cognition.

The historical pillar on which the modern understanding of formalization rests is the monumental two-volume work by Hilbert and Bernays, "Foundations of Mathematics" [34]. This work, which grew out of Hilbert's famous program, is an exhaustively detailed attempt to build the edifice of mathematics on a solid syntactic foundation. Although its style may seem cumbersome today, it is the primary source to which all modern expositions trace back in one way or another.

A pedagogically sound and canonical development of these ideas can be found in the fundamental work of Stephen Kleene, "Introduction to Metamathematics" [41]. It is here that the distinction between object-language and metalanguage is drawn with unparalleled thoroughness, which directly echoes our "Matryoshka principle" and the recursive definition of lexemes. For decades, Kleene's book has been the gold standard in teaching the foundations of mathematical logic.

A completely different perspective, linking formal systems, art, and human cognition, is offered in the Pulitzer Prize-winning work by Douglas Hofstadter, "Gödel, Escher, Bach: an Eternal Golden Braid" [35]. In a uniquely accessible and engaging manner, it explores the deep philosophical and structural aspects of mathematics, focusing on themes of meaning, self-reference, and isomorphism that lie at the heart of any formal language. To understand the origins of the formal approach to syntax, which influenced both logic and computer science, it is necessary to turn to the revolutionary work of the linguist Noam Chomsky, "Syntactic Structures" [12], where the concepts of generative grammar and the hierarchy of formal languages were introduced, which directly correspond to the method of constructing lexemes from graphemes using templates introduced in this chapter and have today become an integral part of Computer Science.

Chapter D2
Logic

Wir müssen wissen — wir werden wissen.
(We must know — we will know.)

— David Hilbert

D2.1 Propositional Logic

Propositional logic (or propositional calculus) is one of the simplest formalisms. Its purpose is to describe the logical connectives between arguments as abstractly as possible, without assuming any dependence among these arguments other than truth-functional dependence. Therefore, the alphabet of the system consists of so-called propositional variables $\mathcal{A}, \mathcal{B}, C, \ldots$, which may be supplemented with indices (so that the number of variables is potentially infinite, allowing for arbitrarily long formulas), symbols for logical connectives $\wedge, \vee, \neg, \rightarrow, \leftrightarrow, \uparrow$, and parentheses. We have immediately added two more connectives to the language ($\leftrightarrow$ and $\uparrow$) to include them in formulas without additional conventions.

The formulas of this theory constitute the minimal set of texts in this alphabet that is closed under the composition of connectives and formulas. We discussed the language itself and the rules for parenthesis reduction in detail in Section B1.1.2.

Propositional logic has several kinds of models, one of which (the standard one) is that of Boolean functions of the form $f : \{0, 1\}^n \rightarrow \{0, 1\}$. By assigning the values 0 or 1 to the propositional variables and computing the

N. I. Kazimirov, *Mathematics as a Foreign Language*, Mathematics in Mind,
https://doi.org/10.1007/978-3-032-30721-7_7

value of the logical connectives according to the truth table D2.1, we obtain a value of 0 or 1 for any formula. Thus, every formula corresponds to a Boolean function (which is unique up to the numbering of its variables).

Table D2.1. Truth table for logical connectives

$\mathcal{A}$	$\mathcal{B}$	$\neg\mathcal{A}$	$\mathcal{A} \wedge \mathcal{B}$	$\mathcal{A} \vee \mathcal{B}$	$\mathcal{A} \to \mathcal{B}$	$\mathcal{A} \leftrightarrow \mathcal{B}$	$\mathcal{A} \uparrow \mathcal{B}$
0	0	1	0	0	1	1	1
1	0	0	0	1	0	0	1
0	1	1	0	1	1	0	1
1	1	0	1	1	1	1	0

Formulas that yield 1 under any valuation of their variables are called *tautologies*. For example, the formulas $\mathcal{A} \vee \neg\mathcal{A}$ and $\mathcal{A} \to \mathcal{A}$ are tautologies. The tautology $\mathcal{A} \vee \neg\mathcal{A}$ is often replaced by the symbol $\top$ (*truth*), while the identically false formula $\mathcal{A} \wedge \neg\mathcal{A}$ is replaced by the symbol $\bot$ (*falsity*); they are used as constant symbols in the language of propositional logic. Formulas are called *equivalent* if they evaluate to the same Boolean functions. To denote the equivalence of formulas, we use the meta-level predicate $\equiv$. Thus, tautologies are all formulas that are equivalent to $\top$. Here are some examples of equivalences:

- law of double negation: $\neg\neg\mathcal{A} \equiv \mathcal{A}$;
- associativity of conjunction and disjunction:

$$((\mathcal{A}\wedge\mathcal{B})\wedge C) \equiv (\mathcal{A}\wedge(\mathcal{B}\wedge C)), \quad ((\mathcal{A}\vee\mathcal{B})\vee C) \equiv (\mathcal{A}\vee(\mathcal{B}\vee C)); \quad \text{(D2.1)}$$

- commutativity of conjunction and disjunction:

$$(\mathcal{A} \wedge \mathcal{B}) \equiv (\mathcal{B} \wedge \mathcal{A}), \quad (\mathcal{A} \vee \mathcal{B}) \equiv (\mathcal{B} \vee \mathcal{A}); \qquad \text{(D2.2)}$$

- distributivity of conjunction and disjunction:

$$\begin{aligned} (\mathcal{A} \wedge (\mathcal{B} \vee C)) &\equiv ((A \wedge B) \vee (A \wedge C)), \\ (A \vee (B \wedge C)) &\equiv ((A \vee B) \wedge (A \vee C)); \end{aligned} \qquad \text{(D2.3)}$$

- equivalences with constants:

$$\mathcal{A} \wedge \top \equiv \mathcal{A}, \quad \mathcal{A} \vee \top \equiv \top, \quad \mathcal{A} \vee \bot \equiv \mathcal{A}, \quad \mathcal{A} \wedge \bot \equiv \bot \qquad \text{(D2.4)}$$

- idempotence[1] of conjunction and disjunction:

$$(\mathcal{A} \wedge \mathcal{A}) \equiv \mathcal{A}, \quad (\mathcal{A} \vee \mathcal{A}) \equiv \mathcal{A};$$

- absorption laws:

$$\mathcal{A} \vee (\mathcal{A} \wedge \mathcal{B}) \equiv \mathcal{A}, \quad \mathcal{A} \wedge (\mathcal{A} \vee \mathcal{B}) \equiv \mathcal{A};$$

- other equivalences:

$$(\mathcal{A} \to \mathcal{B}) \equiv (\neg\mathcal{A} \vee \mathcal{B}), \quad (\mathcal{A} \leftrightarrow \mathcal{B}) \equiv ((\mathcal{A} \to \mathcal{B}) \wedge (\mathcal{B} \to \mathcal{A})), \quad \text{(D2.5)}$$

$$\neg(\mathcal{A} \wedge \mathcal{B}) \equiv \neg\mathcal{A} \vee \neg\mathcal{B},$$
$$\neg(A \vee B) \equiv \neg A \wedge \neg B,$$
$$(A \to B) \equiv (\neg B \to \neg A),$$

$$\mathcal{A} \uparrow \mathcal{B} \equiv \neg(\mathcal{A} \wedge \mathcal{B}), \quad \neg\mathcal{A} \equiv \mathcal{A} \uparrow \mathcal{A},$$
$$\mathcal{A} \vee \mathcal{B} \equiv (\mathcal{A} \uparrow \mathcal{A}) \uparrow (\mathcal{B} \uparrow \mathcal{B}), \quad \mathcal{A} \wedge \mathcal{B} \equiv (\mathcal{A} \uparrow \mathcal{B}) \uparrow (\mathcal{A} \uparrow \mathcal{B}).$$

Furthermore, the following statement is true: $\varphi \leftrightarrow \psi$ is a tautology if and only if $\varphi \equiv \psi$ (or, alternatively: $(\varphi \leftrightarrow \psi) \equiv \top \Leftrightarrow \varphi \equiv \psi$). The principle of substituting an equivalent subformula is also valid: if $\psi_1 \equiv \psi_2$, then $\varphi[\mathcal{A}/\psi_1] \equiv \varphi[\mathcal{A}/\psi_2]$.[2]

From the equivalences provided, it is evident, in particular, that all Boolean connectives can be expressed through a pair of connectives such as $\neg, \wedge$ or $\neg, \vee$ or $\neg, \to$, or even through the single connective $\uparrow$, which is called the *Sheffer stroke*.

Theorem D2.1 (on Functional Completeness).
For every Boolean function $f : \{0,1\}^n \to \{0,1\}$, there exists a formula with n variables whose valuation coincides with this function.

Proof. Let a function $f : \{0,1\}^n \to \{0,1\}$ and propositional variables $\mathcal{A}_1, \ldots, \mathcal{A}_n$ be given. We need to construct a formula φ in these variables such that for any valuation ε, the identity $f \circ \varepsilon = \varepsilon(\varphi)$ holds.

[1] The word means “equal in power” (*Lat. idem potens*), as any power of the operation is equivalent to a single application.

[2] This is quite simple to establish using induction on the construction of formulas.

If f is identically zero, we can take the formulas $\mathcal{A}_k \wedge \neg\mathcal{A}_k$, for $k = 1, \ldots, n$, and combine them, for instance, with disjunction.

Let f take the value 1 on a non-empty set X of tuples $\vec{x}$. Let us define the *literal* of a variable:

$$\mathcal{A}^x = \begin{cases} \mathcal{A}, & x = 1, \\ \neg\mathcal{A}, & x = 0. \end{cases}$$

After which, we consider the formula

$$\varphi \overset{\text{def}}{\leftrightarrow} \bigvee_{\vec{x} \in X} (\mathcal{A}_1^{x_1} \wedge \cdots \wedge \mathcal{A}_n^{x_n}). \tag{D2.6}$$

Let us check that the valuations of this formula coincide with the function f. Let $\varepsilon(\mathcal{A}_k) = x_k$. If $\vec{x} \in X$, i.e., $f(\vec{x}) = 1$, then the valuation of the conjunction corresponding to $\vec{x}$ is

$$\varepsilon(\mathcal{A}_1^{x_1} \wedge \cdots \wedge \mathcal{A}_n^{x_n}) = x_1^{x_1} \wedge \cdots \wedge x_n^{x_n} = 1,$$

from which it follows that the entire disjunction in the definition of φ also evaluates to 1.

If, however, $\vec{x} \notin X$, i.e., $f(\vec{x}) = 0$, then for each $\vec{y} \in X$ there exists some k such that $x_k \neq y_k$, from which $x_k^{y_k} = 0$, and thus $x_1^{y_1} \wedge \cdots \wedge x_n^{y_n} = 0$. This means that all conjunctions in the definition of φ evaluate to zero.

Thus, for any valuation ε, the value of the formula φ coincides with the value of f when the variable $\mathcal{A}_k$ is associated with the k-th argument of f. □

The representation of a formula in the form (D2.6) is called the disjunctive normal form (DNF). It follows from the theorem that any formula of propositional logic has a representation in DNF.

All tautologies of propositional logic can be obtained from the axioms using two rules of inference: substitution (B1.6) and Modus Ponens (B1.7). Recall that the rule of substitution applies only to tautologies, whereas MP can be applied to any hypotheses.

The following list is a classical axiomatization of propositional logic:

A1 introduction of implication: $\mathcal{A} \to (\mathcal{B} \to \mathcal{A})$;
A2 Frege's axiom: $(\mathcal{A} \to (\mathcal{B} \to \mathcal{C})) \to ((\mathcal{A} \to \mathcal{B}) \to (\mathcal{A} \to \mathcal{C}))$;
A3 elimination of conjunction: $(\mathcal{A} \wedge \mathcal{B}) \to \mathcal{A}$, $(\mathcal{A} \wedge \mathcal{B}) \to \mathcal{B}$;
A4 introduction of conjunction: $\mathcal{A} \to (\mathcal{B} \to (\mathcal{A} \wedge \mathcal{B}))$;
A5 introduction of disjunction: $\mathcal{A} \to (\mathcal{A} \vee \mathcal{B})$, $\mathcal{B} \to (\mathcal{A} \vee \mathcal{B})$;
A6 constructive dilemma: $(\mathcal{A} \to \mathcal{C}) \to ((\mathcal{B} \to \mathcal{C}) \to ((\mathcal{A} \vee \mathcal{B}) \to \mathcal{C}))$;

A7 law of Duns Scotus: $\neg\mathcal{A} \to (\mathcal{A} \to \mathcal{B})$ (*ex falso quodlibet*);
A8 introduction of negation: $(\mathcal{A} \to \mathcal{B}) \to ((\mathcal{A} \to \neg\mathcal{B}) \to \neg\mathcal{A})$ (*reductio ad absurdum*);
A9 law of the excluded middle: $\mathcal{A} \vee \neg\mathcal{A}$ (*tertium non datur*).

Accordingly, the theorems of propositional logic are all such formulas that can be derived from these axioms by the rules of inference. The fact that a formula φ is a theorem is denoted as: $\vdash \varphi$.

In modern textbooks, the rule of substitution is not used at all; instead, the axioms listed above are presented not as specific formulas, but as schemas, where the variables $\mathcal{A}, \mathcal{B}, C$ are replaced by non-terminal variables that can be substituted with arbitrary formulas of the language of propositional logic. In essence, this is an implementation of the rule of substitution, but all results of substitutions are included in the list of axioms.

It is worth noting that the axiomatic approach presented here, often called Hilbert-style, is not the only one. There is another, more intuitive way to formalize logical reasoning—the **natural deduction calculus**, proposed by G. Gentzen. In it, as a rule, there are no logical axioms; instead, for each logical connective, there is a pair of rules: a rule for *introducing* that connective into a formula and a rule for its *elimination*. For example, for conjunction, the introduction rule is $\frac{\mathcal{A},\ \mathcal{B}}{\mathcal{A}\wedge\mathcal{B}}$, and the elimination rules are $\frac{\mathcal{A}\wedge\mathcal{B}}{\mathcal{A}}$ and $\frac{\mathcal{A}\wedge\mathcal{B}}{\mathcal{B}}$ (cf. axioms A3 and A4). This approach is much closer to how mathematicians construct proofs informally, operating with hypotheses and conclusions drawn from them. Although both approaches (Hilbert-style and Gentzen-style) are equivalent for classical logic, the second often proves more convenient for analyzing the structure of proofs themselves.

Note also that if not all logical connectives are considered basic, but only, for example, $\neg$ and $\to$, then the number of axioms can be sharply reduced, leaving axioms only for these connectives, and introducing all other connectives as abbreviations, defining by definition that $\mathcal{A} \vee \mathcal{B} = \neg\mathcal{A} \to \mathcal{B}$, $\mathcal{A} \wedge \mathcal{B} = \neg(\mathcal{A} \to \neg\mathcal{B})$. However, with this approach, we are purely syntactically imposing classical behavior on these logical connectives.

Theorem D2.2 (Soundness and Completeness of Derivation).
A formula φ of the language of propositional logic is derivable from the axioms if and only if it is a tautology.

Proof. Necessity ($\Rightarrow$) is not difficult to verify. To begin, we note that all axioms are tautologies. To convince ourselves of this, it is sufficient to check them using a truth table. For example, consider the axiom $\mathcal{A} \to (\mathcal{B} \to \mathcal{A})$:

since it has the form of an implication, it can be false in only one case, when $\mathcal{A} = 1$ and $(\mathcal{B} \to \mathcal{A}) = 0$, i.e., when $(\mathcal{B} \to 1) = 0$. But the latter equality is false for any valuation of $\mathcal{B}$, so this axiom is a tautology.

Next, we need to show that the rules of inference preserve tautologicity (they transform tautologies into tautologies):

$$\frac{\varphi}{\varphi[\mathcal{A}/\psi]}, \quad \frac{\varphi, \quad \varphi \to \psi}{\psi}.$$

Indeed, let $\varphi(\mathcal{A}, \mathcal{B}, \dots)$ be a tautology. What happens if we substitute the formula ψ (about whose tautologicity we know nothing) for the variable $\mathcal{A}$? For this, it is sufficient to note that if the formula $\varphi(\mathcal{A}, \mathcal{B}, \dots)$ corresponds to a Boolean function $f(a, b, \dots)$ (we evaluate propositional variables with corresponding lowercase Latin variables over the set $\{0, 1\}$), and the formula $\psi(a, b, \dots)$ corresponds to a function $g(a, b, \dots)$, then according to the definition of formula valuation, the valuation of the formula $\varphi[\mathcal{A}/\psi]$ will be the composition of functions $f(g(a, b, \dots), b, \dots)$ (for chosen valuations of the variables $a, b, \dots$). But since f is identically one, the composition will also be so. Consequently, $\varphi[\mathcal{A}/\psi]$ is a tautology.

MP rule: assuming that φ and $\varphi \to \psi$ are tautologies, and following the truth table for implication, we notice that ψ is also a tautology, since the equation $(1 \to x) = 1$ has the unique solution $x = 1$.

Thus, when given tautologies as input, the rules of inference always produce tautologies as output, which proves the soundness of the derivability relation.

Sufficiency ($\Leftarrow$). Here the situation is much more complicated, so we will not present the classical proof of completeness, referring instead to the book [16], where it is considered in detail.

Instead, we will provide an alternative way of proving completeness, using some algebraic constructions. We will present it schematically, omitting details but preserving the essence of the reasoning.

Suppose that the formula φ_0 is underivable in propositional logic. Our task in this case is to construct a valuation ε of the propositional variables for which $\varepsilon[\varphi_0] = 0$. This will mean that φ_0 is not a tautology. Thus, if a formula is a tautology, then it is derivable.

Now, for propositional logic, we construct the Lindenbaum algebra, whose elements are classes of provably equivalent formulas. We say that a formula φ is *provably equivalent* to a formula ψ if the equivalence $\varphi \leftrightarrow \psi$ is derivable in propositional logic. Then we set $[\varphi] = \{\psi \mid \vdash \varphi \leftrightarrow \psi\}$. For these classes, operations and constants are introduced in a natural way: $\neg_L[\varphi] = [\neg\varphi]$, $[\varphi] \wedge_L [\psi] = [\varphi \wedge \psi]$, $[\varphi] \vee_L [\psi] = [\varphi \vee \psi]$, $\top_L = [\varphi \vee \neg\varphi]$, $\bot_L = [\varphi \wedge \neg\varphi]$.

The quotient set of formulas $L = \mathbb{L}/_{\leftrightarrow}$ with the operations $\neg_L, \wedge_L, \vee_L$ and constants $\top_L, \bot_L$ is a countable Boolean algebra.[3]

Since the formula φ_0 is not derivable, its class $[\varphi_0] \neq \top_L$, and therefore, $[\neg\varphi_0] \neq \bot_L$.

Next, we construct an ultrafilter on the Lindenbaum algebra containing $[\neg\varphi_0]$. In universal algebra, to prove the existence of such an ultrafilter, we would use Tarski's ultrafilter theorem (the one that is a weakening of the axiom of choice from the list on p. 185), but since our set of formulas is countable (and moreover, the set of provable formulas is enumerable), such an ultrafilter can be constructed by a countable recursion within ZF, without resorting to stronger principles. Let us denote this ultrafilter by U.[4]

Now we can define a valuation function for formulas in the following way:

$$\varepsilon(\mathcal{A}) = \begin{cases} 1, & [\mathcal{A}] \in U, \\ 0, & [\mathcal{A}] \notin U, \end{cases}$$

for propositional variables. Next, we extend the valuation to all formulas, following the usual scheme. Note that this same valuation defines a function $h : L \to \{0, 1\}$ by the rule $h([\varphi]) = \varepsilon(\varphi)$, and h is a homomorphism of Boolean algebras from L to $\mathbf{2}$.[5] In fact, we could have gone the other way: first choose the homomorphism h, defining it as $h([\varphi]) = 1 \Leftrightarrow [\varphi] \in U$. And then extract the valuation of propositional variables from it.

A *homomorphism* of algebras is a map that preserves operations, and our h is such (the key fact here is that we defined it as the characteristic function of an ultrafilter). Moreover, since $h([\neg\varphi_0]) = 1$ and h is a homomorphism, we get that $h([\varphi_0]) = 0$, and therefore the valuation $\varepsilon(\varphi_0) = 0$. That is, φ_0 is not a tautology. □

---

[3] **A Boolean algebra** is a set B on which are defined two binary operations $\vee_B$ (disjunction, join, OR), $\wedge_B$ (conjunction, meet, AND), one unary operation $\neg_B$ (negation, complement, NOT), and two distinct special elements $\bot_B$ (zero, false, bottom, 0) and $\top_B$ (one, truth, top, 1), for which the axioms are the identities (D2.1)–(D2.4), where the sign $\equiv$ must be replaced by $=$, as well as the definitions of the logical constants $\top_B = a \vee_B \neg_B a$, $\bot_B = a \wedge_B \neg_B a$. An inequality is also introduced on a Boolean algebra by the rule: $a \leqslant b \overset{\text{def}}{\leftrightarrow} a \wedge_B b = a$.

[4] **An ultrafilter** on a Boolean algebra B is a subset $U \subseteq B$ such that $\bot_B \notin U$, $\forall x, y \in U : x \wedge_B y \in U$, $\forall x, y \in B : (x \in U) \wedge (x \leqslant_B y) \to (y \in U)$, $\forall x \in B : (x \in U) \vee (\neg_B x \in U)$.

[5] Here, $\mathbf{2}$ is the two-element boolean algebra.

From the proof of the second part of the theorem, it might seem that we do not use the axiomatics of propositional logic and the rules of inference at all. But in fact, we use them in a substantial way. First, without the law of the excluded middle, the Lindenbaum algebra would cease to be a Boolean algebra, as this law holds in a Boolean algebra. This means we would not have been able to construct a homomorphism that "decides" how to evaluate the elements of this algebra using the algebra **2**. Furthermore, our reasoning that if $[\varphi_0] \neq \top_L$, then $[\neg\varphi_0] \neq \bot_L$, would have been incorrect. The other axioms of logic are directly related to the axiomatics of Boolean algebra. Second, the definition of equivalence classes (and the fact that they are indeed equivalence classes) is based directly on derivability, i.e., on the MP rule. If this rule were weakened, then indeed, either the proof itself would be incorrect, or we would prove that in that case, derivability would not be complete.

Both proofs (the classical one and the one via the homomorphism of algebras) are quite non-trivial. In one case (the classical one), it requires the analysis of numerous cases and induction on the complexity of formulas; in the other (as above), it requires bringing in the apparatus of Boolean algebras and the non-trivial construction of an ultrafilter. Therefore, the completeness theorem for propositional logic is a very significant and deep mathematical fact.

The Lindenbaum algebra is an immanent example of a Boolean algebra as a model of propositional calculus. However, it turns out that the entire class of Boolean algebras is precisely the class of models of classical propositional logic.

Theorem D2.3 (on Algebraic Completeness).
A formula φ is derivable in propositional logic if and only if it is true in every Boolean algebra (is valid).

Proof. Here, soundness ($\Rightarrow$) is proven as above: one must show that the axioms are equivalent to $\top$, and that the rules of inference always map $\top$ to $\top$.

Completeness ($\Leftarrow$) is proven by contraposition. Suppose that the formula φ is not derivable. Then we construct an example of a Boolean algebra where it will not be true (not equal to $\top$). Such an example is precisely the Lindenbaum algebra.

Indeed, if the formula φ is not a theorem ($\nvdash \varphi$), then its equivalence class $[\varphi]$ is not equal to the top element $\top_L$ in the Lindenbaum algebra by definition.

We can define an interpretation of the variables in this algebra as follows: the propositional variable $\mathcal{A}$ is mapped to its own equivalence class $[\mathcal{A}]$. Then the value of the formula φ under this interpretation will be equal to $[\varphi]$. Since $[\varphi] \neq \top_L$ in this algebra, we have found a Boolean algebra in which φ is not true, and thus, φ is not valid. $\square$

Continuing the theme of using Boolean algebras, one cannot help but wonder how the derivability relation can be expressed in a Boolean algebra. Let $\Gamma \vdash \varphi$, where Γ is an arbitrary set of formulas. Due to the compactness of derivation, it is actually the case that $\Gamma' \vdash \varphi$, where Γ' is a finite subset of Γ. Moreover, this is equivalent to $\bigwedge \Gamma' \vdash \varphi$, where $\bigwedge \Gamma'$ is the conjunction of all formulas in Γ'. In other words, derivability reduces to a binary relation between formulas, and thus it will be interpreted in the algebra as a binary relation between elements.

Let us set, for $a, b \in B$, where B is a Boolean algebra, $a \leqslant b \overset{\text{def}}{\leftrightarrow} (a \wedge_B b) = a$, i.e., we define the relation $\leqslant$ via the equality relation and the meet operation. Algebraically, it is not difficult to show that $\leqslant$ is a partial (non-strict) order on B, where $\top_B$ is the maximum and $\bot_B$ is the minimum of this relation ($\forall x \in B : \bot_B \leqslant x \leqslant_B \top_B$).

Let's check if it will be a model of the derivability relation on formulas. To do this, we need to verify that under this translation, the rules of inference are true predicates in the algebra.

Rule of substitution $\frac{\varphi}{\varphi[\mathcal{A}/\psi]}$. Let φ be a tautology and ε be an arbitrary valuation of propositional variables in the algebra B. We need to show that $\varepsilon(\varphi) \leqslant \varepsilon(\varphi[\mathcal{A}/\psi])$, i.e., that $(\varepsilon(\varphi) \wedge_B \varepsilon(\varphi[\mathcal{A}/\psi])) = \varepsilon(\varphi)$ for all valuations ε. Since φ is a tautology, we get $\varepsilon(\varphi) = \top_B$ (by soundness). Furthermore, $\varepsilon(\varphi[\mathcal{A}/\psi]) = \varphi(\varepsilon(\psi)) = \top_B$, hence we need to check that $\top_B \wedge_B \top_B = \top_B$, which is an obvious consequence of the axioms of a Boolean algebra.

MP rule: $\frac{\varphi, \quad \varphi \to \psi}{\psi}$. We need to show that for any valuation ε: $\varepsilon(\varphi) \wedge_B \varepsilon(\varphi \to \psi) \leqslant \varepsilon(\psi)$. For this, we expand the implication by definition: $\varepsilon(\varphi \to \psi) = \varepsilon(\neg\varphi \vee \psi) = \neg_B \varepsilon(\varphi) \vee_B \varepsilon(\psi)$. In other words, we need to check that the identity

$$a \wedge_B (\neg_B a \vee_B b) \leqslant b, \text{ where } a = \varepsilon(\varphi), b = \varepsilon(\psi)$$

holds in the algebra B. Following the axioms of Boolean algebra, we easily get that

$$a \wedge_B (\neg_B a \vee_B b) = (a \wedge_B \neg_B a) \vee_B (a \wedge_B b) = \bot_B \vee_B (a \wedge_B b) = a \wedge_B b,$$

so it remains to check that $a \wedge_B b \leqslant b$, which by definition means $(a \wedge_B b) \wedge_B b = a \wedge_B b$, and this is also true by the axioms of a Boolean algebra.

Thus, applying the rules of inference to formulas means moving up the $\leqslant$ scale for the valuations of the formulas. And this means that for the entire derivation $\Gamma' \vdash \varphi$, the inequality $\varepsilon(\bigwedge \Gamma') \leqslant \varepsilon(\varphi)$ holds for any valuation ε.

On the other hand, if for any valuation ε we have $\varepsilon(\bigwedge \Gamma') \leqslant \varepsilon(\varphi)$, then it is also true that $\neg_B \varepsilon(\bigwedge \Gamma') \vee_B \varepsilon(\varphi) = \top_B$, from which, by virtue of completeness, it follows that the formula $\neg \bigwedge \Gamma' \vee \varphi$ is a tautology, i.e., the implication $\bigwedge \Gamma' \to \varphi$ is derivable. Then, by the MP rule, one can show that $\Gamma' \vdash \varphi$.

Thus, $\Gamma' \vdash \varphi$ if and only if for any valuation in any Boolean algebra, $\varepsilon(\bigwedge \Gamma') \leqslant \varepsilon(\varphi)$.

From this, it easily follows

Theorem D2.4 (on Deduction). *$\Gamma, \varphi \vdash \psi$ if and only if $\Gamma \vdash (\varphi \to \psi)$.*

Proof. We can immediately assume that Γ is finite (if not, a finite subset can be chosen instead). Let χ denote the formula $\bigwedge \Gamma$; then the statement of the theorem is as follows:

$$\chi \wedge \varphi \vdash \psi \quad \Leftrightarrow \quad \chi \vdash (\varphi \to \psi),$$

and from the preceding arguments, this is equivalent to a theorem of Boolean algebra

$$(a \wedge_B b \leqslant c) \leftrightarrow (a \leqslant (\neg_B b \vee_B c)),$$

which is equivalent, according to the definition of inequality, to:

$$(a \wedge_B b \wedge_B c = a \wedge_B b) \leftrightarrow (a \wedge_B (\neg_B b \vee_B c) = a). \qquad \text{(D2.7)}$$

There are two ways to prove this: 1) honestly check this statement using the axioms of Boolean algebra, 2) use a "cheat" which consists in the fact that any identity in the language of Boolean algebra is either true or not true in all Boolean algebras simultaneously. This result is a consequence of the fact that from any Boolean algebra one can construct a homomorphism into the algebra **2**. Therefore, if some identity were violated in some Boolean algebra, it would also be violated in **2**.[6]

Consequently, it only remains for us to check that (D2.7) holds in the algebra **2**, which can be done by a simple case analysis, substituting the numbers 0 and 1 for the variables. □

[6] This fundamental fact is closely related to Stone's representation theorem for Boolean algebras, which guarantees a sufficient number of ultrafilters/homomorphisms into **2**.

Now we begin to better understand that logic is not just a game of formulas that tries to model an abstract thought process, but also the internal language of a whole class of algebras, where logical symbols become real operations, and derivability acquires a concrete meaning in the form of an order on the elements of the algebra. From this arises the mathematical discipline called algebraic logic, which allows for the analysis of logic by purely algebraic (traditional for mathematics) methods.[7]

The following examples of Boolean algebras show that even classical propositional calculus has a very diverse class of models: 1) the set $\mathbf{2} = \{0, 1\}$ with the corresponding Boolean functions (the standard model); 2) the powerset $\mathcal{P}(A)$ of any non-empty set A with the operations of complement, intersection, and union; 3) the set of all divisors of a square-free number N with the operations N/a, $\gcd(a, b)$, $\operatorname{lcm}(a, b)$; 4) the algebra of clopen subsets of a topological space (with the operations of complement, intersection, and union).

On the other hand, the commonality of these structures is manifested in the theorem that every Boolean algebra is isomorphic to an algebra that is some (Cartesian) product of copies of the two-element Boolean algebra **2**. This theorem is a consequence of the well-known Stone's representation theorem for Boolean algebras (which we mentioned in the list of consequences of the axiom of choice on p. 185).

D2.2 Predicate Logic

Since predicate logic was covered in sufficient detail in Section B1.2, we will focus here only on some key constructions and theorems.

Let us recall the axioms and inference rules of predicate logic:

PC1 the universal closures of all substitutions of all formulas of the language of predicate logic into all tautologies of propositional logic;

PC2 axioms for quantifiers:

$$(\forall) \quad (\forall x\, \varphi[a/x]) \to \varphi[a/T]; \qquad (\exists) \quad \varphi[a/T] \to \exists x\, \varphi[a/x],$$

[7] This refers, of course, not only and not so much to classical logic and Boolean algebra, as to a large variety of logics that differ from classical and have their own algebraic interpretations.

where φ does not contain the bound variable x, and T is any term of the theory;

PC3 rules of inference:

$$(MP)\quad \frac{\varphi,\quad \varphi \to \psi}{\psi}; \qquad (R1)\quad \frac{\varphi \to \psi}{\varphi \to \forall x\, \psi[a/x]};$$

$$(R1')\quad \frac{\psi}{\forall x\, \psi[a/x]}; \qquad (R2)\quad \frac{\psi \to \varphi}{(\exists x\, \psi[a/x]) \to \varphi},$$

where for rules (R1), (R1'), and (R2), the variable a does not occur in φ nor in any of the active (undischarged) hypotheses on which the derivation of the formula in the premise of these rules depends.

The rules $(R1)$ and $(R2)$ are called **Bernays' rules**. Rule (R1'), called the *generalization rule*, is a derived rule in our inference system, as it can be obtained from rule (R1) by substituting a tautology for the formula φ.

Indeed, if ψ is proven (the premise for (R1')), then it follows that $\top \to \psi$ (where $\top$ is any substitution into a tautology or a proven theorem not containing a). Applying (R1) to $\top \to \psi$, we obtain $\top \to \forall x\, \psi[a/x]$. Since $\top$ is true (provable), by MP we obtain $\forall x\, \psi[a/x]$.

Theorem D2.5 (on Deduction). *Let T be a set of formulas (in a given first-order language), and let φ, ψ be arbitrary formulas. Then if there exists a derivation $T, \varphi \vdash \psi$ in which no application of rule $(R1)$ to formulas that depend on φ in this derivation binds any of the free variables of the formula φ, then $T \vdash (\varphi \to \psi)$.*

The proof of this theorem almost completely reproduces the syntactic proof of the deduction theorem in propositional logic, with an adjustment for checking the application of rules (R1) and (R2). The restriction on the free variables of the formula φ, mentioned in the theorem, is required precisely to correctly check the application of rule (R1). If the formula φ is closed, then the theorem takes the same form as in propositional logic. Moreover, the converse statement is in any case obvious by virtue of the Modus Ponens rule, so for closed formulas we have an equivalence:

$$T, \varphi \vdash \psi \quad \Leftrightarrow \quad T \vdash (\varphi \to \psi).$$

Theorem D2.6. *Let φ be an arbitrary formula with a free variable a (and possibly some other free variables), and let t be an arbitrary term, then*

$$\vdash \varphi[a/t] \leftrightarrow \exists x\ (\varphi[a/x] \wedge (x = t)).$$

Proof. ($\Rightarrow$) Consider the new formula $\psi(a) \overset{\text{def}}{\leftrightarrow} \varphi(a) \wedge (a = t)$. It is obvious that $\varphi[a/t]$ derives $\psi[a/t]$. Furthermore, by the axiom for the existential quantifier, the implication $\psi[a/t] \rightarrow \exists x\ \psi[a/x]$ is valid. From this and the preceding, by the Modus Ponens rule, we obtain that $\exists x\ (\varphi[a/x] \wedge (x = t))$. And by the deduction theorem, we have the required implication.

($\Leftarrow$) It is clear that $\psi(a) \rightarrow \varphi[a/t]$ by virtue of the congruence of equality. Moreover, since a does not occur in the formula $\varphi[a/t]$, one can apply Bernays' rule for introducing an existential quantifier:

$$\frac{\psi(a) \rightarrow \varphi[a/t]}{\exists x\ \psi[a/x] \rightarrow \varphi[a/t]},$$

and this is precisely the converse implication of the theorem. □

The following statement we gave as an example of a valid formula of pure predicate calculus in Section B1.2.8.

Theorem D2.7. *Let φ be an arbitrary formula with free variables a, b (and possibly some other free variables), then:*

$$(\exists x \forall y\ \varphi[a/x, b/y]) \rightarrow \forall y \exists x\ \varphi[a/x, b/y].$$

Proof. First, let us prove the following lemma (in the case of one variable, we do not explicitly write the substitution $[a/x]$):

$$\forall x\ (\psi_1(x, \vec{p}) \rightarrow \psi_2(x, \vec{q})) \rightarrow ((\exists x\ \psi_1(x, \vec{p})) \rightarrow \exists x\ \psi_2(x, \vec{q})), \qquad \text{(D2.8)}$$

where $\vec{p}, \vec{q}$ are sets of free variables (parameters), which we will omit going forward to save space, as they do not affect the derivation. We use the same bound variable under the quantifiers here, since the scopes of these quantifiers do not overlap, and a collision of variables is impossible.

Let us take the formula $\forall x\ (\psi_1(x) \rightarrow \psi_2(x))$ as a hypothesis. Using Axiom ($\forall$) we get the formula $\psi_1(a) \rightarrow \psi_2(a)$ with free a. By the axiom for introducing the existential quantifier, we note that

$$\psi_2(a) \rightarrow \exists x\ \psi_2(x),$$

then using the transitivity of implication we obtain that

$$\psi_1(a) \to \exists x\ \psi_2(x),$$

now let us consider the resulting formula as a formula of the form $\psi_1(a) \to \psi_3$, where the variable a does not occur in the formula ψ_3 (we have already replaced it with x there), and apply Bernays' rule ($R2$) to this implication to introduce an existential quantifier; we will get the formula

$$(\exists x\ \psi_1(x)) \to \exists x\ \psi_2(x).$$

By the deduction theorem, we place our hypothesis in the antecedent of the implication:

$$\forall x\ (\psi_1(x) \to \psi_2(x)) \to ((\exists x\ \psi_1(x)) \to \exists x\ \psi_2(x)),$$

which proves (D2.8). Thus, we can introduce existential quantifiers into both parts of an implication over the same free variable.

In set-theoretic terms, this lemma states that if a non-empty set A is a subset of a set B, then B must also be non-empty. Let us use this to prove the theorem.

By the axiom for eliminating the universal quantifier, the following is valid:

$$(\forall y\ \varphi[a, b/y]) \to \varphi(a, b).$$

Now, viewing this implication as $\psi_1(a) \to \psi_2(a)$, using (D2.8) and the MP rule, we get that:

$$(\exists x \forall y\ \varphi[a/x, b/y]) \to \exists x\ \varphi[a/x, b].$$

The resulting formula has the form $\varphi_1 \to \varphi_2(b)$, where b does not occur in the formula φ_1 (since it has already been replaced by y), therefore, we can apply Bernays' rule ($R1$) to introduce a universal quantifier:

$$(\exists x \forall y\ \varphi[a/x, b/y]) \to \forall y \exists x\ \varphi[a/x, b/y],$$

which completes the proof of the theorem.

Note that the scopes of the quantifiers $\exists x \forall y$ and $\forall y \exists x$ here do not overlap, so we can use the same bound variables under the quantifiers. □

As an example of the application of Theorem D2.7, note that the statement from real analysis “a function uniformly continuous on a set is continuous on it“ reduces to this formula.

D2.2.1 Soundness and Completeness of the Calculus

Now let us give a precise definition of the truth of a formula of predicate logic in a model. Let there be a first-order language with a signature Σ, a non-empty set M, and an interpretation I, which maps each symbol from Σ to a corresponding object (a relation, function, or constant) on M (taking arity into account). Then the structure $\mathcal{M} = \langle M, I[\Sigma] \rangle$ is called a **model** of the language generated by this signature, and the set M is called the *domain* of the interpretation.

A *valuation* (or *assignment*) of free variables is a function ε from the set of free variables to M, i.e., an assignment of (arbitrary) values from M to the variables. The definition of the interpretation of a language and a theory is built around the arbitrariness of the choice of valuation.

Given an interpretation of the signature's symbols in a model, as well as a valuation of the free variables, we can compute the values of terms and the truth values of formulas of the language with signature Σ in this model. For convenience, we will also denote the values of terms under a given valuation of variables as a valuation $\varepsilon(t)$, where t is an arbitrary term.

The values of terms are computed recursively:

- constant symbols are assigned their interpretation: $\varepsilon(c) = I(c)$;
- let the values of the terms $t_1, \ldots, t_n$ be already computed, and let F be an n-ary function symbol of the signature, then $\varepsilon(F(t_1, \ldots, t_n)) = I(F)(\varepsilon(t_1), \ldots, \varepsilon(t_n))$.

Then we begin to assign truth values to formulas, following a similar scheme, but now we define the relation $\mathcal{M} \models$ (*is true in the model*):

- for atomic predicates: $\mathcal{M} \models_\varepsilon P(t_1, \ldots, t_n)$ if and only if $\langle \varepsilon(t_1), \ldots, \varepsilon(t_n) \rangle \in I(P)$;
- for logical connectives: $\mathcal{M} \models_\varepsilon \neg\varphi \Leftrightarrow$ it is not the case that $\mathcal{M} \models_\varepsilon \varphi$;
  $\mathcal{M} \models_\varepsilon \varphi \wedge \psi \Leftrightarrow \mathcal{M} \models_\varepsilon \varphi$ AND $\mathcal{M} \models_\varepsilon \psi$;
  $\mathcal{M} \models_\varepsilon \varphi \vee \psi \Leftrightarrow \mathcal{M} \models_\varepsilon \varphi$ OR $\mathcal{M} \models_\varepsilon \psi$;
  $\mathcal{M} \models_\varepsilon \varphi \rightarrow \psi \Leftrightarrow$ NOT $\mathcal{M} \models_\varepsilon \varphi$ OR $\mathcal{M} \models_\varepsilon \psi$;

- for quantifiers: $\mathcal{M} \models_\varepsilon \forall x\ \varphi(a/x, \vec{p})$ if and only if $\mathcal{M} \models_{\varepsilon[a/m]} \varphi(a, \vec{p})$ for all $m \in M$; $\mathcal{M} \models_\varepsilon \exists x\ \varphi(a/x, \vec{p})$ if and only if $\mathcal{M} \models_{\varepsilon[a/m]} \varphi(a, \vec{p})$ for some $m \in M$, where we have used a modified valuation

$$\varepsilon[a/m](v) = \begin{cases} \varepsilon(v), & v \neq a, \\ m, & v = a, \end{cases} \qquad m \in M.$$

In short, we interpret the logical connectives and quantifiers of our language as the logical connectives and quantifiers of the meta-theory, i.e., of the set theory in which the model $\mathcal{M}$ is defined.

Finally, let us say that $\mathcal{M} \models \varphi$ (*φ is true in the model $\mathcal{M}$*) if for any valuation ε it is true that $\mathcal{M} \models_\varepsilon \varphi$, where φ is an arbitrary formula of the language with signature Σ.

A formula is called **satisfiable** if it is true in some model, and **valid** if it is true in every model. A set of formulas T is true in a model ($\mathcal{M} \models T$, has a model $\mathcal{M}$) if all its formulas are true in that model. Obviously, the relation $\mathcal{M} \models T$ is monotonic in T, i.e., if $\mathcal{M} \models T$ and $T' \subseteq T$, then $\mathcal{M} \models T'$.

Correspondingly, the satisfiability of a set of formulas (in particular, of an axiomatics or a theory) is also defined. A set of formulas that has a model is called **satisfiable**.

Recall that a theory is called **(syntactically) consistent** if a contradiction cannot be derived in it ($T \nvdash \bot$).

The notion of semantic (or logical) consequence is closely related to models. We say that a formula φ is a **semantic** (logical) **consequence** of a set of formulas T if for any model $\mathcal{M}$, from $\mathcal{M} \models T$ it follows that $\mathcal{M} \models \varphi$. Notation: $T \Vdash \varphi$.[8]

As was already noted in Section B1.2.9, it is not difficult to show that predicate logic is *sound* with respect to the semantics, i.e.,

[8] There is often confusion with the notations for truth in a model and logical consequence; for example, the symbol $\models$ or a very similar symbol $\vDash$ is used in both cases. Here we have decided to use a separate symbol $\Vdash$ for logical consequence, although this symbol is also often used for other purposes.

Theorem D2.8 (on the Soundness of Derivation). *If the derivation $T \vdash \varphi$ holds, then the semantic consequence $T \Vdash \varphi$ also holds.*

This follows directly from the fact that we interpret the logic of the language as the logic of the meta-theory, i.e., a formal derivation will correspond to a derivation in the model by virtue of the definition of truth in a model.

A key corollary connects the syntactic and semantic concepts:

Corollary D2.1. *If a theory is satisfiable (has a model), then it is consistent.*

Indeed, suppose that a theory has a model, but is inconsistent. Then $T \vdash \bot$. By soundness, it would follow that $T \Vdash \bot$, meaning a contradiction must be true in the model, which is impossible.

Corollary D2.2.
If the sets T, φ and $T, \neg\varphi$ are both satisfiable, then φ is independent of T.

Recall that the independence of a formula φ from a set of formulas T means that $T \nvdash \varphi$ and $T \nvdash \neg\varphi$. This corollary of soundness is the main tool for demonstrating the independence of certain statements. For example, the independence of the parallel postulate from the other axioms of geometry (the model of $\mathbb{R}^2$ and the Poincaré model), the independence of the axiom of choice and the continuum hypothesis from the axioms of ZF (the models of Gödel and Cohen).

The models we presented in Sections B1.2.7, B1.2.10, C1.4.3 showed us a number of examples of consistent theories, such as the theory of linear orders, the theory of groups, the theory of equality, Peano arithmetic (its model is the ordinal ω), the theory of real closed fields, and some fragments of set theory. Of course, all this is proven under the assumption of the consistency of our meta-theory, which is usually ZF, but for simple cases (like equality, linear orders, and groups) very primitive, finite, almost physical models suffice.

As for ZF itself, by virtue of Gödel's second incompleteness theorem, it cannot justify its own consistency, and to clarify this issue, an even stronger theory would be needed (for example, ZF with the axiom of a strongly inaccessible cardinal number), which requires stronger ontological assumptions.

Nevertheless, as we have already discussed, ZF without the axiom of infinity, or HF, looks like an undoubtedly reliable theory, given the possibility of its interpretation in arithmetic, as well as the bracket model. Therefore, there is currently a consensus among mathematicians that ZF is consistent.

Theorem D2.9 (Gödel's Completeness Theorem for Predicate Logic).

(1) *If a first-order theory is consistent, then it is satisfiable.*
(2) *Equivalent formulation: for any T and formula φ, if $T \Vdash \varphi$, then $T \vdash \varphi$.*

The proof of (1) was first obtained by Gödel in his doctoral dissertation in 1929. The work was published in 1930 under the title „Die Vollständigkeit der Axiome des logischen Funktionenkalküls“ ("The Completeness of the Axioms of the Logical Function Calculus"). Then Henkin (in 1949) found a more transparent proof based on the use of constants that are witnesses for existential formulas. In essence, the model is constructed from the existing constants, new constants extracted from theorems of the theory of the form $\exists x\ \varphi(x)$, and all closed terms that are formed by the rules for term formation from these constants.

In this proof, the ability to extend the theory to a maximal consistent (and, therefore, complete) one is used, by constructing an ultrafilter in the Lindenbaum algebra, which for countable languages is feasible in pure ZF theory (thanks to the enumerability of all formulas), and for uncountable languages requires Tarski's ultrafilter theorem (see p. 185).

Point (1), with the help of the soundness theorem, is generalized as follows: a theory is consistent if and only if it is satisfiable (has a model).

Point (2) is also generalized to an equivalence: $T \Vdash \varphi$ if and only if $T \vdash \varphi$, by virtue of the soundness of first-order predicate logic with respect to semantics.

Proof of the equivalence of the formulations. ($\Rightarrow$) Let (1) hold and $T \Vdash \varphi$. Then, assuming that $T \nvdash \varphi$, we get that $\neg\varphi$ does not contradict T, therefore, the set of formulas $T \cup \{\neg\varphi\}$ is consistent, and thus has a model by virtue of (1). Let us denote this model by $\mathcal{M}$. Then $\mathcal{M} \models T \cup \{\neg\varphi\}$ and, in particular, $\mathcal{M} \models T$. Then, on the one hand, $\mathcal{M} \models \varphi$ by virtue of the assumption $T \Vdash \varphi$, and on the other hand, $\mathcal{M} \models \neg\varphi$ by virtue of the choice of the model $\mathcal{M}$ itself. A contradiction. Consequently, $T \vdash \varphi$, i.e., from (1) follows (2).

($\Leftarrow$) Conversely, let (2) hold and the theory T be consistent. Suppose that it has no model, i.e., for any structure $\mathcal{M}$: $\mathcal{M} \not\models T$. Let us write out condition (2) in full for this theory T:

$$\forall\varphi\ [\forall\mathcal{M}\ (\mathcal{M} \models T)\rightarrow(\mathcal{M} \models \varphi)]\rightarrow(T \vdash \varphi),$$

where red highlights the quantifiers and implication of the meta-theory. As we can see, the premise of the first implication is false for this theory T, which means the entire implication is true. Consequently, from the second implication with a true premise, we get that

$$\forall \varphi \ (T \vdash \varphi),$$

in which case, let us take a false formula as φ and we get that $T \vdash \bot$, which contradicts the fact that T is consistent. Thus, from (2) follows (1).

Using the completeness theorem, one can rewrite the deduction theorem differently:

$$T, \varphi \Vdash \psi \quad \Leftrightarrow \quad T \Vdash (\varphi \to \psi), \tag{D2.9}$$

and this result can already be proven semantically in much the same way as we proved it for propositional logic. The difference is that instead of Boolean algebras, one would need to use cylindrical or polyadic algebras, in which an interpretation of quantifiers is provided.[9] Then the statement (D2.9) would be rewritten in the form of an analogous algebraic theorem:

$$(g \wedge_B a) \leqslant_B b \quad \Leftrightarrow \quad g \leqslant (a \to_B b),$$

the truth of which easily follows from the axioms of the corresponding algebra.

Using the completeness theorem, one can give another (semantic) proof of Theorem D2.7, which now, taking into account the deduction theorem, can be formulated differently:

$$\exists x \forall y \ \varphi[a/x, b/y] \Vdash \forall y \exists x \ \varphi[a/x, b/y].$$

Let an arbitrary model $\mathcal{M}$ be given, consisting of a non-empty set (the domain of interpretation) M and an interpretation of the predicate symbol φ. Suppose that $\mathcal{M} \models \exists x \forall y \ \varphi(x, y)$. This means that in M there exists at least one element, let's call it a, such that $\mathcal{M} \models \forall y \ \varphi(a, y)$. This means that for any element $d \in M$: $\mathcal{M} \models \varphi(a, d)$.

Let us show that under this assumption, $\mathcal{M} \models \forall y \exists x \ \varphi(x, y)$.

Take an arbitrary element $b \in M$. From the preceding, we know that there exists an element $a \in M$ such that $\mathcal{M} \models \varphi(a, d)$ for all $d \in M$. Since this is true for all d, it will also be true for our arbitrarily chosen b. That is, $\mathcal{M} \models \varphi(a, b)$. This means that for our arbitrary b, we have found an element x (namely, a) for which $\varphi(x, b)$ is true.

Since we chose b arbitrarily, this reasoning is valid for any element $y \in M$. Consequently, $\mathcal{M} \models \forall y \exists x \varphi(x, y)$.

Since we did not restrict the choice of the model in any way, the logical consequence $\exists x \forall y \ \varphi[a/x, b/y] \Vdash \forall y \exists x \ \varphi[a/x, b/y]$ is proven.

[9] Cylindrical algebras were introduced by Alfred Tarski and his students (Leon Henkin, Donald Monk). Polyadic algebras were developed by Paul Halmos. See [32, 49].

D2.2.2 Definability

Next, we provide for reference a few more facts about first-order logic (without proof).

We have already briefly spoken about isomorphisms of structures; now let us give a precise definition. Let there be two models $\mathcal{M} = \langle M, \mathcal{P}, \mathcal{F}, C\rangle$ and $\mathcal{N} = \langle N, \mathcal{P}', \mathcal{F}', C'\rangle$, and let there be a one-to-one correspondence I between their signatures such that $I(P) = P'$ for all $P \in \mathcal{P}$, $I(F) = F'$ for all $F \in \mathcal{F}$ and $I(c) = c'$ for all $c \in C$, preserving arity. A **homomorphism** from structure $\mathcal{M}$ to structure $\mathcal{N}$ is a function $g : M \to N$ that preserves predicates and functions, i.e.,

$$\begin{aligned} F(a_1, \ldots, a_n) = b \quad &\Rightarrow \quad F'(g(a_1), \ldots, g(a_n)) = g(b), \\ \langle a_1, \ldots, a_n\rangle \in P \quad &\Rightarrow \quad \langle g(a_1), \ldots, g(a_n)\rangle \in P', \\ g(c) &= c' \end{aligned}$$

If g is a bijection between M and N and in the definition above $\Rightarrow$ can be replaced by $\Leftrightarrow$, then g is called an **isomorphism** of structures $\mathcal{M}$ and $\mathcal{N}$ (an equivalent definition: there exists $g^{-1} : N \to M$, which is also a homomorphism). The existence of an isomorphism of structures is denoted by $\mathcal{M} \cong \mathcal{N}$.

If $\mathcal{M} = \mathcal{N}$ and $I = \mathrm{id}$, then the isomorphism is called an **automorphism** of the structure $\mathcal{M}$. It is easy to check that the automorphisms of a structure form a group with respect to the operation of function composition, which is denoted by $\mathsf{Aut}(\mathcal{M})$.

Now let there be a first-order language with a signature Σ, which we interpret in the same way (up to the correspondence of their signatures) in the models $\mathcal{M}$ and $\mathcal{N}$.

Theorem D2.10. *The truth of formulas is preserved under an isomorphism of models, i.e.,*

$$\textit{if } \mathcal{M} \cong \mathcal{N}, \textit{ then} \qquad \mathcal{M} \models \varphi \quad \Leftrightarrow \quad \mathcal{N} \models \varphi.$$

It is important to emphasize here that the formula φ is given in the same language, which we interpret in the same way in these two models (taking into account the correspondence I between the signatures of the models).

This theorem, when applied to automorphisms, provides a powerful tool for analyzing the definability of model predicates in the language of the theory.

Consider an example: the theory of torsion-free abelian groups with the operation + is interpreted in the model $\mathbb{Z}$ in a natural way. However, in the structure $\mathbb{Z}$ there is a linear order relation $\leqslant$. The question is: can it be expressed through the addition operation in $\mathbb{Z}$? If this is possible, i.e., the predicate $a \leqslant b$ can be written as a formula $\varphi(a, b)$ in the language of this abelian group, then this predicate must preserve its truth under automorphisms in all models, in particular in $\mathbb{Z}$. But in the structure $\mathbb{Z}$ there are two automorphisms that preserve the signature of the theory, i.e., the addition operation: these are $f(x) = x$ and $g(x) = -x$. However, the automorphism g obviously does not preserve the formula $a \leqslant b$, as it reverses the order relation ($-b \leqslant -a$). Consequently, the predicate $a \leqslant b$ is undefinable through + in the structure $\mathbb{Z}$.

Such an argument would not work if we were considering the signature of the theory of rings, since in the ring $\mathbb{Z}$ with the operations of addition and multiplication, the order predicate can already be expressed:

$$(a \leqslant b) \overset{\text{def}}{\leftrightarrow} \exists x_1, x_2, x_3, x_4 : \; a + x_1 \cdot x_1 + x_2 \cdot x_2 + x_3 \cdot x_3 + x_4 \cdot x_4 = b.$$

It is not violated by automorphisms, since in the ring $\mathbb{Z}$ with the operations of addition and multiplication there is exactly one automorphism: id. At the same time, we are not claiming that the predicate so defined will define an order in *all* models of the theory of rings. It is enough to give the example of the ring of Gaussian integers $\mathbb{Z}[i]$, where the specified definition will not define a linear order (both $-1 \leqslant 1$ and $1 \leqslant -1$ will hold simultaneously).

In other words, the theorem on automorphisms is usually useful when one needs to refute the definability of some relation (or function) through the elements of a given signature, rather than to show the possibility of defining it.

One can give a few more examples:

- In the integers, the operation + and the constant 0 cannot be expressed through the order relation alone.
- In the language of Tarski's elementary planimetry, the following are undefinable:
  - any specific figure other than the entire plane;
  - the unit of length, i.e., the predicate "the length of the segment AB is equal to 1";
  - orientation, i.e., the predicate "the vertices of the triangle ABC are traversed counterclockwise".

► In the language of the theory of real closed fields, the following are undefinable:

- the predicate “to be an integer”;
- the predicate “to be a rational number”;
- predicates expressing elementary functions $y = \sin(x)$, $y = e^x$, etc.;
- the numbers π and e.

D2.2.3 Categoricity

Theorem D2.11 (Löwenheim-Skolem Downward Theorem).
If a first-order theory in a countable language has an infinite model, then it has a countable model.

Remark: this theorem can be generalized to talk about uncountable languages and models, in which case the lower bound for the cardinality of the model will be different, but since in this book we consider only countable languages, we will omit such details.

Almost all the theories familiar to us from this course satisfy the conditions of this theorem. For example, the theory of equality without axioms about a finite domain, the theory of linear orders, the theory of groups, Peano arithmetic, the theory of real closed fields, Zermelo-Fraenkel set theory...

Doubts may arise about the last two examples, especially if one recalls that in set theory there is an axiom that allows one to increase the cardinality of a set for as long as one likes by applying the operation $\mathcal{P}(X)$.

However, one must understand here that the capabilities of a theory are often much weaker, both in terms of provability and in terms of expressibility, than the capabilities of the meta-theory in which we construct the models. If a theory says that one can build an ascending chain of powersets, this does not yet mean that in its model these provable powersets will correspond to “real” powersets. In particular, a countable model of set theory ZF is constructed on the so-called Gödel’s constructible universe L, in which all sets are definable, i.e., they are only such collections that can be defined by formulas. More precisely, each subsequent constructible universe is built as the set of all *definable* subsets of the previous one, which means that in reality (from the point of view of the meta-theory) the cardinalities of all these universes are no more than countable.

How do things stand there with Cantor's theorem, which states that there is no bijection between X and $\mathcal{P}(X)$? Very simply, since this bijection must also be a constructible set. And when the theory proves to us that no bijection exists, it tells us only about the bijections that are sets within this model, while in reality such a bijection exists, but this bijection will not be definable in the model.

Thus, although the model is countable in the metatheory, the theory still treats many sets as uncountable: internally, countability means having a bijection that is itself a set of the model, and a metatheoretic enumeration of the same objects need not furnish such a set.

Perhaps this can be compared to the tunnel effect in race car drivers. Under normal conditions, they see much more than in a race situation, where their field of vision narrows to the minimum necessary to reach the goal. Or with the observer effect in relativity theory: from the outside, the observer sees things completely differently than the observer inside the system (the theory).

Similarly, the theory of real closed fields has as its model, for example, the countable set of real algebraic numbers, since the most it can require is the existence of real roots of odd-degree polynomials, and these are algebraic numbers, the set of which can be enumerated by relying on the enumeration of polynomials.

On the other hand, it is important to understand that the existence of a countable model does not mean its uniqueness. Thus, Peano arithmetic has non-standard models that include the standard model $\mathbb{N}$ only as an initial segment. In these models, there are so-called galaxies of non-standard numbers, which are copies of $\mathbb{Z}$. And again, it may seem that we are encountering some contradiction, since in $\mathbb{Z}$ there are infinitely decreasing sequences, while in Peano arithmetic they cannot exist due to the principle of induction. And here we again understand that from the point of view of an external observer, such sequences exist, but Peano arithmetic is neither able to prove their existence nor even to define them, i.e., it simply "does not see" them as a single object (the truth domain of some of its formulas). And here it turns out that the model as a structure in the meta-theory is much richer than the theory being modeled.

Moreover, countable non-standard models of arithmetic are not isomorphic to the standard model. That is, the class of countable models can be quite diverse, no matter what the theory itself thinks about them.

Let us record one standard method of producing such non-standard models, widely known as the *Compactness Theorem* for first-order logic.

Theorem D2.12 (Compactness, Gödel–Maltsev). *A set Γ of closed first-order formulas has a model if and only if every finite subset $\Gamma' \subseteq \Gamma$ has a model.*

This is the semantic counterpart of the (trivial) compactness of derivability already noted among the basic properties of $\vdash$ in B1.3, and is obtained from Soundness and Completeness in one step: if every finite $\Gamma' \subseteq \Gamma$ has a model, then by Soundness each such Γ' is consistent; hence Γ itself is consistent (otherwise $\Gamma \vdash \bot$ would already hold on a finite fragment, by B1.3); and by Completeness Γ has a model. The converse direction is immediate.

A non-standard model of PA is then obtained by adjoining to its axioms a fresh constant c together with the infinite schema $\{c \neq S^n0 \mid n \in \mathbb{N}\}$, where S^n0 denotes the term 0 preceded by n symbols of the successor function S. Every finite fragment of this schema is satisfied by the standard model $\mathbb{N}$ interpret c by a sufficiently large natural number), hence by the Compactness Theorem the whole extension has a model — necessarily containing an element greater than S^n0 for every $n \in \mathbb{N}$. The explicit description of such models lies beyond the scope of this book (see [39]).

This distinction between standard and non-standard models of arithmetic runs much deeper than a simple lack of isomorphism. It concerns their very computational nature, which is elegantly demonstrated by Tennenbaum's Theorem. It establishes a fundamental connection between axiomatics and computability theory, granting the standard model a special, "constructive" status.

Theorem D2.13 (Tennenbaum, 1959). *The only countable computable model of Peano arithmetic (PA) is its standard model on the ordinal ω. In any countable non-standard model of PA, at least one of the operations—addition or multiplication—is not a computable function.*

Thus, although first-order axioms are incapable of distinguishing the standard model from non-standard ones, the requirement of operation *computability* instantly eliminates all "pathological" variants. From the point of view of the meta-theory, we can construct a multitude of strange models, but as soon as we require these models to be algorithmically realizable, only one remains—the natural numbers we are accustomed to.

Henceforth, by the *cardinality of a set* we will mean the cardinal number, assuming that we are working in ZFC. Recall that a **cardinal number** (a cardinal) is the smallest among equinumerous ordinals. From the axiom of choice, it follows that for any set, there exists a unique equinumerous cardinal. The countable cardinal is denoted either by ω or, following the general numbering of cardinals, by $\aleph_0$. The next cardinal after ω is denoted by ω_1 or $\aleph_1$. The cardinality of the continuum is often denoted by $\mathfrak{c}$ or $2^{\aleph_0}$.

In general, for any proper subclass of the class of ordinals, there exists a unique strictly increasing one-to-one enumeration of the elements of this subclass by all ordinals. That is, if there is a proper class $\mathcal{K} \subseteq \mathrm{Ord}$, then there exists a unique definable mapping $F : \mathrm{Ord} \to \mathcal{K}$ such that $F(\alpha) < F(\beta) \leftrightarrow \alpha < \beta$ and $\forall \alpha \in \mathcal{K} : \exists \beta \in \mathrm{Ord} : F(\beta) = \alpha$. This is how all infinite cardinals $\aleph_\alpha$ are numbered, or, say, the beth-sequence:

$$\beth_0 = \omega, \quad \beth_{\alpha+1} = 2^{\beth_\alpha}, \quad \beth_\lambda = \sup\{\beth_\beta \mid \beta < \lambda\}.$$

We will talk more about cardinal numbers in Section D4.3.4.

Theorem D2.14 (Löwenheim-Skolem Upward Theorem).
If a theory T has an infinite model of cardinality κ, then it has a model of any cardinality $\tau > \kappa$.

Combining the two Löwenheim-Skolem theorems, one can say that if a first-order theory in a countable language has a model of at least one infinite cardinality, then it has a model of any infinite cardinality. And thus we notice a surprising fact that first-order theories in countable languages are poor at distinguishing infinities from the point of view of an external observer (the meta-theory).

This is also confirmed by the following theorem, for the formulation of which we need to give a definition. A theory T is called κ-categorical (categorical in power κ), where κ is a cardinal, if all models of this theory whose carrier has cardinality κ are isomorphic. For example, the theory DLO_∞ (dense linear orders without maximum and minimum) is categorical in countable power, and the theory DTFA of divisible torsion-free abelian groups is categorical in power $\mathfrak{c}$ (and any other uncountable power).

Theorem D2.15 (Morley's Categoricity Theorem). *Let T be a complete first-order theory in a countable language. If there exists an uncountable cardinality κ such that T is κ-categorical, then T is λ-categorical for any uncountable cardinality λ.*

Thus, we see that complete categorical (first-order) theories can only distinguish between two infinities—countable and uncountable. Moreover, as we will see later, the requirement of completeness here is superfluous, as categorical theories turn out to be complete automatically.

D2.2.4 Completeness of a Theory

Since we have covered the topic of completeness of calculi in such detail, we must say at least a few words about the completeness of theories, although we have already discussed this in sufficient detail in Section B1.2.9.

We have already seen several examples of independent statements, such as the Fifth Postulate, the Axiom of Choice, and Goodstein's theorem, which means that for theories there is a certain gray area regarding derivability: some formulas we can derive from the axioms (theorems), some we can refute (antitheorems), and about some our theory knows nothing. The desire arises to supplement the axiomatics so that the gray area disappears.

A theory is called **complete** if every closed formula of the language of this theory is either provable or refutable.

It turns out that sometimes a theory can be completed with a perfectly *enumerable* set of axioms so that it becomes complete, and sometimes such a step-by-step completion is fundamentally impossible; a certain transcendental leap is required to obtain a complete theory (as in the aforementioned construct with an ultrafilter on the Lindenbaum algebra). Between "cooperative" (constructively completable) and "stubborn" theories, there is a clear watershed called Gödel's incompleteness theorem (the first). We have already cited it in Section B1.2.9, but such a theorem is worth repeating several times.

Theorem D2.16 (Gödel's First Incompleteness Theorem).
If a first-order theory

(G1) *is consistent;*
(G2) *is enumerable (i.e., there exists an algorithm that can enumerate all its theorems);*
(G3) *can interpret first-order arithmetic,*

then it is incomplete, i.e., there exists a sentence in the language of this theory that can neither be proved nor disproved within its system of axioms.

Remark. This general version, known as the Rosser form, strengthens Gödel's original 1931 result by requiring only simple consistency rather than ω-consistency. Since this chapter focuses on model-theoretic aspects, we provide the theorem here as a fundamental fact about the limits of formal systems. A detailed proof based on the syntax of arithmetization and the diagonal lemma will be given in Section D3.5, Theorem D3.11.

Theories that satisfy conditions (G1)–(G3) are often called *Gödelian*.

The requirement of consistency is absolute, since in a contradictory theory everything is derivable. Enumerability is a certain constructive limitation that forces our theory to be more "cooperative"; such a theory is easier to analyze algorithmically, its axioms and theorems are constructed in some regular way much like the grammar of a language is constructed, i.e., "regularity" is a guarantee of "cooperativeness". Condition (G3), on the contrary, pulls our theory towards "stubbornness"; it is this condition that allows one to express within the theory itself the predicate $\mathrm{Pr}(\ulcorner\varphi\urcorner)$, which says that the formula φ is provable in this theory ($\ulcorner\varphi\urcorner$ means taking the Gödel number of the formula φ). This is done using a special encoding of formulas and proofs (Gödel numbering) by natural numbers, which allows this predicate to be written as an arithmetical formula. Of course, strictly speaking the theory proves only arithmetical facts about natural numbers; it does not quantify over formulas or formal derivations in its own language—those are objects of the metatheory, and assigning Gödel codes to them is an interpretation we supply from outside. Even this is quite sufficient to present an example of a formula G such that $G \leftrightarrow \neg\mathrm{Pr}(\ulcorner G\urcorner)$, i.e., the truth of the formula is equivalent to its unprovability. Moreover, for his second incompleteness theorem, Gödel showed that the consistency statement itself, $\neg\mathrm{Pr}(\ulcorner\bot\urcorner)$, could serve as this unprovable formula G. In other words, a Gödelian theory is unable to either prove or refute its own consistency.

Therefore, all Gödelian theories are "stubborn," but there are many enumerable theories that are unable to interpret arithmetic, and therefore for the most part are "cooperative," i.e., they can be relatively easily completed to be complete. Such theories include, for example,

- the theory of linear orders (completed to the theory of dense linear orders without a maximum and minimum);
- Tarski's geometry (complete);
- the theory of real closed fields (complete);
- the theory of algebraically closed fields of characteristic 0 (complete);
- the theory of divisible torsion-free abelian groups (complete).

However, as we will see shortly, the expressive power of such theories leaves much to be desired.

On the other hand, theories capable of expressing a great deal, such as arithmetic and set theory, have the "stubbornness gene" (G3), i.e., they are constructively and fundamentally incompletable: by virtue of Gödel's incompleteness theorem, no matter how many new independent axioms we add to their axiomatics, the theory will never become complete. A qualitative transcendental leap occurs here only if we extend such a "stubborn" theory to the set of all formulas true in its standard model (for example, for Peano arithmetic, this would be the elementary theory $Th(\mathbb{N})$). But then the theory will become non-enumerable, i.e., it will lose the "cooperativeness gene" (G2).

Thus, Gödelian theories are a kind of boundary between completable and expressive first-order theories. In the context of studying mathematics as a foreign language, it might be appropriate to compare these two types of theories with spoken and academic languages in ordinary language, or, say, with Vulgar and High Latin of the Middle Ages.

Theorem D2.17 (Łoś–Vaught test). *If two conditions are met:*

(1) *a first-order formal theory has at least one infinite model, but no finite models;*
(2) *the theory is κ-categorical for some infinite cardinal number κ,*

then this theory is complete.[10]

With the help of this theorem, one can prove, for example, the following facts: **1)** the theory of equality with axioms excluding a finite domain of interpretation (see Section B1.2.10) is complete, since it is categorical in countable cardinality; **2)** the theory DLO_∞ of dense linear orders without a first and last element (see Section B1.2.7) is complete, since it is categorical in countable cardinality; **3)** the theory DTFA of divisible torsion-free abelian groups (see Section B1.2.7) is complete, since it is categorical in uncountable cardinality.

From the Łoś-Vaught test and Morley's theorem, we get a simply formulated fact: a first-order theory categorical in some infinite cardinality is not Gödelian. Simply because such theories are complete, and Gödelian theories are not.

A theory T is said to *admit quantifier elimination* if for every formula $\varphi(\vec{p})$ in the language of this theory, there exists a quantifier-free formula $\varphi_0(\vec{p})$ in the language of this theory such that $T \vdash \forall\vec{p}\ (\varphi \leftrightarrow \varphi_0)$. A simple example: the formula $(a \neq 0) \wedge \exists x\ (ax^2 + bx + c = 0)$ is equivalent in RCF to the formula $(a \neq 0) \wedge (b^2 - 4ac \geqslant 0)$. Note that the condition of quantifier elimination is essential only for sentences from the "gray zone," i.e., those that are not theorems or anti-theorems of the theory, as well as for formulas with free variables, since in any theory every theorem is provably equivalent to the quantifier-free formula $\top$, and every anti-theorem to the formula $\bot$.

---

[10] This theorem was proven by Jerzy Łoś in 1954 and independently by Robert L. Vaught, also in 1954.

Theorem D2.18 (Tarski-Seidenberg).
The theory RCF *admits quantifier elimination.*[11]

The following theorem provides another criterion for the completeness of a theory, using quantifier elimination.

Theorem D2.19. *Let T be a theory in some first-order language L. If*

(1) *the theory T admits quantifier elimination;*
(2) *for any closed quantifier-free sentence φ of the language L, the theory T proves either φ or $\neg\varphi$,*

then the theory T is complete.

Remark: A closed quantifier-free sentence is constructed from atomic formulas of the form $P(t_1, \ldots, t_n)$, where t_k are terms without variables (i.e., constants or combinations of constants and function symbols), and P is a predicate symbol, using logical connectives $(\neg, \wedge, \vee, \rightarrow, \leftrightarrow)$. For example, if the language has constants $0, 1$ and a predicate $<$, then $0 < 1$ and $\neg(1 = 0)$ are closed quantifier-free sentences.

If a theory T admits quantifier elimination, then any closed sentence σ is equivalent in T to some closed quantifier-free sentence σ', i.e., $T \vdash (\sigma \leftrightarrow \sigma')$. If the second condition of the theorem is also met—the theory T decides the truth of all closed quantifier-free sentences (i.e., for each such σ', either $T \vdash \sigma'$ or $T \vdash \neg\sigma'$)—then it automatically decides the truth of all sentences. Consequently, T will be complete.

Special case: a language without constants. If the language L does not contain constant symbols, then there are no non-trivial closed atomic (or quantifier-free) sentences in it, other than those equivalent to $\top$ (truth) or $\bot$ (falsity). Thus, *if a theory T in a language without constants is consistent and admits quantifier elimination, then it is automatically complete.*

For example, the theory of dense linear orders without a minimal and maximal element (DLO_∞) is given in a language without constants with one predicate symbol $<$ (besides $=$) and admits quantifier elimination. Indeed, to show the possibility of quantifier elimination, it is sufficient to show that one existential quantifier can be eliminated. Any formula describing the properties of

[11] The main work was done by Alfred Tarski. He published his result on quantifier elimination for real closed fields (and, as a consequence, the decidability of elementary algebra and geometry) in 1948 in the work "A Decision Method for Elementary Algebra and Geometry". Abraham Seidenberg in 1954 proposed another, more algebraic and, in the opinion of some, simpler and more natural proof of this result.

x reduces to one of three cases:[12] 1) $x = p$; 2) $x < p$; 3) $q < x$, where p, q are parameters (other free variables).

The formula $\exists x(x = p)$ is equivalent to the quantifier-free formula $\top$ (its witness is the parameter p).

The formula $\exists x((q < x) \wedge (x < p))$ is equivalent to the formula $(q < p)$ by virtue of the axiom of density of the order.

The formulas $\exists x(q < x)$ and $\exists x(x < p)$ are equivalent to the formula $\top$ by virtue of the axiom of the absence of a maximum and minimum.

So, DLO_∞ admits quantifier elimination, and since its language has no constants, it is complete.

The theory of real closed fields (RCF) admits quantifier elimination by the Tarski-Seidenberg theorem. Closed quantifier-free sentences (order and equality relations between the results of polynomial expressions of constants) are decided by the axioms of RCF. From this follows its completeness.

D2.2.5 Summary of Results

Let's gather the material we have studied into a short reference table classifying theories by such features as: Gödelian or non-Gödelian (separately highlighting the non-enumerable case), completeness (completability) or incompleteness (essential), categoricity (in countable or uncountable cardinality), decidability. We did not analyze the last property of theories, but for reference, we note that this is a very useful property of a theory, meaning that there exists an algorithm that in a finite number of steps for any closed formula of the language of the theory can report whether this formula is a theorem of this theory or not (the second is also important to be able to determine in a finite number of steps) (Table D2.2).

[12] Here we will refer to the known fact that any formula can be brought to prenex normal form (all quantifiers at the front), and then its quantifier-free part can be brought to DNF, and also the quantifier $\forall$ can be expressed through $\exists$ and $\neg$, after which it only remains to push $\exists$ into disjunctions of atomic conjunctions, which are precisely the cases presented here. Note also that due to the connexity of the order, the negations of these cases reduce to the same cases, i.e., $\neg$ is also eliminable.

Table D2.2. Properties of some first-order theories

Theory	Gödelian	Completeness	κ-Categ.	Decidable
Pure theory of equality	non-Gödelian	completable	$\checkmark(\kappa \geqslant \omega)$	✓
= with infinite domain	non-Gödelian	✓	$\checkmark(\kappa \geqslant \omega)$	✓
DLO_∞	non-Gödelian	✓	$\checkmark(\omega)$	✓
Group theory	non-Gödelian	x	x	x
DTFA	non-Gödelian	✓	$\kappa > \omega$	✓
ACF *	non-Gödelian	completable	$\kappa > \omega$	✓
Tarski's geometry	non-Gödelian	✓	x	✓
RCF	non-Gödelian	✓	x	✓
Presburger arithmetic	non-Gödelian	✓	x	✓
Q **	Gödelian	essentially incomplete***	x	x
PA	Gödelian	essentially incomplete	x	x
ZF/ZFC	Gödelian	essentially incomplete	x	x
Th($\mathbb{N}$), Th($\mathbb{Z}$), Th($\mathbb{Q}$)	non-enumerable	✓	x	x

* the theory ACF of algebraically closed fields is completed to a complete theory by means of an axiom that specifies the characteristic of the field—0 or a prime p.
** the axiomatics of Q we will give in the next chapter.
*** essential incompleteness means that this theory cannot be extended to a complete one while preserving the recursive enumerability of the axioms.

Exercises

D2.1 Let P be the statement "I am in Paris", F — "I am in France", L — "I am in London", E — "I am in England". We accept the following geographic facts as true premises:

$$(P \to F) \wedge (L \to E).$$

Prove that from these premises logically follows the statement:

$$(P \to E) \vee (L \to F)$$

("Either 'if I am in Paris, then I am in England' is true, or 'if I am in London, then I am in France' is true").

D2.2 Derive the rule of *Existential Introduction* in the consequent:

$$\varphi[a/t] \vdash \exists x \varphi[a/x],$$

where t is a term.

D2.3 Using the axioms of predicate logic and the Modus Ponens rule (and the Deduction Theorem), derive the rule of *Universal Instantiation* combined with implication:

$$\forall x(\varphi(x) \to \psi(x)) \vdash \forall x \varphi(x) \to \forall x \psi(x).$$

D2.4 Prove the following logical equivalence (distribution of universal quantifier over conjunction):

$$\vdash \forall x(\varphi(x) \land \psi(x)) \leftrightarrow (\forall x \varphi(x) \land \forall x \psi(x)).$$

D2.5 Prove that the existential quantifier distributes over disjunction:

$$\vdash \exists x(\varphi(x) \lor \psi(x)) \leftrightarrow (\exists x \varphi(x) \lor \exists x \psi(x)).$$

D2.6 Prove the "Drinker Principle" formula (classically valid, though intuitively paradoxical):

$$\vdash \exists x(\varphi(x) \to \forall y \varphi(y)).$$

(Hint: Use the Law of Excluded Middle for the formula $\forall y \varphi(y)$).

D2.7 Using the Deduction Theorem, show that if the variable a does not occur free in φ, then:

$$\vdash (\varphi \to \forall x \psi[a/x]) \leftrightarrow \forall x(\varphi \to \psi[a/x]).$$

D2.8 Derive the rule of *Generalization of the Consequent*:

$$\Gamma \vdash \varphi \to \psi[a/x] \implies \Gamma \vdash \varphi \to \forall x \psi[a/x],$$

provided a is not free in φ and Γ.

D2.9 Prove semantically that from $\exists x \forall y \varphi(x, y)$ follows $\forall y \exists x \varphi(x, y)$.

D2.10 Let $\mathcal{M}_1 = \langle \mathbb{Q}, < \rangle$ and $\mathcal{M}_2 = \langle \mathbb{Z}, < \rangle$. Find a sentence in the language of order that is true in $\mathcal{M}_1$ but false in $\mathcal{M}_2$.

D2.11 Consider the language containing only the addition symbol $+$.

1. Show that in the structure $\langle \mathbb{N}, + \rangle$ (natural numbers with zero), the strict order relation $x < y$ is definable. Write the corresponding formula.
2. Prove that in the structure $\langle \mathbb{Z}, + \rangle$ (integers), the relation $x < y$ is NOT definable. (Hint: Use the automorphism $x \mapsto -x$).

D2.12 Consider the structure $\langle \mathbb{R}, \cdot \rangle$ (multiplicative group of real numbers).

1. Show that the set of non-negative numbers $\{x \in \mathbb{R} \mid x \geqslant 0\}$ is definable in this structure.
2. Show that the standard order relation $x \leqslant y$ is definable in the field of real numbers $\langle \mathbb{R}, +, \cdot \rangle$ using only algebra operations.

D2.13 Consider the structure $\langle \mathbb{R}, < \rangle$. Determine which of the following sets are definable in this structure:

1. The set $\{0\}$;
2. The set of all numbers greater than 5;
3. The interval $(0, 1)$.

D2.14 Prove that the structures $\langle \mathbb{Q}, < \rangle$ and $\langle \mathbb{Q} \cup \{\pi\}, < \rangle$ (where π is treated just as another point in the order) are isomorphic.

D2.15 1. Describe a countable model for the theory ACF_0 (Algebraically Closed Fields of characteristic 0). (Hint: Think about the closure of $\mathbb{Q}$).

2. Prove the existence of an uncountable model of Peano Arithmetic.

D2.16 Prove that the theory of the structure $\langle \mathbb{N}, < \rangle$ is not $\aleph_0$-categorical. (Hint: Consider a model consisting of two copies of $\mathbb{N}$, one following the other).

D2.17 In the theory of Dense Linear Orders (DLO), eliminate the quantifiers from the following formula (find an equivalent quantifier-free formula):

$$\exists x(a < x \wedge x < b).$$

D2.18 In the theory DLO, eliminate the quantifiers from:

$$\forall x(a < x \to b < x).$$

D2.19 In the Theory of Equality (with infinite domain), eliminate the quantifier from the formula:

$$\exists x(x \neq a \wedge x \neq b).$$

D2.20 In the theory of Real Closed Fields (RCF), determine the condition on coefficients a, b for the formula:

$$\exists x(ax + b > 0).$$

(Write an equivalent quantifier-free formula involving a and b).

D2.21 In the theory RCF, eliminate the quantifier from the formula

$$\exists x(ax^2 + bx + c = 0).$$

D2.22 Let L be a language with a single binary predicate symbol $\sim$. Let T be a theory asserting that:

- $\sim$ is an equivalence relation;
- There are exactly two equivalence classes;
- Each equivalence class is infinite.

Use the Łoś–Vaught test to prove that the theory T is complete.

Recommended Reading

In this chapter, we have looked into the "engine room" of mathematics—the theory of logical inference. The literature here reflects different approaches to constructing this mechanism.

The most direct and, perhaps, the most intuitive for beginners is the axiomatic (Hilbert-style) approach, which was detailed in this chapter. The classic textbook that guides one along this path with exemplary clarity is "Introduction to Mathematical Logic" by E. Mendelson [46]. In it, step by step, from the simplest axioms, the entire structure of the calculus of propositions and predicates is built, and key metatheorems are proven.

For a more contemporary undergraduate perspective, H. Enderton's "A Mathematical Introduction to Logic" [17] is widely considered the standard in American universities. Additionally, the Springer undergraduate text by

H.-D. Ebbinghaus, J. Flum, and W. Thomas [16] offers a rigorous yet accessible mathematical treatment. For those studying independently, D. Goldrei's [27] structural approach is particularly helpful and pedagogically close to the spirit of this book.

A more modern and powerful perspective on the structure of proofs is offered by structural proof theory, or sequent calculus, which originates from the work of G. Gentzen [24]. This approach makes it possible to analyze not just the result, but the very "fabric" of a proof. An excellent introduction to this world is the classic textbook "Basic Proof Theory" by A. S. Troelstra and H. Schwichtenberg [61]. For those ready for a deep dive into the essence of proof theory, the classic, though challenging, work by K. Schütte [52] is the definitive resource.

Finally, to see how the syntactic constructions of logic "come to life" in mathematical structures (models), one should turn to the textbook by J. R. Shoenfield [54]. It offers one of the most elegant and concise paths to the central results of model theory—the theorems on completeness, compactness, and Löwenheim-Skolem, which connect the worlds of syntax and semantics, discussed in detail in the second half of this chapter.

Chapter D3
Arithmetic

Mathematics is the queen of the sciences and arithmetic the queen of mathematics.

— Carl Friedrich Gauß

Recall that first-order Peano Arithmetic is formulated in the language of first-order predicate logic with the signature (in addition to logical connectives and quantifiers) $\Sigma_{\mathsf{PA}} \overset{\text{def}}{\rightleftharpoons} \langle =, 0, S, +, \cdot, < \rangle$. The predicate symbol for inequality, $<$, can either be included in the signature or added later using its direct definition in terms of addition (see (D3.2) below). Here, we will opt for the version with a recursive definition of inequality, and thus $<$ is included in the signature from the outset.

Let us once again provide the complete list of axioms for first-order arithmetic:

PA1 $a = a$; $(a = b) \rightarrow (b = a)$; $(a = b) \wedge (b = c) \rightarrow (a = c)$;

PA2 $a = b \rightarrow S(a) = S(b)$;

PA3 $S(a) = S(b) \rightarrow (a = b)$;

PA4 $S(a) \neq 0$;

PA5 $a + 0 = a$; $a + Sb = S(a + b)$;

PA6 $a \cdot 0 = 0$; $a \cdot Sb = ab + a$;

PA7 $\neg(a < 0)$; $(a < Sb) \leftrightarrow (a < b) \vee (a = b)$;

PA8 $\varphi[a/0] \wedge \forall x(\varphi[a/x] \rightarrow \varphi[a/Sx]) \rightarrow \forall y\, \varphi[a/y]$,

as well as the axioms and inference rules of predicate logic:

N. I. Kazimirov, *Mathematics as a Foreign Language*, Mathematics in Mind,
https://doi.org/10.1007/978-3-032-30721-7_8

PC1 the universal closures of all formulas obtained by substituting any formula of PA into any tautology of propositional logic;

PC2 axioms for quantifiers:

$$(\forall)\quad (\forall x\ \varphi[a/x]) \to \varphi[a/T]; \qquad (\exists)\quad \varphi[a/T] \to \exists x\ \varphi[a/x],$$

where φ does not contain the bound variable x, and T is any term of the theory;

PC3 rules of inference:

$$(MP)\quad \frac{\varphi,\quad \varphi \to \psi}{\psi}; \qquad (R1)\quad \frac{\varphi \to \psi}{\varphi \to \forall x\ \psi[a/x]};$$

$$(R1')\quad \frac{\psi}{\forall x\ \psi[a/x]}; \qquad (R2)\quad \frac{\psi \to \varphi}{(\exists x\ \psi[a/x]) \to \varphi},$$

where for rules (R1), (R1'), and (R2), the variable a does not occur in φ nor in any of the active (undischarged) hypotheses upon which the derivation of the formula in the premise of these rules depends. The rules $(R1)$ and $(R2)$ are called **Bernays' rules**. Rule (R1'), called the *generalization rule*, is a derived rule in our inference system, as it can be obtained from rule (R1) by substituting a tautology for the formula φ. It is listed here for convenience of reference.

Let us introduce additional predicates by definition:

$$(a \leqslant b) \overset{\text{def}}{\leftrightarrow} (a < b) \vee (a = b), \quad (a > b) \overset{\text{def}}{\leftrightarrow} (b < a), \quad (a \geqslant b) \overset{\text{def}}{\leftrightarrow} (b \leqslant a).$$

D3.1 Basic Properties

Lemma D3.1. $a \neq 0 \to \exists x\ a = Sx$.

Proof. Let $\varphi(a) \overset{\text{def}}{\leftrightarrow} (a \neq 0 \to \exists x\ a = Sx)$ and prove this formula by induction on the variable a. For the substitution $[a/0]$, we must prove $0 \neq 0 \to \exists x\ 0 = Sx$. This formula is true because the antecedent of the implication is false (by virtue of PA1). Next, we introduce the hypothesis $\varphi(a)$, from which we need to derive $\varphi[a/Sa]$, i.e., the formula $Sa \neq 0 \to \exists x\ Sa = Sx$. It is sufficient to show that the consequent $\exists x\ Sa = Sx$ is true. To do this, consider the formula $\psi(a, b) \overset{\text{def}}{\leftrightarrow} (Sa = Sb)$ and note that the substitution $\psi[b/a]$ is true, since $Sa =$

Sa by the first axiom PA1, in which we performed the substitution $[a/Sa]$. Furthermore, by rule ($\exists$), understanding the variable a in it as the variable b, and the term T as the variable a, we obtain the truth of the implication $\psi[b/a] \to \exists x\ \psi[b/x]$. Now, by the MP rule, from the formulas $\psi[b/a]$ and $\psi[b/a] \to \exists x\ \psi[b/x]$, we arrive at the formula $\exists x\ \psi[b/x]$, i.e., $\exists x\ Sa = Sx$, as required. Next, using the tautology $\mathcal{A} \to (\mathcal{B} \to \mathcal{A})$ and the MP rule, substituting the formula $\exists x\ Sa = Sx$ for $\mathcal{A}$ and the formula $Sa \neq 0$ for $\mathcal{B}$, we obtain the implication $Sa \neq 0 \to \exists x\ Sa = Sx$, i.e., the formula $\varphi[a/Sa]$.

Thus, even without using the hypothesis $\varphi(a)$, we have managed to derive $\varphi[a/Sa]$. Consequently, again by virtue of the tautology $\mathcal{A} \to (\mathcal{B} \to \mathcal{A})$ and the MP rule, the implication $\varphi(a) \to \varphi[a/Sa]$ is true, whence by the generalization rule (R1') we obtain that

$$\forall y\ \varphi[a/y] \to \varphi[a/Sy].$$

Now, together with the formula $\varphi[a/0]$ proven above, by the induction axiom PA8 and the MP rule, we arrive at the conclusion $\forall y\ \varphi[a/y]$. Removing the quantifier by rule ($\forall$), we obtain the statement of the lemma. □

Remark: The induction hypothesis $\varphi(a)$ was not actually used in the proof, and therefore, using induction to prove the lemma might seem redundant. However, in this case, induction essentially allows us to use the fact that any number is either zero or a successor. This is embedded in the premise of induction. Not yet having the lemma's statement at our disposal, only with the help of induction can we indirectly use this fact and thereby prove the lemma. In other words, the axiom of induction (and other similar axiom schemas) is not always applied, so to speak, for its direct purpose. Having Lemma D3.1, we can now in many cases avoid using induction in a proof, simply by considering two possible cases: a is zero or a is a successor. Nevertheless, Lemma D3.1 is not equivalent to induction; it is much weaker. For example, it does not allow for the proof of the commutativity of addition in the absence of induction.

Arithmetic with axioms PA1–PA6 plus Lemma D3.1 is called **Robinson's** arithmetic and is denoted by Q. Inequality in Q is usually defined in the strong sense (see below). In this arithmetic, the commutativity and associativity of addition and multiplication, and many other useful properties of natural numbers, are unprovable, which makes such an arithmetic quite weak. However, Robinson's arithmetic is good because it is finitely axiomatizable (it has no axiom schema) and one can still implement the Gödel numbering of formulas, which makes it a (minimal) Gödelian theory.

Another testament to the gap between induction and Lemma D3.1 is the so-called *strong axiom of inequality*, which we will introduce below, in the course of proving the following main theorem.

Theorem D3.1. *The following theorems hold in the theory* PA *(with universal closure of the formulas understood):*

1° $a = b \to a + c = b + c$ [congruence of = with respect to addition];
2° $a + b = b + a$ [commutativity of addition];
3° $a = b \to ac = bc$ [congruence of = with respect to multiplication];
4° $a + (b + c) = (a + b) + c$ [associativity of addition];
5° $ab = ba$ [commutativity of multiplication];
6° $a(b + c) = ab + ac$ [distributivity];
7° $a(bc) = (ab)c$ [associativity of multiplication];
8° $a = b \to (a < c) \leftrightarrow (b < c)$ [congruence of = with respect to inequality];
9° $a = b \to (c < a) \leftrightarrow (c < b)$ [congruence of = with respect to inequality];
10° $(a < b) \leftrightarrow (Sa < Sb)$ [compatibility of order with S];
11° $(a + c = b + c) \to (a = b)$ [cancellation law for +];
12° $(a < b) \leftrightarrow (a + c < b + c)$ [compatibility of order with +];
13° $<$ is a linear order;
14° $(c \neq 0) \to [(a < b) \leftrightarrow (ac < bc)]$ [compatibility of order with $\cdot$];
15° $(ac = bc) \wedge (c \neq 0) \to (a = b)$ [cancellation law for $\cdot$];
16° $(a \leqslant b) \wedge (b < Sa) \to (b = a)$ [discreteness of order];
17° $a \cdot S0 = a$ [identity element for multiplication];
18° $(a \cdot b = 0) \to (a = 0 \vee b = 0)$ [no zero divisors].

Proof. Here, with each step, we will appeal less and less to the use of logical axioms and inference rules, assuming that their application in the previous steps was sufficiently convincing. This will allow us to concentrate on the essence of the proofs and not get bogged down in a sea of formal transformations.

Congruence of equality with respect to addition will be proven by induction on c. Let $\varphi(c) \overset{\text{def}}{\leftrightarrow} \forall x, y\ (x = y) \to (x + c = y + c)$. For the substitution $[c/0]$, we need to show that $\forall x, y\ (x = y) \to (x + 0 = y + 0)$. We will instead prove the implication $(a = b) \to (a + 0 = b + 0)$ with free variables a, b. And to prove the implication, we will use the Deduction Theorem.

Let us introduce the hypothesis $(a = b)$. By virtue of PA5, we have $a+0 = a$ (more formally: in closed form, PA5 is $\forall x\ (a + 0 = a)[a/x]$, whence by removing the quantifier via axiom $(\forall)$ we get $a + 0 = a$), from which, by the transitivity of equality, we conclude that $a + 0 = b$. Substituting b for a in PA5 (which is also done using axiom $(\forall)$), we see that $b + 0 = b$. Then, by the symmetry and transitivity of equality, we arrive at $a + 0 = b + 0$.

Thus, by the Deduction Theorem, we have proven the implication $(a = b) \to (a + 0 = b + 0)$, after which we can apply the universal quantifier twice using the generalization rule (R1'). As a result, we obtain $\forall x, y\ (x = y) \to (x + 0 = y + 0)$, i.e., $\varphi[c/0]$.

Next, we need to prove that $\forall z\ \varphi(z) \to \varphi(Sz)$. We proceed similarly: we will prove $\varphi(c) \to \varphi(Sc)$ and then apply the universal quantifier using the generalization rule. Let us introduce the induction hypothesis $\varphi(c)$, i.e., $\forall x, y\ (x = y) \to (x + c = y + c)$. Here, we temporarily remove the universal quantifiers and consider the formula $(a = b) \to (a+c = b+c)$ as a hypothesis, where the variables a, b are fresh, and the variable c is used in the hypothesis.

Let us introduce another hypothesis: $(a = b)$, then within the hypotheses, by the MP rule, we obtain the formula $a + c = b + c$. Next, $a + Sc = S(a + c)$ by axiom PA5 (again, we use it in closed form and remove the quantifier, replacing the variables as needed). Next, in axiom PA2, using it in closed form and removing the universal quantifier by axiom $(\forall)$ with the substitution $[a/a+c; b/b+c]$, we obtain the formula $(a+c = b+c) \to S(a+c) = S(b+c)$, which, together with the already proven equality $a + c = b + c$, gives us $S(a + c) = S(b+c)$ by the MP rule. Now, by the transitivity of equality, from the formulas proven above, we obtain that $a + Sc = S(b + c)$. Finally, by virtue of PA5 and the symmetry of equality, we have $S(b+c) = b+Sc$, whence from the previous equality and by the transitivity of equality, we obtain that $a + Sc = b + Sc$.

By the Deduction Theorem, we conclude that within the hypothesis $(a = b) \to (a + c = b + c)$, we have proven the implication $(a = b) \to (a + Sc = b + Sc)$, which means, again by virtue of the Deduction Theorem, we have proven the implication

$$[(a = b) \to (a + c = b + c)] \to [(a = b) \to (a + Sc = b + Sc)].$$

Next, to transition to the formula $\varphi(c)$, we first use axiom $(\forall)$:

$$[\forall x, y\ (x = y) \to (x + c = y + c)] \to [(a = b) \to (a + c = b + c)],$$

whence from the preceding

$$[\forall x, y\ (x = y) \to (x + c = y + c)] \to [(a = b) \to (a + Sc = b + Sc)],$$

and since the antecedent of the implication on the left has no free variables a, b, we can apply Bernays' rule (R1) and introduce a universal quantifier in the consequent:

$$[\forall x, y\ (x = y) \to (x + c = y + c)] \to [\forall x, y\ (x = y) \to (x + Sc = y + Sc)],$$

which corresponds to the implication $\varphi(c) \to \varphi(Sc)$, and this is true without restrictions on the variable c, and thus, by the generalization rule: $\forall z\ \varphi(z) \to \varphi(Sz)$.

Thus, we have derived the formulas $\varphi(0)$ and $\forall z\ \varphi(z) \to \varphi(Sz)$. To combine them with a conjunction, we use the tautology $\mathcal{A} \to (\mathcal{B} \to (\mathcal{A} \wedge \mathcal{B}))$, where we substitute $\varphi(0)$ for $\mathcal{A}$ and $\forall z\ \varphi(z) \to \varphi(Sz)$ for $\mathcal{B}$, and apply MP twice, after which we obtain the formula $\varphi(0) \wedge \forall z(\varphi(z) \to \varphi(Sz))$, which is the antecedent of the induction axiom PA8 up to the choice of bound variables.

Applying the MP rule, we obtain $\forall z\ \varphi(z)$, as required.

Commutativity of addition $(a + b = b + a)$ will be proven by induction on b, or more precisely, we will conduct induction on the formula $\varphi(b) \overset{\text{def}}{\leftrightarrow} \forall x\ x + b = b + x$. The base case of this induction, $\varphi(0)$, was already proven on p. 135. So we have the identity $\forall x\ x + 0 = 0 + x$. Let us introduce the hypothesis $(\Gamma 1)$ $\forall x\ x + b = b + x$ and from this prove that $\forall x\ x + Sb = Sb + x$. We will also prove this formula by induction on the formula $\psi(c) \overset{\text{def}}{\leftrightarrow} c + Sb = Sb + c$.

Obviously, $\psi(0)$ holds, as it is the base case of the first induction (in the formula $a + 0 = 0 + a$, we substitute Sb for a, using axiom $(\forall)$ and MP). Let us introduce the hypothesis $(\Gamma 2)$ $c + Sb = Sb + c$ and prove that $Sc + Sb = Sb + Sc$:

$$Sc + Sb \overset{\mathsf{PA5}}{=\!=} S(Sc + b) \overset{\Gamma 1,\mathsf{PA2}}{=\!=\!=} S(b + Sc) \overset{\mathsf{PA5},\mathsf{PA2}}{=\!=\!=} SS(b + c) \overset{\Gamma 1,\mathsf{PA2}}{=\!=\!=}$$
$$SS(c + b) \overset{\mathsf{PA5},\mathsf{PA2}}{=\!=\!=} S(c + Sb) \overset{\Gamma 2,\mathsf{PA2}}{=\!=\!=} S(Sb + c) \overset{\mathsf{PA5}}{=\!=} Sb + Sc.$$

Note that we are almost always writing the use of PA2, which allows for the substitution of equals for equals under the successor operator S, but in fact, this substitution also requires the use of axiom $(\forall)$, since we take the axioms of PA in closed form and remove the quantifier $\forall$, replacing the variables with the terms we need, and then we also use the MP rule to move from the implication in axiom PA2 to the identity from its consequent.

Thus, we have shown the transition from formula $\psi(c)$ to formula $\psi(Sc)$ within the hypothesis $(\Gamma 1)$, where c is a free variable not occurring in the formula of the outer induction on b, i.e., it is in no way related to hypothesis $(\Gamma 1)$, which means,

$$(\Gamma 1) \vdash \psi(0) \wedge \forall x\ (\psi(x) \rightarrow \psi(Sx)),$$

whence by induction and by virtue of MP, we conclude that

$$(\Gamma 1) \vdash \forall x\ \psi(x),$$

i.e., within the hypothesis $\forall x\ x + b = b + x$, we have proven $\forall x\ x + Sb = Sb + x$, i.e., we now have

$$\varphi(0) \wedge \forall y\ (\varphi(y) \rightarrow \varphi(Sy)),$$

and now from this, using induction and the MP rule, we finally obtain the formula $\forall y\ \varphi(y)$, i.e., $\forall y \forall x\ x + y = y + x$, which completes the proof of the commutativity of addition.

From the commutativity of addition, the congruence of equality with respect to addition for the second argument easily follows:

$$(a = b) \rightarrow (c + a = c + b),$$

so that in the future, we can apply the congruence of equality with respect to addition to either argument.

Congruence of equality with respect to multiplication is also proven by induction on c. Let $\varphi(c) \overset{\text{def}}{\leftrightarrow} \forall x, y\ (x = y) \rightarrow (xc = yc)$. For $c = 0$, we need to show that $\forall x, y\ (x = y) \rightarrow (x \cdot 0 = y \cdot 0)$. We will instead prove the implication $(a = b) \rightarrow (a \cdot 0 = b \cdot 0)$ with free variables a, b. And to prove the implication, we will use the Deduction Theorem.

Let us introduce the hypothesis $(a = b)$. By virtue of PA6, we have $a \cdot 0 = 0$. Substituting b for a in PA6, we also see that $b \cdot 0 = 0$. Then, by the symmetry and transitivity of equality, we arrive at $a \cdot 0 = b \cdot 0$.

Thus, by the Deduction Theorem, we have proven the implication $(a = b) \rightarrow (a \cdot 0 = b \cdot 0)$, after which we can apply the universal quantifier twice using the generalization rule. As a result, we obtain $\forall x, y\ (x = y) \rightarrow (x \cdot 0 = y \cdot 0)$, i.e., $\varphi(0)$.

Next, we lower the degree of formality and proceed in broader steps. We need to show, within the hypothesis $(a = b) \rightarrow (ac = bc)$, that $(a = b) \rightarrow (a \cdot Sc = b \cdot Sc)$. We introduce the hypothesis $a = b$, then compute according to the axioms:

$$a \cdot Sc \xlongequal{\mathsf{PA6}} ac + a \xlongequal{Cong^+} bc + a \xlongequal{Cong^+} bc + b \xlongequal{\mathsf{PA6}} b \cdot Sc,$$

where $Cong^+$ denotes the application of the congruence of addition (to either argument).

We have shown the inductive step for the formula $\varphi(c)$ (taking into account the application of quantifiers as we did when proving the congruence of addition).

Thus, by induction and the MP rule, we conclude that $\forall z\ \varphi(z)$, as required.

Associativity of addition is proven by induction on c. Let

$$\varphi(c) \overset{\text{def}}{\leftrightarrow} \forall x, y\ (x + (y + c) = (x + y) + c).$$

For $c = 0$, we need to obtain the equality $(a + (b + 0) = (a + b) + 0)$ and then close it with quantifiers over the free variables.

$$a + (b + 0) \xlongequal{\mathsf{PA5}, Cong^+} a + b \xlongequal{\mathsf{PA5}} (a + b) + 0.$$

The congruence of equality with respect to addition was needed here to derive the equality $a + (b + 0) = a + b$ from the equality $b + 0 = b$.

To prove the inductive step $\varphi(c) \rightarrow \varphi(Sc)$, we will first prove it in quantifier-free form, introducing the hypothesis (Γ) $(a+(b+c) = (a+b)+c)$:

$$\begin{aligned} a + (b + Sc) &\xlongequal{\mathsf{PA5}, Cong^+} a + S(b + c) \xlongequal{\mathsf{PA5}} S(a + (b + c)) \\ &\xlongequal{\mathsf{PA2}, \Gamma} S((a + b) + c) \xlongequal{\mathsf{PA5}} (a + b) + Sc, \end{aligned}$$

where, as above, we do not indicate the use of axiom $(\forall)$ when applying the axioms of PA. Thus, by virtue of the Deduction Theorem, the implication is true:

$$[(a + (b + c) = (a + b) + c)] \rightarrow [(a + (b + Sc) = (a + b) + Sc)].$$

Next, reasoning as in the proof of the congruence of addition, we move to the implication with quantifiers and obtain the derivation $\varphi(c) \rightarrow \varphi(Sc)$. And then, by the generalization rule, we obtain $\forall z\ \varphi(z) \rightarrow \varphi(Sz)$. It remains to apply induction and obtain the final conclusion $\forall z\ \varphi(z)$.

Commutativity of multiplication $ab = ba$ will be proven by two nested inductions, similar to the proof of the commutativity of addition. The outer induction is on b. But first, we need to prove that $0 \cdot a = 0$ for any a. This is done by induction on a. Indeed, $0 \cdot 0 = 0$ (PA6), and within the hypothesis $0 \cdot a = 0$, we have

$$0 \cdot Sa \overset{\text{PA6}}{=\!=} (0 \cdot a) + 0 \overset{\text{PA5}}{=\!=} 0 \cdot a \overset{\text{hypothesis}}{=\!=\!=} 0,$$

so that by virtue of induction, $\forall x\ 0 \cdot x = 0$.

Now, the equality $ab = ba$ for $b = 0$ is obvious, since $a \cdot 0 = 0$ by axiom PA6 and $0 \cdot a = 0$ by what has been proven.

Let's check the inductive step. We introduce the hypothesis ($\Gamma 1$) $\forall x\ xb = bx$. We prove that $\forall x\ x \cdot Sb = (Sb) \cdot x$. We will prove this formula by induction within the hypothesis ($\Gamma 1$). For $x = 0$, the equality holds, as it is a consequence of the equality obtained above for $b = 0$. We now proceed without quantifiers for convenience. Let us introduce the hypothesis ($\Gamma 2$) $c \cdot Sb = (Sb) \cdot c$ and check the equality for Sc:

$$Sc \cdot Sb \overset{\text{PA6}}{=\!=} (Sc)b + Sc \overset{\Gamma 1, Cong^{+}}{=\!=\!=} bSc + Sc \overset{\text{PA5}}{=\!=}$$
$$S(bSc + c) \overset{\text{PA6}, Cong^{+}, \text{PA2}}{=\!=\!=\!=} S((bc + b) + c);$$

$$Sb \cdot Sc \overset{\text{PA6}}{=\!=} (Sb)c + Sb \overset{\Gamma 2, Cong^{+}}{=\!=\!=} cSb + Sb \overset{\text{PA5}}{=\!=}$$
$$S(cSb + b) \overset{\text{PA6}, Cong^{+}, \text{PA2}}{=\!=\!=\!=} S((cb + c) + b).$$

On the right, we have obtained two terms whose equality is established using PA2, hypothesis $\Gamma 1$, the congruence of addition, the commutativity of addition, and the associativity of addition. Thus, by the transitivity of equality, we obtain $Sc \cdot Sb = Sb \cdot Sc$, which completes the inductive step for the formula $\forall x\ x \cdot Sb = (Sb) \cdot x$, and with this, the inductive step for the variable b for the formula $\forall x\ xb = bx$ is completed, so that we finally obtain $\forall x, y\ (xy = yx)$.

From the congruence of equality with respect to multiplication for the first argument proven above and the commutativity of multiplication, the congruence of equality with respect to multiplication for the second argument also follows.

Distributivity of addition with multiplication is proven by induction on c. Let $c = 0$, then

$$a(b + 0) = ab + a \cdot 0$$

by virtue of PA5, PA6, the congruence of addition and multiplication. Assume that $a(b + c) = ab + ac$, then

$$a(b + Sc) = a \cdot S(b + c) = a(b + c) + a = ab + (ac + a) = ab + aSc,$$

where we have used axioms PA5, PA6, the hypothesis, the associativity of addition, and the congruence of equality with respect to addition and multiplication. Thus, the inductive step is obtained, and distributivity is proven.

Associativity of multiplication is proven by induction on c. For $c = 0$, we have

$$a(b \cdot 0) = a \cdot 0 = 0, \quad (ab) \cdot 0 = 0.$$

Introducing the hypothesis $a(bc) = (ab)c$, we obtain that

$$a(bSc) = a(bc + b) = a(bc) + ab = (ab)c + ab = (ab)Sc,$$

which gives the inductive step and completes the proof of the associativity of multiplication.

Now, having the associativity and commutativity of addition and multiplication, we can automatically arrange parentheses and swap operands as is convenient for further calculations. We will also no longer refer to the properties of distributivity and the congruence of equality with respect to addition and multiplication, as they are familiar to the point of automaticity since childhood. It is important that we have already proven them in our system of axioms.

Let's prove an auxiliary statement that will greatly simplify working with inequality.

Lemma D3.2 (strong definition of inequality).

$$\text{PA7'} \qquad (a < b) \leftrightarrow \exists x\ b = a + Sx.$$

Proof. We will use induction on the variable b.

For $b = 0$, we need to derive the formula $(a < 0) \leftrightarrow \exists x\ 0 = a + Sx$. On the left is the false formula $a < 0$ (axiom PA7). On the right, using PA5 with the substitution $[a/a, b/x]$, we get $\exists x\ 0 = S(a + x)$, which is false by virtue of PA4. So we have false formulas in both parts of the equivalence, and the entire equivalence is true for any a. The base case of the induction is obtained.

Let us introduce the induction hypothesis $(a < b) \leftrightarrow \exists x\ b = a + Sx$ and from this prove the formula $(a < Sb) \leftrightarrow \exists x\ Sb = a + Sx$.

Let's prove from left to right. Let $a < Sb$, then $a = b$ or $a < b$ (PA7). Next, using the tautology $((\mathcal{A} \to C) \wedge (\mathcal{B} \to C)) \to ((\mathcal{A} \vee \mathcal{B}) \to C)$, we will derive the formula $\exists x\ Sb = a + Sx$ by case analysis, i.e., we will first show two implications

$$(a = b) \to \exists x\ Sb = a + Sx, \quad (a < b) \to \exists x\ Sb = a + Sx, \tag{D3.1}$$

then combine them with a conjunction, using the tautology $\mathcal{A} \to (\mathcal{B} \to (\mathcal{A} \wedge \mathcal{B}))$, and then by the MP rule we will get the implication $(a = b) \vee (a < b) \to \exists x\ Sb = a + Sx$, whence by the Deduction Theorem we obtain $(a < Sb) \to \exists x\ Sb = a + Sx$.

Let's prove the first implication in (D3.1). Let $a = b$. From axiom PA2, we have $(a = b) \to (Sa = Sb)$, therefore, $Sa = Sb$. Furthermore, $Sa = a + S0$ (PA2, PA5). Then $Sb = a + S0$, whence we obtain the formula $\exists x\ Sb = a + Sx$. Using the Deduction Theorem and the hypothesis $a = b$, we derive the first implication in (D3.1).

Let's prove the second implication in (D3.1). Let us introduce the hypothesis $a < b$. From the induction hypothesis, we obtain the formula $\exists x\ b = a + Sx$. The formula under the quantifier is equivalent to $Sb = a + S(Sx)$ (PA5, PA2, PA3), therefore, $\exists y\ Sb = a + Sy$ (where $y = Sx$). Using the Deduction Theorem and the hypothesis $a < b$, we derive the second implication in (D3.1).

From this, as already mentioned, $(a = b) \vee (a < b) \to \exists x\ Sb = a + Sx$ is derived, and as a consequence, $(a < Sb) \to \exists x\ Sb = a + Sx$.

Conversely, let $\exists x\ Sb = a + Sx$. Since $Sb = a + Sx$ is equivalent to $b = a + x$ (PA5, PA3, PA2), let's consider the alternative (Lemma D3.1): $(x = 0) \vee (x = Sy)$. We then proceed by case analysis.

Let $x = 0$, then $b = a + 0 = a$. And from $a = b$, by axiom PA7, it follows that $a < Sb$.

Let $x = Sy$. Then from $b = a + x$, we get that $b = a + Sy$, whence by the induction hypothesis, we get $a < b$, from which, by axiom PA7, we arrive at the formula $a < Sb$.

Thus, by the rule of case analysis and the Deduction Theorem, we arrive at the implication $(\exists x\ Sb = a + Sx) \to (a < Sb)$.

Combining both derived implications into a conjunction, we obtain the formula $(a < Sb) \leftrightarrow \exists x\ Sb = a + Sx$, and with this, the induction on the variable b is complete. □

The strong definition of inequality allows for the simplification of a number of proofs for inequality, which we will consider below. Note that it is called strong because to obtain it from the weak definition (PA7), induction is required (as we saw in the proof of the lemma). At the same time, if inequality is defined by formula PA7', then the proof of PA7 can be done without induction, if one uses the statement of Lemma D3.1.

Indeed, let $<$ be defined by formula PA7'; let us prove that PA7 holds for it. First, we need to check that $\neg(a < 0)$. If we assume the opposite, $a < 0$, then by virtue of PA7', we have $\exists x\ 0 = a + Sx$, but by virtue of PA5, under the quantifier we have $0 = S(a + x)$, which contradicts PA4.

Let's prove the recursive part of PA7: $(a < Sb) \leftrightarrow (a < b) \vee (a = b)$:

$$(a < Sb) \leftrightarrow \exists x\ Sb = a + Sx \leftrightarrow \exists x\ Sb = S(a + x) \leftrightarrow \exists x\ (b = a + x),$$

where we used PA5, PA3, PA2, PA1. Now, by virtue of Lemma D3.1, there are two alternatives for x: $x = 0$ or $x = Sy$ for some y. In the first case, the equality $b = a+x$ is equivalent to $b = a$ (PA5). In the second case, $b = a+Sx$, which, by definition PA7', is equivalent to $a < b$. Thus, $a < Sb$ is equivalent to $(a = b) \vee (a < b)$.

With this, we have completed the proof of Theorem B2.3, given in Part I.

In Robinson's arithmetic Q, where Lemma D3.1 is used instead of induction, our two definitions of inequality are not equivalent (the weak one is derivable from the strong one, but not vice versa).

Congruence of equality with respect to inequality for the first argument: $a = b \to (a < c) \leftrightarrow (b < c)$. We will use the strong definition of inequality. Since the inequality $a < c$ is equivalent to the formula $\exists x\ c = a + Sx$, within the hypothesis $a = b$, this formula is equivalent to $\exists x\ c = b + Sx$, and the latter formula is equivalent to the inequality $b < c$.

Similarly, we obtain the congruence of equality with respect to inequality for the second argument: $a = b \to (c < a) \leftrightarrow (c < b)$.

$$(c < a) \leftrightarrow \exists x\ a = c + Sx \leftrightarrow \exists x\ b = c + Sx \leftrightarrow (c < b)$$

within the hypothesis $a = b$.

Thus, we have the congruence of equality with respect to all operations and predicates of our arithmetic signature.

Let's prove that the **order is compatible with the operation** S, i.e., $(a < b) \leftrightarrow (Sa < Sb)$. Having the strong definition of inequality at our disposal, we can do without induction. Indeed,

$$(a < b) \leftrightarrow \exists x\ b = a + Sx \leftrightarrow \exists x\ Sb = S(a + Sx)$$
$$\leftrightarrow \exists x\ Sb = Sa + Sx \leftrightarrow (Sa < Sb),$$

where we have used a series of identities:

$$S(a + Sx) = S(Sx + a) = Sx + Sa = Sa + Sx,$$

which follow from the properties of addition proven above.

Let's prove the **cancellation law for addition**: $a + c = b + c \to a = b$. We proceed by induction on c. For $c = 0$, it is tautological. Let us accept the hypothesis $a + c = b + c \to a = b$. Then

$$(a + Sc = b + Sc) \to S(a + c) = S(b + c) \to a + c = b + c \to a = b.$$

The inductive step is proven. Together with congruence, this gives us the equivalence

$$(a = b) \leftrightarrow (a + c = b + c)$$

for any a, b, c.

Let's prove that the **order is compatible with the operation** $+$, i.e., $(a < b) \leftrightarrow (a + c < b + c)$.

$$(a < b) \leftrightarrow \exists x\ b = a + Sx \leftrightarrow \exists x\ b + c = a + Sx + c \leftrightarrow$$
$$\exists x\ b + c = (a + c) + Sx \leftrightarrow (a + c < b + c).$$

Let's prove that the relation $<$ is a strict linear order, i.e., it is irreflexive, transitive, and connex.

Irreflexivity $\neg(a < a)$: if $a < a$, then $\exists x\ a + 0 = a + Sx$, where by the cancellation law we would get $Sx = 0$, which contradicts PA4.

Transitivity: $((a < b) \wedge (b < c)) \to (a < c)$. From $b = a + Sx$ and $c = b + Sy$, it follows that $c = a + Sx + Sy = a + S(Sx + y)$, whence $c = a + Sz$, where $z = Sx + y$, i.e., $a < c$.

Connexity: $(a < b) \vee (a = b) \vee (b < a)$. We will prove this by induction on a. For $a = 0$, we need to check the formula $(0 < b) \vee (0 = b) \vee (b < 0)$. By Lemma D3.1, either $b = 0$ or $b = Sx$ for some x. In the first case, the target formula is true because of the equality; in the second case, we rewrite the equality $b = Sx$ as follows: $b = 0 + Sx$, whence by the strong definition of inequality, we get $0 < b$, and the target formula is again true.

Induction hypothesis: $(a < b) \vee (a = b) \vee (b < a)$ for any b. We need to prove that $(Sa < b) \vee (Sa = b) \vee (b < Sa)$ for any b. If $(a = b) \vee (b < a)$ is true, then by axiom PA7, we have $b < Sa$. If $a < b$ is true, then $b = a + Sx = S(a + x) = S(x + a) = x + Sa = Sa + x$. Then if $x = 0$, then $Sa = b$, and if $x = Sy$, then $b = Sa + Sy$, i.e., by axiom PA7, we have $Sa < b$. In all cases, we obtain $(Sa < b) \vee (Sa = b) \vee (b < Sa)$ for any b. The induction is complete.

Furthermore, the property of *antisymmetry* is also true: $(a < b) \to \neg(b < a)$. Indeed, if $(a < b) \wedge (b < a)$, then by transitivity we get $a < a$, which contradicts irreflexivity. Antisymmetry turns the property of connexity into the law of trichotomy: one and only one of the alternatives $(a < b)$, $(a = b)$, $(b < a)$ is true.

Let's prove that the **order is compatible with the operation** $\cdot$, i.e., $(c \neq 0) \to [(a < b) \leftrightarrow (ac < bc)]$. Let $c \neq 0$, then $c = Sx$ by Lemma D3.1. Let also $a < b$, then $b = a + Sy$ for some y, then $bc = ac + cSy = ac + (yc + c) = ac + (yc + Sx) = ac + S(yc + x)$, whence by the strong definition of inequality, $ac < bc$. Thus, the implication in one direction is proven.

It remains for us to show that if $c \neq 0$ and $ac < bc$, then $a < b$. Suppose this is not true; then by the trichotomy property of the order, we have two alternatives: $a = b$ or $b < a$. In the first case, we would have $ac = bc$ by virtue of the congruence of equality with respect to multiplication, and in the second case, we would get $bc < ac$ by the first part of the proof of the compatibility of order with multiplication (within the hypothesis $c \neq 0$). Then, taking into account the hypothesis $ac < bc$ and the transitivity of inequality, we get that $ac < ac$, which contradicts the irreflexivity of inequality. Consequently, the assumption is not true, and $a < b$.

Let's prove the **cancellation law for multiplication**: $(ac = bc) \wedge (c \neq 0) \to (a = b)$. Suppose this is not so, i.e., $(ac = bc) \wedge (c \neq 0) \wedge (a \neq b)$, then by virtue of the trichotomy of inequality, either $a < b$ or $b < a$ is true. Let it be the former. Then by what was proven above, $ac < bc$, which excludes the equality $ac = bc$. The case $b < a$ is analogous. Consequently, our assumption is not true.

Let's prove the **discreteness of order**: $(a \leqslant b) \wedge (b < Sa) \to (a = b)$. Let $(a \leqslant b) \wedge (b < Sa)$; then we need to exclude the case $a < b$. Suppose that $a < b$. From $b < Sa$, by axiom PA7, we have $b < a$ or $b = a$, which together with the assumption gives $a < a$. Contradiction.

This property shows that between a and Sa on the scale of natural numbers, there are no intermediate numbers.

Identity element for multiplication $a \cdot S0 = a$ follows from the axioms and already proven properties: $a \cdot S0 = a \cdot 0 + a = a$.

No zero divisors $(ab = 0) \to (a = 0 \vee b = 0)$. Suppose that $(a \neq 0) \wedge (b \neq 0)$, then by Lemma D3.1, $a = Sx$ and $b = Sy$, whence

$$ab = Sx \cdot Sy = Sx \cdot y + Sx = S(Sx \cdot y + x) \neq 0.$$

The theorem is fully proven. □

The collection of proven properties tells us that the natural numbers in Peano's axiomatics form a discrete ordered commutative semiring with a unit.

We now have to completely close the question of the congruence of equality by proving the general formulas (B1.9) we gave earlier.

Theorem D3.2 (Congruence). *Let T, T_1, T_2 be arbitrary terms of the language of* PA*, and φ be an arbitrary formula of the language of* PA*. Then*

$$(T_1 = T_2) \to T[a/T_1] = T[a/T_2]; \tag{D3.2}$$

$$(T_1 = T_2) \to (\varphi[a/T_1] \leftrightarrow \varphi[a/T_2]), \tag{D3.3}$$

and these formulas are also valid for any other free variable (not just a).

Proof. The theorem is proven by induction on the construction of terms and formulas, using as a base case the properties of the congruence of equality proven in Theorem D3.1.

First, let us introduce the notion of the *height of a term* $H[T]$ and the *height of a formula* $H[\varphi]$, which is defined recursively, following the rules for constructing terms and formulas:

1) for free variables and constants: $H[a] = 0$, $H[0] = 0$ (in PA there is only one constant, 0);

2) for function symbols: $H[S(T)] = 1 + H[T]$, $H[T_1 + T_2] = 1 + \max(H[T_1], H[T_2])$, $H[T_1 \cdot T_2] = 1 + \max(H[T_1], H[T_2])$;

3) for atomic formulas: $H[T_1 = T_2] = 0$ (we exclude the predicate $a < b$ from the signature, considering it to be introduced by definition PA7'; this makes the language of PA simpler to analyze);

4) for logical connectives and quantifiers: $H[\neg\varphi] = 1 + H[\varphi]$, $H[\varphi \wedge \psi] = 1 + \max(H[\varphi], H[\psi])$, $H[\exists x\ \varphi[a/x]] = 1 + H[\varphi]$ (we also reduce the signature of the language of predicate logic to the symbols $\neg, \wedge, \exists$, since the other logical connectives and the universal quantifier are expressible through them).

For example, $H[\underbrace{S \ldots S}_{n} 0] = n$.

The proof of (D3.2) and (D3.3) is carried out by *complete* induction (see Section B2.1.3) on the height of terms and formulas. Recall that in complete induction, a separate check for the case $n = 0$ is not formally required; however, in this case, the inductive step from a height less than $n = 0$ to a height of $n = 0$ has a vacuously true premise for the induction hypothesis, so the check of the inductive step for $n = 0$ must still be checked separately.

Proof for terms. Let $n = 0$, i.e., the term T is atomic, meaning it is either a variable or a constant. There are three possible cases here:

1) $T \overset{\text{def}}{\rightleftharpoons} a$, i.e., the term T is the variable for which the substitution is made, then (D3.2) takes the form: $(T_1 = T_2) \rightarrow (T_1 = T_2)$, which is a tautology;
2) $T \overset{\text{def}}{\rightleftharpoons} b$, i.e., the term T is a variable different from the substitution variable, then the substitution changes nothing, and (D3.2) takes the form: $(T_1 = T_2) \rightarrow (b = b)$, which is true because the consequent is true;
3) $T \overset{\text{def}}{\rightleftharpoons} 0$ — here the situation is analogous: $(T_1 = T_2) \rightarrow (0 = 0)$.

Next, the inductive step for $n > 0$ is proven. Assume that (D3.2) holds for terms T' whose height is less than n. We need to show congruence for a term T whose parse tree has height n.

For example, let $T \overset{\text{def}}{\rightleftharpoons} S(T')$. Within the induction hypothesis, the implication $(T_1 = T_2) \rightarrow (T'[a/T_1] = T'[a/T_2])$ holds. Then we take axiom PA2 and substitute the terms $T'[a/T_1]$ and $T'[a/T_2]$ for the variables a, b, obtaining the implication $(T'[a/T_1] = T'[a/T_2]) \rightarrow S(T'[a/T_1]) = S(T'[a/T_2])$, whence by the transitivity of implication, we arrive at the formula $(T_1 = T_2) \rightarrow S(T'[a/T_1]) = S(T'[a/T_2])$. But the terms in the last equality are graphically (i.e., as text strings) the same terms that are obtained directly from the term T: $S(T'[a/T_1]) \equiv T[a/T_1]$ and $S(T'[a/T_2]) \equiv T[a/T_2]$, where $\equiv$ denotes the graphical identity of terms. Thus, the implication $(T_1 = T_2) \rightarrow T[a/T_1] = T[a/T_2]$ is true.

We proceed similarly for the other two cases of term construction for T — via addition and multiplication, applying the congruence proven in Theorem D3.1. As a result, the inductive step is proven, and the congruence of equality with respect to terms is proven.

Proof for formulas. Let the height of the formula φ be zero; then it is a formula of the form $T' = T''$, where the substitution variable a may or may not be present in the terms T', T'', and they may also be closed. We need to show the implication

$$(T_1 = T_2) \to (T'[a/T_1] = T''[a/T_1]) \leftrightarrow (T'[a/T_2] = T''[a/T_2]).$$

Let $T_1 = T_2$, and for brevity, let us denote $L_1 \equiv T'[a/T_1]$, $L_2 \equiv T'[a/T_2]$, $R_1 \equiv T''[a/T_1]$, $R_2 \equiv T''[a/T_2]$. From what was proven for terms, we have: $T'[a/T_1] = T'[a/T_2]$, i.e., $L_1 = L_2$, and similarly $T''[a/T_1] = T''[a/T_2]$, i.e., $R_1 = R_2$. Then by the properties of equality (the corresponding substitutions into axiom PA1), we derive that $(L_1 = R_1) \leftrightarrow (L_2 = R_2)$, which is the required equivalence.

Now let $n > 0$; then the formula φ has one of three forms: $\neg\psi$, or $\psi_1 \wedge \psi_2$, or $\exists x\ \psi[b/x]$, where the height of the formulas ψ, ψ_1, ψ_2 is less than n, i.e., the induction hypothesis (D3.3) holds for them. Here, the variable b is bound by the quantifier, so that a remains as the substitution variable (although it may not be present in the formula).

Let $\varphi \overset{\text{def}}{\leftrightarrow} \neg\psi$ and let $T_1 = T_2$. By the induction hypothesis, we have $(\psi[a/T_1] \leftrightarrow \psi[a/T_2])$. Here, we can apply negation to both sides of the equivalence by the rules of propositional calculus, so that we obtain $\neg\psi[a/T_1] \leftrightarrow \neg\psi[a/T_2]$, whence by the Deduction Theorem, we get the implication $(T_1 = T_2) \to (\varphi[a/T_1] \leftrightarrow \varphi[a/T_2])$.

We proceed similarly for the case $\varphi \overset{\text{def}}{\leftrightarrow} \psi_1 \wedge \psi_2$.

Let's consider the case $\varphi \overset{\text{def}}{\leftrightarrow} \exists x\ \psi[b/x]$. Let $T_1 = T_2$; then by the induction hypothesis, $\psi[a/T_1] \leftrightarrow \psi[a/T_2]$. From this, we can move to the equivalence $(\exists x\ \psi[a/T_1, b/x] \leftrightarrow \exists x\ \psi[a/T_2, b/x])$, using Lemma D2.8. And thus we obtain the equivalence $\varphi[a/T_1] \leftrightarrow \varphi[a/T_2]$.

Thus, the inductive step is proven, and we have the congruence of equality with respect to all formulas of the language of PA.

Remark: We have not explicitly indicated the presence of free parameters in the formulas and terms here to save space, although in fact, all these arguments should be carried out assuming that the formulas and terms may contain arbitrary free parameters $p_1, \ldots, p_k$. This is necessary in order to be able to transition to formulas with quantifiers, in which the number of parameters is reduced. □

This theorem is formulated as a schema of theorems involving non-terminal symbols, and therefore it is not a theorem of PA in the proper sense.

Although the formulation and proof of this theorem belong to the metatheory, the implementation of such a schema for specific terms and formulas is provable by means of PA itself. Thus, for example, if we substitute the elementary term $a + c$ for the term T, and the variables a and b for the terms T_1 and T_2, then the substitution $(T_1 = T_2) \to T[a/T_1] = T[a/T_2]$ turns into point 1° of Theorem D3.1. Indeed, having a general meta-theoretical proof, it is not difficult to "extract" from it a proof for each specific formula, since the main idea of this meta-proof is that obtaining congruence for a more complex formula or term reduces to obtaining congruence for simpler formulas and terms with the help of the axioms of PA, and for the simplest formulas and terms, we already have Theorem D3.1.

The Congruence Theorem, in essence, legitimizes the fundamental mathematical technique — the substitution of equals for equals — within the formal system of PA. This is not just a technical detail, but a meta-mathematical justification for the correctness of algebraic transformations.

D3.2 Induction

Recall the notations introduced in Section B2.1.3:

$$\begin{aligned} Ind_a[\varphi] &\overset{\text{def}}{\leftrightarrow} \varphi[a/0] \wedge \forall x\,(\varphi[a/x] \to \varphi[a/Sx]); \\ Prog_a[\varphi] &\overset{\text{def}}{\leftrightarrow} \forall x\,(\forall y < x :\ \varphi[a/y]) \to \varphi[a/x]; \\ Tot_a[\varphi] &\overset{\text{def}}{\leftrightarrow} \forall y\ \varphi[a/y], \end{aligned}$$

where the formula φ may contain parameters, i.e., free variables other than a. As usual, we do not specify them for the sake of brevity.

In these notations, axiom PA8 is written as $Ind_a[\varphi] \to Tot_a[\varphi]$ and is called *local induction*. Furthermore, we defined the schema of formulas

$$Prog_a[\varphi] \to Tot_a[\varphi] \tag{D3.4}$$

as *complete induction* (sometimes called *course-of-values induction*).

Theorem D3.3. *The schema of complete induction is equivalent to the schema of local induction over Robinson's arithmetic Q (with congruence axioms).*

Proof. Let us first show that over Q, the schema of complete induction implies the schema of local induction. To do this, we begin by proving that $Ind_a[\varphi] \to Prog_a[\varphi]$. Let us introduce the hypotheses: $Ind_a[\varphi]$ and the antecedent of the implication in the progressiveness formula, $\forall y < a :\ \varphi(y))$. We need to verify its consequent, i.e., the formula $\varphi(a)$. By Lemma D3.1 (which is an axiom of Q), there is an alternative for a: $a = 0$ or $a = Sz$ for some z.

Let $a = 0$; then $\varphi(a)$ is true by virtue of the hypothesis $Ind_a[\varphi]$.

Let $a = Sz$; then $z < a$ by axiom PA7 ($z < Sz$) and the congruence of inequality, and thus from the hypothesis $\forall y < a : \ \varphi(y))$, by axiom ($\forall$), it follows that $\varphi(z)$, whence by virtue of the hypothesis $Ind_a[\varphi]$, we obtain $\varphi(Sz)$, i.e., $\varphi(a)$.

Thus we have shown $Ind_a[\varphi] \to Prog_a[\varphi]$ (by the generalization rule and the Deduction Theorem).

Suppose that the schema of complete induction is true, i.e., $Prog_a[\varphi] \to Tot_a[\varphi]$ for any formula φ. From this, by the proven implication $Ind_a[\varphi] \to Prog_a[\varphi]$ and the transitivity of implication, we obtain that $Ind_a[\varphi] \to Tot_a[\varphi]$, i.e., local induction holds.

Note that in this case, for each formula φ, local induction follows from complete induction. To prove the converse, we will have to prove that from the entire schema of local induction follows the schema of complete induction, i.e., for a single, separately taken formula, this does not work.

Let us show that local induction implies complete induction. Let $Ind_a[\psi] \to Tot_a[\psi]$ for any formula ψ. Let us introduce the hypothesis $Prog_a[\varphi]$. Let also $\psi(b) \overset{\text{def}}{\leftrightarrow} \forall y < b : \varphi[a/y]$. We apply local induction to the formula $\psi(b)$ within the hypothesis.

$\psi(0)$ is true, since $\psi(0) \leftrightarrow \forall y < 0 : \varphi(y)$. Expanding the bounded quantifier by definition (B2.2), we get $\psi(0) \leftrightarrow \forall y\ (y < 0) \to \varphi(y)$. The implication under the quantifier is true because $y < 0$ is false (PA7); therefore, $\psi(0)$.

Let us show the inductive step $b \to Sb$. Suppose that $\psi(b)$, i.e., $\forall y < b : \varphi(y)$. Then by virtue of the hypothesis $Prog_a[\varphi]$, we obtain $\varphi(b)$. Further, $\psi(b)$ and $\varphi(b)$ together imply $\forall y \leqslant b : \varphi(y)$, and the latter is equivalent to the formula $\forall y < Sb : \varphi(y)$, since $(y \leqslant b) \leftrightarrow (y < Sb)$ (PA7), and the latter formula is by definition $\psi(Sb)$. Thus $\psi(b) \to \psi(Sb)$, and the inductive step is proven.

Applying local induction to the formula ψ, we obtain its totality: $\forall z\ \psi(z)$. Now from this, by axiom ($\forall$), replacing z with Sa, we obtain $\psi(Sa)$, where the variable a is not bound by any hypothesis. Expanding $\psi(Sa)$ by definition, we get $\forall y < Sa : \ \varphi(y)$, whence again by axiom ($\forall$), replacing y with a, we get $(a < Sa) \to \varphi(a)$. The antecedent of the last implication is true (PA7), whence we derive $\varphi(a)$.

Now, since the variable a is free and not bound by hypotheses, by the generalization rule we arrive at the formula $Tot_a[\varphi]$, and the derivation of the formula of complete induction for φ is obtained. □

Remark: The fact that this equivalence is provable over a theory as weak as Q is a typical result in the field of Reverse Mathematics. This branch of logic seeks to determine the minimal axiomatic system required to prove a given theorem. In this case, the equivalence between the different forms of induction does not require the full power of PA, but only the axioms of Robinson's arithmetic (with congruence).

Corollary D3.1. *The following schema, which is called the well-foundedness of the natural numbers,*

$$(WOP) \qquad (\exists y\ \varphi(y)) \to \exists x\ \varphi(x) \land \forall y < x:\ \neg\varphi(y),$$

is equivalent to the schema of local induction. The formula φ may contain parameters.

In Section B2.1.3, we showed that the WOP schema is equivalent to the schema of complete induction in pure predicate calculus (without the involvement of PA axioms), therefore, taking into account Theorem D3.3, the WOP schema is equivalent to the schema of local induction over the base theory Q (with the congruence axioms (D3.2) and (D3.3)).

Corollary D3.2. *Fermat's principle of infinite descent*

$$(\exists y\ \varphi(y)) \to \neg(\forall x\ (\varphi(x) \to \exists y < x:\ \varphi(y)))$$

is equivalent to the schema of induction. The formula φ may contain parameters.

This principle, as was shown in the same place, is also equivalent to the schema of complete induction in pure predicate calculus, and thus is equivalent to local induction over the base theory Q (with the congruence axioms (D3.2) and (D3.3)).

Thus, we can conclude that over the base theory Q (+congruence), the following principles are equivalent:

- the schema of local induction;
- the schema of complete induction;
- the well-foundedness of the natural numbers;
- the principle of infinite descent,

with the last three being equivalent in pure predicate calculus.

D3.3 Elementary Theory of Divisibility

Let's show how some simple functions can be defined in PA. For example, the predecessor function $P(a)$ is often used. This is a fresh function symbol that requires its own axiom:

$$b = P(a) \overset{\text{def}}{\leftrightarrow} (a = 0 \to b = 0) \wedge (a \neq 0 \to S(b) = a).$$

From Lemma D3.1, it follows that this definition correctly defines a total function, i.e., for each a, the number b is defined and is unique. Furthermore, the congruence of equality works for it, i.e., $(T_1 = T_2) \to P(T_1) = P(T_2)$, which follows from the fact that instead of the equality $P(T_1) = P(T_2)$, one can write the formula $\exists x\ x = P(T_1) \wedge x = P(T_2)$, which allows for the elimination of the symbol P by its definition.[1]

In fact, one can prove a general meta-theorem stating that if a fresh function symbol F is introduced by a formula in the "old" signature, for which the congruence of equality already holds, then congruence extends to the symbol F and to all terms and formulas containing it. The easiest way to prove such a theorem is by external induction (i.e., in the meta-theory) on the complexity of formulas, where complexity is defined by the number of occurrences of the symbol F in the given formula. Each expansion of one occurrence of F reduces the complexity of the formula in this sense (although it increases its complexity in the sense of $H[\varphi]$, which we used earlier) and ultimately reduces it to formulas without the symbol F, for which everything has already been proven.

Next, let's define the **monus** function (truncated subtraction), which is defined at the meta-level as follows:

$$a \dot{-} b = \begin{cases} c, & (b + c = a) \wedge (b < a), \\ 0, & b \geqslant a, \end{cases}$$

and in the language of Peano arithmetic, this is written as an axiom:

[1] It is understood that the terms T_1, T_2 may also contain symbols P, but their number is finite, and their substitution into any term can also be expanded by the definition. Thus, in a finite number of steps, we can expand a formula involving symbols P into a longer formula without them. For example, if we encounter the expression $a = b + P(c)$, we replace it with the formula $\exists x\ (a = b + x) \wedge (x = P(c))$, after which we expand the equality $x = P(c)$ by its definition.

$$(a \dot{-} b = c) \overset{\text{def}}{\leftrightarrow} (b < a \wedge b + c = a) \vee (b \geqslant a \wedge c = 0).$$

For this definition, one can also prove congruence of equality and correctness. Monus has the following properties, which we leave as an exercise:

1° $a \dot{-} b \leqslant a$, and $(a \dot{-} b = a) \leftrightarrow (a = 0 \vee b = 0)$;
2° $(a \dot{-} b = 0) \leftrightarrow (a \leqslant b)$;
3° $(a = (a \dot{-} b) + b) \leftrightarrow (b \leqslant a)$;
4° $(a + b) \dot{-} a = b$;
5° $a \dot{-} (b + c) = (a \dot{-} b) \dot{-} c$;
6° $c \cdot (a \dot{-} b) = (c \cdot a) \dot{-} (c \cdot b)$;
7° $(a \leqslant b) \rightarrow (a \dot{-} c \leqslant b \dot{-} c)$;
8° $(a \leqslant b) \rightarrow (c \dot{-} b \leqslant c \dot{-} a)$.

Furthermore, monus can be used to define certain functions:

$$\max(a, b) \overset{\text{def}}{\rightleftharpoons} a + (b \dot{-} a), \qquad \min(a, b) \overset{\text{def}}{\rightleftharpoons} a \dot{-} (a \dot{-} b),$$
$$|a - b| \overset{\text{def}}{\rightleftharpoons} (a \dot{-} b) + (b \dot{-} a), \qquad [a = b] \overset{\text{def}}{\rightleftharpoons} 1 \dot{-} (|a - b|),$$

in the last example, the equality should not be perceived as an arithmetic predicate; this notation should be considered in its entirety, including the square brackets, as a designation for a function equal to 1 when $a = b$, and 0 otherwise (such a predicate is called an *Iverson bracket*). And a more complex definition:

$$if(a = b, c, d) \overset{\text{def}}{\rightleftharpoons} ([a = b] \cdot c) + ((1 \dot{-} [a = b]) \cdot d),$$

which expresses a function equal to c when $a = b$, and d otherwise.

Theorem D3.4 (The Division Algorithm). *For any natural numbers a (the dividend) and $b \neq 0$ (the divisor), there exist unique natural numbers q (the quotient) and r (the remainder) such that $a = bq + r$ and $q \leqslant a,\ r < b$.*

Proof. Given numbers $a \geqslant 0$ and $b > 0$. Let us define the formula

$$\varphi(a, b, k) \overset{\text{def}}{\leftrightarrow} \exists x\ (a = bx + k).$$

This formula is true for all such k that are obtained by subtracting multiples of b from a.

The formula φ is true for some k, namely, for $k = a$ (one must take $x = 0$ as a witness); therefore, by the well-ordering principle (WOP), there exists a smallest k that satisfies this formula. Let us denote this smallest k by r.

Since $\varphi(a, b, r)$ holds, there exists a q such that $a = bq+r$. It is also obvious that $r < b$, because otherwise the number $r' = r \dot{-} b$ would also satisfy the formula φ (with $x = q + 1$) in contradiction to the fact that r is the smallest such number. Furthermore, $q \leqslant a$, because otherwise $(q > a)$ we would have $bq + r > a$ for any r, since $b \geqslant 1$.

It remains to prove the uniqueness of the pair of numbers q, r. Suppose that the pair q', r' also satisfies the equality $a = bq' + r'$ and the condition $r' < b$. Then from $a = bq + r$, we get that $bq + r = bq' + r'$.

Next, we use the trichotomy of inequality: $(r' < r) \vee (r' = r) \vee (r' > r)$. It is easy to see that

$$(r = r') \leftrightarrow (bq = bq') \leftrightarrow (q = q'),$$

where we used the cancellation laws for addition and multiplication.

Let $r' < r$, then $r = (r \dot{-} r') + r'$, whence from the equality $bq+r = bq'+r'$, by substituting for r and canceling r', we get that $bq' + (r \dot{-} r') = bq$, whence by the strong definition of inequality, we have $bq' \leqslant bq$, from which, by canceling $b \neq 0$, we arrive at $q' \leqslant q$. But from the preceding, $q' = q$ is impossible under the condition $r' < r$, which means $q' < q$.

Then $q = q' + Sx$ for some x, whence $bq = bq' + bx + b$ and, consequently, $bx + b = r \dot{-} r'$. And here we obtain a contradiction, since $r \dot{-} r' \leqslant r < b$, whereas $bx + b \geqslant b$.

Consequently, the case $r' < r$ is impossible. The case $r' > r$ is excluded analogously. Thus, the uniqueness of the quotient and remainder is proven. □

The proof provided is already almost in the usual mathematical style, and therefore takes up only half a page. If we had written it in as much detail as, for example, the proof of the commutativity of addition, it would have been several times longer and much more difficult to comprehend. Therefore, of course, such thorough proofs as in Theorem D3.1 are needed only when we are faced with the task of confirming the ability of a formal system to express and prove the arsenal of tools known to us and to ensure that any of our general mathematical proofs can be reduced to such a "mathematical assembler" as is the formal language of predicate logic.

Using the Division Algorithm (Theorem D3.4), it is not difficult to define the ternary predicate $r = \mathrm{rem}(a, b)$, which expresses that the number r is the remainder of the division of a by b:

$$r = \text{rem}(a, b) \overset{\text{def}}{\leftrightarrow} (r < b) \wedge \exists q \leqslant a : (bq + r = a),$$

which is also a Δ_0-formula, i.e., it contains no unbounded quantifiers. The fact that this formula defines the graph of a function (i.e., that for given $a, b \neq 0$, the remainder r is uniquely determined) follows directly from the uniqueness part of the Division Algorithm.

Note that we can extend the signature of the language of PA with a binary function symbol rem, introducing this formula for it as an axiom. In this case, as has been noted before, such a definition extends the congruence of equality to the new function symbol and all terms and formulas containing it. Moreover, all theorems that can be proven in the extended language are equivalent to the corresponding theorems in the original language if they are expanded using this definition. Therefore, if we want to remain in the original language, we can consider that we have only added a shorthand for the formula, and if it is more convenient for us to understand rem as a function, we can consider that we are operating in the extended language. In the latter case, we should then define $\text{rem}(a, b)$ for the case $b = 0$, for example, as zero, since we can substitute any term into a function symbol, including the constant 0. However, as a rule, the first approach is sufficient.

We say that a number b is a **divisor** of a number a (synonym: a is divisible by b) if $\text{rem}(a, b) = 0$, which is written more concisely as: $b \mid a$. When giving definitions, it is always useful to check the edge cases. In this case, we have: $d \mid 0$ for any d, including $0 \mid 0$, but $0 \nmid a$ for $a > 0$.

Let us define such a concept as the *greatest common divisor* (gcd):

$$d = \text{gcd}(a, b) \overset{\text{def}}{\leftrightarrow} (d \mid a) \wedge (d \mid b) \wedge \forall x\ ((x \mid a) \wedge (x \mid b) \to (x \leqslant d)).$$

From the point of view of studying mathematics as a language, it is very useful from time to time to write down such definitions. Note that if $a = b = 0$, then the greatest common divisor does not exist; the formula $d = \text{gcd}(0, 0)$ is false for any d. In all other cases, $\text{gcd}(a, b)$ exists. For example, if $a > 0, b = 0$, then $\text{gcd}(a, b) = a$.

Next, we can define the predicate

$$a \perp b \overset{\text{def}}{\leftrightarrow} 1 = \text{gcd}(a, b),$$

which expresses the fact that the numbers a and b are **coprime**. Note that $a \perp 1$ for any a, including $a = 0$.

Similarly, we define the predicate "to be a prime number"

$$Prime(a) \overset{\text{def}}{\leftrightarrow} (a > 1) \wedge \forall x\ ((x \mid a) \to (x = 1 \vee x = a)).$$

And here we have immediately forbidden the number a to be 0 or 1. We discussed the concept of a prime number in detail in Section A2.1, including clarifying the need for such a prohibition.

Lemma D3.3 (on the prime divisor). *If a number a has a non-trivial divisor* (> 1), *then it also has a prime divisor.*

Proof. We apply complete induction on the number a. For the cases $a = 0$ and $a = 1$, the check is trivial: for zero, all prime numbers are divisors, and one has no non-trivial divisors. For $a = 2$, the answer is obvious, as 2 is a prime number. We can then apply the inductive step, assuming that the statement is true for all numbers less than a. If a has only two divisors (1 and a), then it is already prime. If a has some other divisor b, then $a = bq$ for some q (by the Division Algorithm), where $1 < b, q < a$, and for them the induction hypothesis works, so there exists a prime $p \mid b$, and thus $p \mid a$. The induction is complete. □

Let's show a few more simple facts from number theory.

Lemma D3.4 (Bézout's identity).
For any a and b, not both zero, there exist s and t such that $\gcd(a, b) = |as - bt|$.

Proof. For $a = 0$ or $b = 0$ (provided they are not both zero), the proof is obvious. Therefore, let $a, b > 0$. Consider the formula:

$$\varphi(a, b, d) \overset{\text{def}}{\leftrightarrow} (d > 0) \wedge \exists s, t\ (d = |as - bt|),$$

i.e., this formula is true for those and only those d that are linear combinations of a and b. This formula has witnesses, for example, $d = a$, since in this case $d = |a \cdot 1 - b \cdot 0|$. Consequently, applying WOP, we find the smallest such d, which we denote by d_0; furthermore, $d_0 = |as - bt|$ for some s, t.

Let us show that $d_0 \mid a$. By the Division Algorithm, $a = qd_0 + r$ or $r = a \dot{-} qd_0$, where $r < d_0$. For definiteness, let us assume that $as \geqslant bt$, then $r = a \dot{-} q(as \dot{-} bt)$, from which it is clear that r is a linear combination of a and b, i.e., $\varphi(a, b, r)$ holds, and at the same time $r < d_0$. Since d_0 is the minimal witness of φ, r cannot be positive, which means $r = 0$, from which it follows that $d_0 \mid a$. It is proven analogously that $d_0 \mid b$. Thus, d_0 is a divisor of a and b.

Now let some d also be their divisor, i.e., $a = kd$ and $b = ld$. Then we substitute these expressions into the representation of d_0, and we get: $d_0 = |kds - ldt|$, from which it will follow that[2] $d \mid d_0$, and thus, $d \leqslant d_0$. □

Thus, the greatest common divisor of a, b is their linear combination.

Lemma D3.5 (Euclid's Lemma).
If p is a prime number and $p \mid ab$, then $(p \mid a) \vee (p \mid b)$.

Proof. Suppose that $p \nmid a$, then $\gcd(a, p) = 1$, and by virtue of Bézout's identity, we have $1 = |as - pt|$ for some s, t. Let's multiply both sides of the equality by b: $b = |abs - ptb|$. The difference is divisible by p, since ab is divisible by p, consequently, $p \mid b$.

Analogously, if $p \nmid b$, then we arrive at $p \mid a$. □

Euclid's Lemma is a fundamental property of the integers, and its generalization to arbitrary rings allows for the consideration of so-called *unique factorization domains*, which have properties very similar to those of the ring of integers.

Lemma D3.6 (Gauss's Lemma). If $c \mid ab$ and $c \perp b$, then $c \mid a$.

Proof. Since $c \perp b$, by Bézout's identity $1 = |cs - bt|$ and, consequently, $a = |acs - abt|$, from which it is clear that $c \mid a$. □

The proven lemmas lead us directly to the **Fundamental Theorem of Arithmetic** (FTA): *every natural number $n > 1$ is a product of prime numbers, and this factorization is unique up to the order of the factors.* And here again we run into the problem that without the ability to talk about arbitrary finite sets of numbers, we are not even in a position to formulate the FTA!

The following analogy with spoken language is appropriate. Imagine that we are trying to tell a story in a language that has only singular pronouns (he, she, it) and proper names, but no nouns for groups (family, crowd, team). We can say: "He loves her." But we cannot say: "The team won the match."

The language of PA without coding is like a spoken language without nouns for groups. It can only talk about individual numbers. But the surprising power

[2] A divisor of one of the factors divides the product, and a divisor of all terms divides the sum. We will not prove the elementary properties of divisibility here.

of PA is that in this language, we can (with the help of Gödel's brilliant solution) reason about the codes of arbitrary finite sequences. A code number becomes a name for an entire sequence $\langle x_1, \dots, x_n \rangle$. And as soon as we have this name, we can make statements about the sequence itself: what its length is, what it consists of, whether it is a permutation of another sequence, etc. Including formulating and proving the FTA and many other theorems of number theory.

This, at first glance, purely technical trick of coding has enormous philosophical and practical significance. The ability of the language of arithmetic to describe its own syntactic constructs (formulas, proofs) and operations on them (derivation) by means of code numbers lies at the heart of computability theory. It is Gödel numbering that allows for the proof of the existence of a universal computable function—the theoretical analogue of a modern computer, capable of executing any algorithm if it is given the code number of that algorithm as input.

Thus, the "language" in which any artificial intelligence operates ultimately reduces to arithmetic. The entire complexity of its programs, neural network architectures, and learning algorithms can be encoded by natural numbers and described by formulas in the language of PA. This is the very "architecture of mathematical thought" mentioned in the subtitle of the book, embodied in silicon.

D3.4 Sequence Coding

We will consider here the classical method of sequence coding, the idea of which is due to Gödel. In ordinary mathematics this method is naturally explained through the Chinese Remainder Theorem, which allows for the magical coding of arbitrary sequences of natural numbers using just pairs of specially chosen numbers. Inside first-order arithmetic, however, we cannot start by invoking the full CRT as a theorem about arbitrary finite families: the very phrase "arbitrary finite family" already presupposes a way of coding sequences. Therefore, below we use only the elementary divisibility facts proved above and build the restricted CRT argument directly into the appending lemma.

But first, let us obtain another important result.

Theorem D3.5. *Let $a \perp b$. Then for any c and any $r < b$, there exists an $i < b$ such that* $r = \mathrm{rem}(c + ai, b)$.

In essence, the theorem states that the mapping $i \mapsto \mathrm{rem}(c + ai)$ is a surjection (and also an injection) as a function whose arguments and values lie in the set $\{0, \ldots, b - 1\}$.

Proof. First, let's consider the edge cases where b is zero or one. It is clear that for $b = 0$, the theorem is vacuously true due to the false premise. For $b = 1$ and any a, the only option for r and i is $r = 0 = i$, and this is obviously true, since $\forall x \ \mathrm{rem}(x, 1) = 0$.

Now let $b > 1$ (from which it automatically follows that $a > 0$). Note that if we prove the theorem for $c = 0$, it can be easily proven for other c. Indeed, suppose we need to obtain $r = \mathrm{rem}(c + ai, b)$, i.e., $c + ai = bq + r$ for some q. At the same time, $c = bq' + r'$, where $r' < b$. From this, it is easy to see that $q' \leqslant q$ (otherwise we get $c \geqslant bq + b + r' > bq + r = c + ai \geqslant c$), then $ai = b(q \dot{-} q') + (r \dot{-} r')$ if $r \geqslant r'$, and $ai = b(q \dot{-} q' \dot{-} 1) + (b \dot{-} r') + r$ if $r < r'$ (in this case, the equality $q = q'$ is excluded). In both cases, we know what $\mathrm{rem}(ai, b)$ must be equal to—it is either $r \dot{-} \mathrm{rem}(c, b)$ or $(b \dot{-} \mathrm{rem}(c, b)) + r$.

It remains to prove the theorem for $c = 0$. This case can also be simplified, namely, by proving it only for $r = 1$. Indeed, if we have found an i such that $\mathrm{rem}(ai, b) = 1$, then for any $r < b$, one should take $\mathrm{rem}(ri, b)$ instead of i, since if $ai = bq + 1$, then $a\,\mathrm{rem}(ri, b) = a(ri \dot{-} bq') = ari \dot{-} abq' = (bq + 1)r \dot{-} abq' = bq'' + r$.

It remains to prove the theorem for $c = 0$ and $r = 1$, i.e., that there exists an i such that $\mathrm{rem}(ai, b) = 1$. By Bézout's identity, $1 = |as - bt|$ for some s, t, since $a \perp b$. Then, as is easy to see, $\mathrm{rem}(as, b) = 1$, i.e., s is the required i. □

Theorem D3.6 (Chinese Remainder Theorem (CRT), informal form). *Let $m_0, \ldots, m_n$ be pairwise coprime numbers. Then for any natural numbers $x_i < m_i$, there exists a number N such that $x_i = \mathrm{rem}(N, m_i)$ for all $i \leqslant n$. Furthermore, if numbers N_1 and N_2 satisfy these conditions, then $|N_1 - N_2|$ is divisible by $m_0 \cdot \ldots \cdot m_n$.*

Imagine you have several alarm clocks. The first rings every m_0 minutes, the second every m_1 minutes, and so on. If you know how much time has passed since the last ring of each alarm clock $(x_0, x_1, \ldots, x_n)$, and the periods m_i are pairwise coprime, then you can uniquely determine the moment in time modulo $m_0 \cdot m_1 \cdot \ldots \cdot m_n$.

This is the usual external mathematical formulation, not yet a closed formula of the language of PA. It requires quantification over two sequences, m_i and x_i, which is impossible before we have learned to encode arbitrary finite sequences by one (or two) natural numbers. Thus, the standard internal formulation of the CRT will become available only after the appending lemma. The apparent circle is broken because the proof of the appending lemma needs only a restricted CRT construction: the moduli are given by the explicit term $1 + (i + 1)d'$, and the remainders are given by a bounded formula with parameters. We will carry out precisely this restricted construction inside the proof.

Next, we define Gödel's β-function:

$$b = \beta(c, d, i) \overset{\text{def}}{\leftrightarrow} b = \mathrm{rem}(c, 1 + (i + 1) \cdot d),$$

or in functional notation, $\beta(c, d, i) = \mathrm{rem}(c, 1 + (i + 1) \cdot d)$. Note that this definition is also a Δ_0-formula. The substitution $i = 0$ is harmless: the divisor then becomes $1 + d$, which is always non-zero (and equals 1 when $d = 0$).

The idea of the encoding is that an arbitrary finite sequence $\langle x_0, x_1, \ldots, x_{n-1} \rangle$ is encoded not by one number, but by a pair of numbers $\langle c, d \rangle$. This pair can then be "packed" into a single number a using Cantor's standard diagonal pairing function: $a = 1/2(c + d)(c + d + 1) + d$, which is easily definable in PA.[3] The number d is chosen such that the moduli $m_i = 1 + (i + 1)d$ for $i < n$ are pairwise coprime, which ensures the possibility of applying the CRT to find a c such that the system of congruences is satisfied:

$$(x_0 = \beta(c, d, 0)) \wedge (x_1 = \beta(c, d, 1)) \wedge \cdots \wedge (x_{n-1} = \beta(c, d, n - 1)), \quad \text{(D3.5)}$$

meaning that we have successfully encoded the sequence $\langle x_0, x_1, \ldots, x_{n-1} \rangle$ with two numbers. In some sources, the sequence $\langle n, x_0, x_1, \ldots, x_{n-1} \rangle$ is encoded, with its length in the first position. In that case, we can immediately extract the length of the encoded sequence as $\beta(c, d, 0)$ to use this bound later.

It is worth emphasizing that the β-function itself is total, i.e., its value is correctly defined for any natural numbers c, d, i, including zero values. This means that, in principle, any pair of numbers c, d (or a single number, if we use Cantor's pairing for pairs) defines some infinite sequence, and thus also some finite sequence of any given length. If $d > 0$, this infinite sequence eventually

---

[3] We recommend verifying independently that Cantor's pairing function is a bijection between pairs of natural numbers and the natural numbers. The beauty of Cantor's pairing function is that its definition uses nothing but basic arithmetic operations.

stabilizes at c, since the divisor $1 + (i + 1)d$ eventually exceeds c; if $d = 0$, it is constantly equal to 0, because all remainders are taken modulo 1.

The merit of the Chinese Remainder Theorem is that the converse is also true: if we already have a finite sequence (of arbitrarily chosen length) given in some way, then it necessarily corresponds to some code $\langle c, d\rangle$, and in fact, there are infinitely many such codes, since c is determined up to a term that is a multiple of the product of all moduli $m_i = 1 + (i + 1)d$ (for the given sequence length).

Next, we will prove not a direct theorem on coding, which states that for every sequence there exists a code in the form of a pair of numbers c, d, since we cannot yet quantify over sequences, but a recursive theorem on appending. It states that, having a pair c, d whose first n decoded values are already fixed, we can construct a new pair c', d' with the same first n values and with one prescribed additional value in position n. This is the exact strength needed fo recursive definitions: induction supplies the already constructed initial segment, and the appending lemma extends it by one term.

Theorem D3.7 (on appending a term to a sequence). *Given numbers $c, d \geqslant 0$ and $n \geqslant 0$. Then for any x, there exist numbers $c' \geqslant 0$ and $d' > 0$ such that*

$$(\forall i < n : \beta(c, d, i) = \beta(c', d', i)) \wedge (x = \beta(c', d', n)),$$

Proof. We introduce local notations:

$$m_i \overset{\text{def}}{\rightleftharpoons} 1 + (i + 1)d', \qquad x_i \overset{\text{def}}{\rightleftharpoons} \begin{cases} \beta(c, d, i), & i < n, \\ x, & i = n. \end{cases}$$

Step 1: choosing d'. We need d' to guarantee two conditions: **a)** the moduli must be greater than the numbers being encoded, i.e., for any $i \leqslant n$, it must be that $x_i < m_i$. **b)** the moduli m_i for $i \leqslant n$ must be pairwise coprime.

Both conditions can be satisfied in one fell swoop. Let's take a number M that is greater than all terms of the sequence x_i and its length $n + 1$. In fact, we need to take $\max(n, x_0, \ldots, x_n) + 1$, but formally in PA, it is sufficient to show that such an M exists. For this, we prove by induction the formula

$$\varphi(c, d, k) \overset{\text{def}}{\leftrightarrow} (k \leqslant n) \to \exists M\ \forall i \leqslant k : (M > x_i) \wedge (M > k),$$

where induction is on the variable k. Indeed, for $k = 0$, the witness is $M = \max(x_0, 1) + 1$. Next, if $\varphi(c, d, k)$ is true, i.e., there is a witness M for the given

k, then for $k + 1$ there are two options: $k + 1 \leqslant n$ and $k + 1 > n$. In the second case, the formula is true due to the false antecedent of the implication, and in the first case, one can take $\max(M, x_{k+1}) + 1$ as a witness. This completes the induction on k, from which it follows that $\forall k\ \varphi(c, d, k)$, and thus, by the rule of specialization, $\varphi(c, d, n)$, which means the existence of the required M.

Now, we need to choose a $d' > 0$ that satisfies a specific property: it must be a multiple of every integer from 1 to M. The existence of such a number is guaranteed by the following theorem of PA (provable by induction on M): $\forall M \exists x > 0 : \forall i\ (1 \leqslant i \leqslant M \rightarrow i \mid x)$. By the Well-Ordering Principle, the set of all such common multiples x has a minimal element. We can choose d' to be this minimal element, which is known as the least common multiple (LCM) of the numbers from 1 to M.

Checking condition **a)**: it is obvious that $x_i < M \leqslant d' < 1 + (i+1)d' = m_i$. The condition is satisfied.

Checking condition **b)**: suppose that the moduli $m_i = 1 + (i + 1)d'$ and $m_j = 1 + (j + 1)d'$, where $i \neq j$, are not coprime. This means they have a common non-trivial divisor, and thus a common prime divisor p. Then p divides their difference: $p \mid |m_j - m_i|$, i.e., $p \mid d'|j - i|$. Consequently, p divides either $|j - i|$ or d' (Euclid's Lemma). But p cannot divide d'. If $p \mid d'$, then since $m_i = 1 + (i + 1)d'$, the remainder of dividing m_i by p would be 1. But we assumed that $p \mid m_i$, i.e., the remainder is 0. A contradiction. Thus, p must divide $|j - i|$. But $|j - i| \leqslant n < M$. Any number less than M, and any of its prime divisors (including p), are divisors of the number d' by its construction. Thus, from $p \mid |j - i|$ it follows that $p \mid d'$. And this, as we just saw, is impossible. Conclusion: no two distinct moduli have a common non-trivial divisor, i.e., the moduli m_i are pairwise coprime.

Step 2: the restricted CRT construction. Now that we have found d' and obtained a set of pairwise coprime moduli $m_0, \ldots, m_n$, defined by the formula $1 + (i + 1)d'$ $(i \leqslant n)$, we construct a number c' such that $x_i = \mathrm{rem}(c', m_i)$ for $i \leqslant n$. This is exactly the CRT argument, but carried out only for this definable family of moduli and remainders.

Consider the formula within the already proven properties of x_i and m_i:

$$\varphi(c, d, d', n, x, k) \overset{\text{def}}{\leftrightarrow} (k \leqslant n) \rightarrow \exists C\ \forall i \leqslant k : x_i = \beta(C, d', i),$$

where $x_i = \beta(c, d, i)$ for $i < n$ and $x_n = x$. We prove it by induction on k.

For $k = 0$, it is required to check that $\exists C\ x_0 = \beta(C, d', 0)$, i.e., that $x_0 = \mathrm{rem}(C, 1 + d')$ for some C. Since $x_0 < 1 + d'$, one can take C to be x_0 itself.

Induction hypothesis: $\varphi(c, d, d', n, x, k)$. We need to prove $\varphi(c, d, d', n, x, k+1)$. From the hypothesis, it follows that there exists a C_k such that for all $i \leqslant k$, we have $x_i = \beta(C_k, d', i)$, i.e., $x_i = \mathrm{rem}(C_k, m_i)$. We need to construct C_{k+1} such that for all $i \leqslant k+1$, it is true that $x_i = \mathrm{rem}(C_{k+1}, m_i)$.

It is not difficult to show by induction that for any $i \leqslant k$, there exists a $D_i > 0$ that is a multiple of all $m_0, \dots, m_i$ ($\forall j \leqslant i : m_j \mid D_i$), then to specialize i by substituting k for it, and by the WOP principle (it is equivalent to induction) to choose the smallest of such D_k. In fact, this means that D_k is the least common multiple of $m_0, \dots, m_k$. And since all m_i are pairwise coprime, then[4] $D_k = m_0 \cdot \ldots \cdot m_k$.

Note that $D_k \perp m_{k+1}$, since if a prime $p \mid D_k$, then there exists an $i \leqslant k$ such that $p \mid m_i$. Indeed, if $p \mid D_k$ and $\forall i \leqslant k : p \nmid m_i$, then $D_k = pq$, while $m_i \mid D_k$ and $\forall i \leqslant k : m_i \perp p$, whence by Gauss's Lemma $\forall i \leqslant k : m_i \mid q$ in contradiction to the minimality of D_k. Consequently, if $p \mid D_k$, then $\exists i \leqslant k : p \mid m_i$. But then this m_i is not coprime with m_{k+1} in contradiction to their construction.

We have: $D_k \perp m_{k+1}, x_{k+1} < m_{k+1}$, then by Theorem D3.5, there exists a $j < m_{k+1}$ such that $x_{k+1} = \mathrm{rem}(C_k + D_k \cdot j, m_{k+1})$. Let us set $C_{k+1} \overset{\mathrm{def}}{\rightleftharpoons} C_k + D_k \cdot j$. We then note that for $i \leqslant k$, we have $\mathrm{rem}(C_k + D_k \cdot j, m_i) = \mathrm{rem}(C_k, m_i) = x_i$, since $m_i \mid D_k$.

Thus, for $C = C_{k+1}$, we obtain that $\forall i \leqslant k+1 : x_i = \mathrm{rem}(C, m_i) = \beta(C, d', i)$, i.e., the formula $\varphi(c, d, d', n, x, k+1)$ is true. With this, the induction is complete.

By making the substitution $[k/n]$ in the formula φ, we arrive at the fact that $\exists C \; \forall i \leqslant n : x_i = \beta(C, d', i)$. By choosing this C as the desired c', we complete the proof of the theorem. Notice that no positivity condition on c' is needed; if one wanted $c' > 0$, one could replace C by $C + D_n$, where D_n is a common multiple of the moduli, without changing any of the remainders.

Thus, the target number c' is the sum of an initial value (e.g., $C_0 = x_0$) and successive products of the moduli ($D_k = m_0 \cdot \ldots \cdot m_k$), each multiplied by a specific coefficient. □

With this theorem, we can now construct the finite initial segments that occur in recursive definitions. For example, we want to define the formula for the graph of the exponential function with base b:

[4] Again, we cannot directly define such a product, so we use induction.

$$y = b^n \overset{\text{def}}{\leftrightarrow} \exists c, d\ (\beta(c, d, 0) = 1) \wedge$$
$$(\beta(c, d, n) = y) \wedge \forall i < n : (\beta(c, d, i + 1) = b \cdot \beta(c, d, i)). \quad \text{(D3.6)}$$

In fact, this formula states the following: there exists a code c, d of a sequence $\langle x_0, \ldots, x_n \rangle$ such that $x_0 = 1$, for each $i < n$, $x_{i+1} = b \cdot x_i$, and $x_n = y$, i.e., it asserts not just the existence of the number b^n, but of the entire segment of the exponential function from 0 to n. Moreover, having the code name c, d, we can restore the entire sequence up to the n-th term (what happens after n is not specified by the formula).

To prove the correctness of the definition and the totality of the exponential function, we need the previous theorem. Correctness means that such c, d with the specified properties indeed exist for each n. Totality is the formula $\forall n \exists y\ (y = b^n)$, which follows from correctness. We prove this by induction on n. For $n = 0$, it is required to prove that there exist y, c, d such that $\beta(c, d, 0) = y = 1$. Since $\beta(c, d, 0) = \text{rem}(c, 1 + d)$, it is sufficient to take $c = 1$ and $d = 1$. Let such a y exist for n, i.e., there are the necessary c, d that define the graph of the exponential function from 0 to n. Apply Theorem D3.7 with the parameter $n + 1$ and with the appended value $x = b \cdot y$. Then there exist c', d' such that

$$b \cdot y = \beta(c', d', n + 1) = b \cdot \beta(c, d, n) \wedge \forall i \leqslant n : \beta(c', d', i) = \beta(c, d, i),$$

i.e., as the value of y corresponding to $n + 1$, we take $b \cdot y$, where y corresponds to n. The induction step is complete.

To prove the functionality of this correspondence, i.e., the condition that if $y = b^n$ and $y' = b^n$, then $y = y'$, one can take one pair c, d and a second pair c', d' — witnesses of the formula (D3.6), and verify by complete induction that they define the same sequence for $i = 0, \ldots, n$.

Thus, we have a formula of the language of PA that defines the function $y = b^n$, and this formula belongs to the class of Σ_1-formulas, i.e., it is a formula with bounded quantifiers, preceded by several unbounded existential quantifiers.

Thus the appending lemma is sufficient for the first non-trivial recursive functions, in particular for the function $y = 2^n$ (take $b = 2$). This already gives enough coding power for many later syntactic constructions: bounded products, finite lists of symbols, and ultimately the formulations of the CRT and the FTA as genuine formulas about sequence codes.

At this point a natural question arises: why then do people distinguish arithmetic with exponentiation from arithmetic without exponentiation? Theanswer depends on what exactly is meant. If we are talking about full PA, then adding a new function symbol for exponentiation together with its defining axioms gives only a definitional extension: it makes formulas shorter and proofs more readable, but does not add new arithmetical consequences, because PA already proves that the graph of exponentiation is total and functional. In weak fragments of arithmetic, however, the situation changes. For example, in theories with only bounded induction, the totality of exponentiation may no longer be provable; adding it as an axiom then genuinely increases the strength of the theory. Thus, in full PA exponentiation is hidden inside the theory, while in weaker arithmetics it marks a real boundary of what the theory can prove.

Here we have come very close to the central result of proof theory, which characterizes the class of all computable (partial recursive) functions.

Theorem D3.8 (Main Theorem of Formal Arithmetic). *The graph of any (partial) recursive function can be defined by a Σ_1-formula of* PA.

The proof of this theorem is beyond the scope of this book, but from the example with the definition of the exponential function, it is already clear how this can be done in the general case, starting from the recursive definition of a function.

Finally, let us answer the main question of this section — *how can we now define quantification over arbitrary finite sequences?* The answer has already been demonstrated in the example (D3.6) of the definition of the exponential function: we need to introduce new quantifiers using sequence codes.

Thus, if we want to say that for any (or some) sequence $\langle x_0, \ldots, x_n \rangle$, some property φ holds, then we will have to write $\forall c, d$ (or, respectively, $\exists c, d$) and then in the property φ, replace every occurrence of x_i with $\beta(c, d, i)$, using bounded quantifiers on the variable i ($\forall i \leqslant n$ or $\exists i \leqslant n$). Although not every pair c, d is a meaningful code, the predicate φ is constructed to be so strict that it will be true only for those codes that decode into sequences with the required properties, and false for all other codes.

Now, after the appending lemma has given us sequence codes, the Chinese Remainder Theorem can be formulated internally. Here is its general form (the first part, about the existence of N):

$$\begin{aligned}\forall n\, \forall c, d\ [\forall i, j \leqslant n : (i \neq j) \to \beta(c, d, i) \perp \beta(c, d, j)] \to \\ \forall c', d' \Big([\forall i \leqslant n : \beta(c', d', i) < \beta(c, d, i)] \to \\ [\exists N\ \forall i \leqslant n : \beta(c', d', i) = \operatorname{rem}(N, \beta(c, d, i))] \Big).\end{aligned}$$

The premise states that all terms of the sequence $\langle m_0, \ldots, m_n \rangle$ (where $m_i = \beta(c, d, i)$) are pairwise coprime, and the conclusion states for an arbitrary sequence $\langle x_0, \ldots, x_n \rangle$ that if all $x_i < m_i$ (where $x_i = \beta(c', d', i)$), then there exists an N such that $x_i = \mathrm{rem}(N, m_i)$ for all $i \leqslant n$. As we can see, we have managed with a closed formula of the language of PA with added predicates expressing the remainder function and Gödel's β-function.

Similarly, one can now express, for example, the **Fundamental Theorem of Arithmetic**. Before sequence coding this theorem could only be stated informally, because "factorization into finitely many primes" is a statement about a finite sequence. But for a more compact and understandable notation, let's introduce a few more auxiliary predicates:

$$\begin{aligned} a = Prod(c, d, m) \overset{\text{def}}{\leftrightarrow} \exists c', d' \ (a = \beta(c', d', m)) \wedge \beta(c', d', 0) = \beta(c, d, 0) \wedge \\ \forall i < m : \beta(c', d', i+1) = \beta(c', d', i) \cdot \beta(c, d, i+1) \end{aligned}$$

expresses that a is the product of all the first $m + 1$ terms of the sequence encoded by c, d;

$$\begin{aligned} &Factor(c, d, n, m) \overset{\text{def}}{\leftrightarrow} (n = Prod(c, d, m)) \wedge \\ &\quad (\forall i \leqslant m : Prime(\beta(c, d, i))) \wedge (\forall i < m : \beta(c, d, i) \leqslant \beta(c, d, i+1)) \end{aligned}$$

which expresses the fact that c, d encode an increasing factorization sequence of the number n, including exactly $m + 1$ prime factors.

Then the formulation of the FTA takes the form:

Theorem D3.9 (Fundamental Theorem of Arithmetic). *For every natural number $n > 1$, there exists a unique factorization into prime numbers:*

$$\begin{aligned} &\forall n > 1 : [\exists m \exists c, d \ Factor(c, d, n, m)] \wedge \forall c', c'', d', d'', m', m'' \\ &\qquad [Factor(c', d', n, m') \wedge Factor(c'', d'', n, m'') \rightarrow \\ &\qquad\qquad (m' = m'') \wedge \forall i \leqslant m' : \beta(c', d', i) = \beta(c'', d'', i)], \end{aligned}$$

i.e., for any n, first, there exists its increasing factorization sequence, and second, it is unique.

The fact that the decomposition of a number into prime factors is determined up to the order of the factors, we have expressed by the fact that the sequence of factors, sorted in increasing order, is unique.

Note that the coding method presented here is far from being the only one. Moreover, depending on the chosen axiomatics, one can use the most convenient coding methods. For example, if we immediately include the definition of the exponential function and the axiom of its totality in the list of axioms, then coding becomes something like a walk in the park. If, on the contrary, we strive for minimalism like Robinson's arithmetic, then internal total coding by means of arithmetic becomes impossible, although one can still code sequences by meta-theoretical methods, and for each specific case of a code, one can find its implementation in Q, which allows for the proof of Gödel's incompleteness theorem even for Q. Intermediate variants use, for example, multiple applications of Cantor's diagonal coding for sequences of a given finite length, coding by a polynomial method (in a positional numeral system) for sequences with uniformly bounded values, etc.

D3.5 Toward Gödel's and Tarski's theorems

We now have all the ingredients needed to compress finite lists of natural numbers into single quantifiable objects of PA: Gödel's β-function from Section D3.4, Cantor pairing if desired, and concrete illustrations such as the exponential (D3.6), together with internal formulations of the CRT and of the FTA above. Fix an injective assignment of numbers to each symbol of the logical alphabet and of the arithmetical signature—the same idea as attaching code points to characters in Unicode. Any formula is a finite string of symbols; hence it determines a finite sequence of numeric codes. As in Section D3.4, we encode that sequence by a pair $\langle c, d\rangle$ using Gödel's β-function, then combine c and d into a single natural number by Cantor's diagonal pairing. This number—once the assignment of codes to symbols is fixed—is what we call the *Gödel number* $\ulcorner\gamma\urcorner$ of an expression γ, consistently with Chapter D2. Formal proofs, once represented as finite structured lists of formulas and inference steps, receive numeric codes by the same mechanism.

We denote by $\mathcal{L}(\mathsf{PA})$ the set of all formulas of this language. As in Section C1.4.3, we write $\mathbb{N}$ for the standard model of PA constructed there. Recall from Section D2.2.1 that first-order derivation is sound. Consequently, PA is *sound* for its standard model $\mathbb{N}$: every theorem of PA is true in $\mathbb{N}$. This semantic connection is crucial for the proofs of incompleteness that follow.

For any natural number n, we denote by $\underline{n}$ the corresponding numeral in PA, defined as $\underbrace{S \ldots S}_{n} 0$.

Theorem D3.10 (Diagonal lemma). *For every formula $\Phi(a)$ of $\mathcal{L}(PA)$ with at most the variable a free, there exists a sentence θ such that*

$$PA \vdash \theta \leftrightarrow \Phi(\underline{\ulcorner \theta \urcorner}).$$

Proof. At the meta-level, define a total primitive recursive function $F : \omega \to \omega$ as follows. If n is the Gödel number of some formula $\psi(a)$ with at most a free, put $F(n) \stackrel{\text{def}}{\rightleftharpoons} \ulcorner \psi(\underline{n}) \urcorner$, where $\psi(\underline{n})$ denotes the result of substituting the numeral $\underline{n}$ for every free occurrence of a in ψ (if a does not occur free in ψ, the string is unchanged). Otherwise set $F(n) \stackrel{\text{def}}{\rightleftharpoons} 0$. Then for every $\psi(a) \in \mathcal{L}(\mathsf{PA})$,

$$F(\ulcorner \psi(a) \urcorner) = \ulcorner \psi(\underline{\ulcorner \psi(a) \urcorner}) \urcorner.$$

By Theorem D3.8, choose a Σ_1-formula $\delta(a, b)$ that strongly represents F: graph correctness in $\mathbb{N}$, and

$$\mathsf{PA} \vdash \forall y \left[\delta(\underline{n}, y) \leftrightarrow y = \underline{F(n)}\right] \quad \text{for each } n \in \omega.$$

Hence, for all $n, m \in \omega$ we have:[5]

$$\mathbb{N} \models \delta(\underline{n}, \underline{m}) \iff F(n) = m.$$

Given $\Phi(a)$, define $\xi(a) \stackrel{\text{def}}{\leftrightarrow} \exists y \, [\delta(a, y) \wedge \Phi(y)]$. Set $\theta \stackrel{\text{def}}{\leftrightarrow} \xi(\underline{\ulcorner \xi(a) \urcorner})$. Then in PA,

$$\begin{aligned}
\theta &\leftrightarrow \exists y \, [\delta(\underline{\ulcorner \xi(a) \urcorner}, y) \wedge \Phi(y)] \\
&\leftrightarrow \exists y \, [y = \underline{F(\ulcorner \xi(a) \urcorner)} \wedge \Phi(y)] \\
&\leftrightarrow \Phi(\underline{F(\ulcorner \xi(a) \urcorner)}) \\
&\leftrightarrow \Phi(\underline{\ulcorner \xi(\underline{\ulcorner \xi(a) \urcorner}) \urcorner}) \\
&\leftrightarrow \Phi(\underline{\ulcorner \theta \urcorner}).
\end{aligned}$$

□

[5] Here $\mathbb{N} \models (\ldots)$ is satisfaction in the standard model introduced in Section D2.2.1.

Having coded formulas, one codes proofs as primitive recursive finite objects as well: given n, one can decide in finitely many steps whether n is the code of a formal proof in PA and, if so, read off the Gödel number of its last formula.

Accordingly, there exists a formula $\mathrm{Pr}(u, v) \in \mathcal{L}(\mathsf{PA})$ such that $\mathbb{N} \models \mathrm{Pr}(\underline{k}, \underline{\ell})$ holds if and only if ℓ codes a proof of the formula with Gödel number k. Moreover, Pr can be chosen Δ_0 (under any standard arithmetization); hence $\exists v\, \mathrm{Pr}(u, v)$ is Σ_1.

Theorem D3.11 (Gödel's first incompleteness theorem).
If PA *is consistent, then there exists a sentence* $G \in \mathcal{L}(\mathsf{PA})$ *such that*

$$\mathsf{PA} \nvdash G \quad \text{and} \quad \mathsf{PA} \nvdash \neg G.$$

Proof. Let G be a diagonal fixed point for the formula $\neg\exists x\, \mathrm{Pr}(a, x)$ supplied by Theorem D3.10, so that

$$\mathsf{PA} \vdash G \leftrightarrow \neg\exists x\, \mathrm{Pr}(\underline{\ulcorner G \urcorner}, x).$$

If $\mathsf{PA} \vdash G$, let p be the code of some formal proof of G. Then $\mathbb{N} \models \mathrm{Pr}(\underline{\ulcorner G \urcorner}, \underline{p})$. By the very meaning of Pr (a formalized, primitive recursive verification that p is a correct derivation with last line coded by $\ulcorner G \urcorner$), this gives $\mathsf{PA} \vdash \mathrm{Pr}(\underline{\ulcorner G \urcorner}, \underline{p})$, and hence $\mathsf{PA} \vdash \exists x\, \mathrm{Pr}(\underline{\ulcorner G \urcorner}, x)$. Together with the equivalence for G, this yields a contradiction in PA.

If $\mathsf{PA} \vdash \neg G$, then $\mathsf{PA} \vdash \exists x\, \mathrm{Pr}(\underline{\ulcorner G \urcorner}, x)$. This last sentence is Σ_1, hence true in $\mathbb{N}$ by soundness. Therefore there exists a (standard) proof of G, i.e., $\mathsf{PA} \vdash G$. Together with $\mathsf{PA} \vdash \neg G$, we again contradict the consistency of PA.

Thus neither G nor $\neg G$ is derivable. □

Theorem D3.12 (Gödel's second incompleteness theorem).
Let $\mathrm{Con}(\mathsf{PA}) \overset{def}{\leftrightarrow} \neg\exists x\, \mathrm{Pr}(\underline{\ulcorner 0{=}1 \urcorner}, x)$. *If* PA *is consistent, then* $\mathsf{PA} \nvdash \mathrm{Con}(\mathsf{PA})$.

Proof. Keep the same G. Formalizing the proof of Theorem D3.11 inside PA gives

$$\mathsf{PA} \vdash \mathrm{Con}(\mathsf{PA}) \to \neg\exists x\, \mathrm{Pr}(\underline{\ulcorner G \urcorner}, x).$$

Combining this with the diagonal equivalence for G yields $\mathsf{PA} \vdash \mathrm{Con}(\mathsf{PA}) \to G$. Hence if $\mathsf{PA} \vdash \mathrm{Con}(\mathsf{PA})$, then $\mathsf{PA} \vdash G$, contradicting Theorem D3.11. □

Theorem D3.13 (Tarski's undefinability theorem). *There is no formula $T(a)$ of $\mathcal{L}(\mathsf{PA})$ with only a free such that for every sentence $\varphi \in \mathcal{L}(\mathsf{PA})$,*

$$\mathbb{N} \models T(\underline{\ulcorner\varphi\urcorner}) \quad \Longleftrightarrow \quad \mathbb{N} \models \varphi.$$

Proof. Suppose such a $T(a)$ existed. Apply Theorem D3.10 to $\neg T(a)$ and obtain a sentence λ with

$$\mathsf{PA} \vdash \lambda \leftrightarrow \neg T(\underline{\ulcorner\lambda\urcorner}).$$

By soundness,

$$\mathbb{N} \models \lambda \quad \Longleftrightarrow \quad \mathbb{N} \models \neg T(\underline{\ulcorner\lambda\urcorner}).$$

But the defining property of T gives $\mathbb{N} \models T(\underline{\ulcorner\lambda\urcorner}) \leftrightarrow \mathbb{N} \models \lambda$. Hence $\mathbb{N} \models \lambda \leftrightarrow \neg\lambda$, impossible. □

Let's summarize the chapter on arithmetic. Starting with the axiomatics, we proved a number of standard algebraic properties of natural numbers and a general statement about the congruence of equality. Then we dealt with principles equivalent to induction, in particular, the well-ordering principle (WOP). Next, we defined several useful arithmetic operations and discussed two equivalent approaches to their definition—by enriching the signature with function symbols and by directly defining a formula describing the graph of a function.

Then we presented and proved several simple facts from number theory related to the properties of divisibility. All the proofs provided were carried out strictly within the framework of first-order arithmetic without using quantification over sequences or sets. This allowed us to then prove the main theorem on appending to a sequence, which makes it possible to give recursive definitions of sequences, encoding them "on the fly" with a pair of suitable numbers (or one number, if using Cantor's numbering).

Having such a tool, we can now talk about sequences as objects of the theory PA, since we can encode them in a certain way. This brings first-order Peano arithmetic to a new level in terms of its proof strength, as it allows for operations on arbitrary finite sets, and thus on arbitrary relations and functions on finite sets, which is much more convenient than just operating on numbers and predicates.

The power of such a "leap" is manifested, for example, in the fact that in PA it becomes possible to talk not only about integers and rational numbers, but also about so-called computable real numbers and functions. Indeed, being able to encode sequences of arbitrary length, we can with one formula $\varphi(n, q)$, where q is a rational number, encode a convergent sequence of rational numbers. Real numbers that are the limits of such sequences are called **computable**. The set of computable numbers contains all algebraic numbers, as well as many others, while remaining a countable set.

Being able to encode computable numbers, the set of which is dense in $\mathbb{R}$, we can encode many real functions, defining their computable values at computable points. For computable functions, a number of usual properties known from analysis can be proven in PA, for example, the intermediate value theorem, the extreme value theorem for a continuous function on a closed interval, the identity $e^{x+y} = e^x \cdot e^y$ for computable x, y, the trigonometric identity $\sin^2(x) + \cos^2(x) = 1$ for computable x, etc.

All the properties of computable numbers and functions are grouped under the general name "computable analysis", which can be implemented (by means of sequence coding) in PA. Thus, PA, being a first-order theory in a countable language, already gives us almost the entire apparatus of real analysis, at least that which is used in computational mathematics, computer systems, and most applications.

Nevertheless, some things remain beyond the reach of PA. The main example is the completeness axiom of the real numbers. It is unprovable in PA because it requires the consideration of arbitrary, not necessarily computably defined, sets of real numbers. This is the territory of stronger theories, such as set theory ZFC.

Exercises

D3.1 Construct a formal derivation for $a + S(0) = S(a)$.

D3.2 Prove the property $\forall x(x \cdot S(0) = x)$ (Multiplicative Identity).

D3.3 Prove the property of discreteness: if $a < b$, then $S(a) \leqslant b$, using the strong definition of inequality ($a < b \leftrightarrow \exists x(b = a + Sx)$) and Lemma D3.1.

D3.4 Using the axioms of Peano Arithmetic (specifically PA5), prove the following property of addition:

$$S(a) + b = S(a + b).$$

Hint: Use induction on b.

D3.5 Prove the properties of monus inequality: $a \dot{-} b \leqslant a$, and $(a \dot{-} b = a) \leftrightarrow (a = 0 \vee b = 0)$.

D3.6 Prove the reconstruction property: $(a = (a \dot{-} b) + b) \leftrightarrow (b \leqslant a)$.

D3.7 Prove the inverse operation property: $(a + b) \dot{-} a = b$.

D3.8 Prove the associativity-like property: $a \dot{-} (b + c) = (a \dot{-} b) \dot{-} c$.

D3.9 Prove the distributivity of multiplication over monus: $c \cdot (a \dot{-} b) = (c \cdot a) \dot{-} (c \cdot b)$.

D3.10 Prove the monotonicity of the minuend: $(a \leqslant b) \rightarrow (a \dot{-} c \leqslant b \dot{-} c)$.

D3.11 Prove the anti-monotonicity of the subtrahend: $(a \leqslant b) \rightarrow (c \dot{-} b \leqslant c \dot{-} a)$.

D3.12 Formally define the "Remainder" relation $R(a, b, r)$ (r is the remainder of a divided by b).

D3.13 Prove that if $a \mid b$ and $b \mid a$, then $a = b$.

D3.14 Find natural numbers s, t satisfying Bézout's identity for $a = 12, b = 18$ (note: in natural numbers, $gcd(a, b) = (a \cdot s) \dot{-} (b \cdot t)$ or vice versa).

D3.15 Calculate the value of of Gödel's β-function $\beta(c, d, i) = \text{rem}(c, 1 + (i + 1)d)$:

1. $\beta(10, 3, 0)$.
2. $\beta(10, 3, 1)$.
3. $\beta(10, 3, 2)$.

D3.16 Calculate the value of Gödel's β-function $\beta(c, d, i)$ for the parameters $c = 13, d = 2$ and indices $i = 0, 1$. What two-element sequence does this encode?

D3.17 Find a pair (c, d) that encodes the sequence $\langle 1, 2 \rangle$ of length 2.

D3.18 Prove that for any sequence $a_0, \ldots, a_n$, there exist infinitely many pairs (c, d) encoding it.

D3.19 Using the definition of the set of integers $Z = (\omega \times \{0\}) \cup (\{0\} \times \omega)$ from Section C1.4.3, write down the pairs representing the numbers 5 and -3.

D3.20 Calculate the sum $\langle 0,5\rangle +_{\mathbb{Z}} \langle 3,0\rangle$ (corresponding to $5 + (-3)$) using the definition of addition for this model.

D3.21 Calculate the sum $\langle 0,2\rangle +_{\mathbb{Z}} \langle 5,0\rangle$ (corresponding to $2 + (-5)$) using the definition of addition.

D3.22 Calculate the product of two negative numbers $\langle 2,0\rangle \cdot_{\mathbb{Z}} \langle 3,0\rangle$ using the definition of multiplication.

D3.23 Calculate the product $\langle 0,3\rangle \cdot_{\mathbb{Z}} \langle 2,0\rangle$ (corresponding to $3 \cdot (-2)$) and verify that the result is correct.

D3.24 Let $\varphi(a)$ be a formula of PA with one free variable. Find a fixed point θ (such that $\mathsf{PA} \vdash \theta \leftrightarrow \varphi(\underline{\ulcorner\theta\urcorner})$) in each of the following cases, under any valid Gödel numbering:

1. $\varphi(a)$ is identically false (e.g., $a \neq a$);
2. $\varphi(a)$ is identically true (e.g., $a = a$).

D3.25 For this and the following problem, assume a simplified concatenation-based Gödel numbering where formulas are encoded directly by replacing symbols with specific digits in base-10. Let $\varphi(a)$ express the fact that the number a ends with the digit 6.

1. Write down a formula in the language of PA expressing this property $\varphi(a)$.
2. Suppose the numbering is arranged such that the closing parenthesis ')' has the code 6. Find a *false* fixed point for the formula $\varphi(a)$.
3. Under the same conditions, find a *true* fixed point for the formula $\varphi(a)$.

D3.26 Assume a concatenation-based coding where specific symbols have the following exact digit codes: $\ulcorner 0\urcorner \overset{\text{def}}{\rightleftharpoons} 1, \ulcorner S\urcorner \overset{\text{def}}{\rightleftharpoons} 2, \ulcorner +\urcorner \overset{\text{def}}{\rightleftharpoons} 3, \ulcorner =\urcorner \overset{\text{def}}{\rightleftharpoons} 4$ (the codes of other symbols are irrelevant for this task).

1. Write down a formula $\varphi(a)$ in PA expressing "a is not divisible by 3".
2. Find a fixed point for the formula $\varphi(a)$. (*Hint:* look for the simplest atomic formulas like $0 = 0$ or $0 + 0 = 0$, compute their base-10 code, and check if it is divisible by 3).

Recommended Reading

Arithmetic is not only the first serious mathematical theory with which we become acquainted, but also the arena in which the greatest dramas of the foundations of mathematics of the 20th century were played out. The literature on this topic is vast, but one can single out several key areas directly related to the material of this chapter.

The central plot, of course, is Gödel's incompleteness theorems. Although their proof can be found in almost any logic textbook, no exposition compares in depth and completeness to the corresponding chapters of Kleene's "Introduction to Metamathematics" [41], which we have already mentioned. It is there that the concepts of recursive functions are introduced with the utmost thoroughness and the entire procedure of Gödel numbering is carried out, the syntactic foundations of which were laid in this chapter through sequence coding. An alternative, more modern exposition, closely linking logic and the theory of algorithms, can be found in the excellent monograph by N. K. Vereshchagin and A. Shen [62]. For a distinctively clear and pedagogical guide specifically focused on these theorems, Peter Smith's "An Introduction to Gödel's Theorems" [55] is highly recommended.

After realizing the fact of incompleteness, the natural question arises: which meaningful statements is arithmetic unable to prove? The answer to this question opened up a whole area of research. The article by Kirby and Paris [40] became a landmark, presenting the first "natural" combinatorial statement (a variation of Ramsey's theorem) independent of Peano's axioms. For those who wish to professionally understand the zoo of models of arithmetic and the subtleties of independence, the specialized work by Richard Kaye, "Models of Peano Arithmetic" [39], should be a go-to book. It details the non-standard models, the existence of which is a consequence of the compactness and Löwenheim-Skolem theorems, and which provide a semantic explanation for the phenomenon of incompleteness.

Recommended Reading

Chapter D4
Set Theory

The essence of mathematics lies in its freedom.

— Georg Cantor

Recall that first-order set theory ZF is formulated in the language of first-order predicate logic with the signature (in addition to logical connectives and quantifiers) $\Sigma_{\mathsf{ZF}} \stackrel{\text{def}}{\rightleftharpoons} \langle =, \in \rangle$. Latin letters (possibly with indices and/or primes) are used as variables, and Greek letters can also be used in quantifiers restricted to classes.

Let us provide the list of axioms for ZF, formulated in the original language, i.e., without applying abbreviations:

ZF1 Congruence: $(a = b) \to ((a \in M) \leftrightarrow (b \in M))$.

ZF2 Extensionality: $(A = B) \leftrightarrow (\forall x\ (x \in A) \leftrightarrow (x \in B))$.

ZF3 Power Set: $\exists P\ \forall x\ (x \in P) \leftrightarrow \forall y (y \in x \to y \in A)$.

ZF4 Union: $\exists U\ \forall x\ (x \in U) \leftrightarrow (\exists a \in A :\ x \in a)$.

ZF5 Regularity: $(\exists x \in A) \to \exists X \in A :\ \forall y \in X :\ (y \notin A)$.

ZF6 Replacement $[\varphi]$:

$$(\forall x \in A : \exists! y : \varphi(x, y, \vec{p})) \to (\exists Y\ \forall y\ (y \in Y) \leftrightarrow \exists x \in A :\ \varphi(x, y, \vec{p})).$$

ZF7 Infinity:

$$\exists X\ (\exists z \in X : \forall y (y \notin z)) \wedge \\ (\forall x \in X : \exists y \in X :\ \forall z (z \in y \leftrightarrow z \in x \vee z = x)),$$

N. I. Kazimirov, *Mathematics as a Foreign Language*, Mathematics in Mind,
https://doi.org/10.1007/978-3-032-30721-7_9

as well as the axioms and inference rules of predicate logic:

PC1 the universal closures of all substitutions of all formulas of the language ZF into all tautologies of propositional logic;

PC2 axioms for quantifiers:

$$(\forall) \quad (\forall x\, \varphi[a/x]) \rightarrow \varphi[a/T]; \qquad (\exists) \quad \varphi[a/T] \rightarrow \exists x\, \varphi[a/x],$$

where φ does not contain the bound variable x, and T is any term of the theory;

PC3 rules of inference:

$$(MP) \quad \frac{\varphi, \quad \varphi \rightarrow \psi}{\psi}; \qquad (R1) \quad \frac{\varphi \rightarrow \psi}{\varphi \rightarrow \forall x\, \psi[a/x]};$$

$$(R1') \quad \frac{\psi}{\forall x\, \psi[a/x]}; \qquad (R2) \quad \frac{\psi \rightarrow \varphi}{(\exists x\, \psi[a/x]) \rightarrow \varphi},$$

where for rules (R1), (R1'), and (R2), the variable a does not occur in φ nor in any of the active (undischarged) hypotheses upon which the derivation of the formula in the premise of these rules depends. The rules $(R1)$ and $(R2)$ are called **Bernays' rules**. Rule (R1'), called the *generalization rule*, is a derived rule in our inference system, as it can be obtained from rule (R1) by substituting a tautology for the formula φ. It is listed here for convenience of reference.

D4.1 Basic Properties

The first thing to do, while our language is as minimal as possible, is to prove the congruence of equality with respect to terms and formulas, i.e., to derive the substitution laws (B1.9).

First, we need to obtain the standard properties of equality: reflexivity, symmetry, and transitivity. These properties follow directly from the Axiom of Extensionality, for example,

$$(A = B) \leftrightarrow [\forall x (x \in A) \leftrightarrow (x \in B)] \leftrightarrow [\forall x (x \in B) \leftrightarrow (x \in A)] \leftrightarrow (B = A),$$

which is true by the properties of symmetry of equivalence.

Second, note that in the pure language of set theory, terms are precisely the variables (due to the absence of constants and function symbols). Therefore, checking the congruence of equality with respect to terms is trivial:

$$(T_1 = T_2) \to (T_1 = T_2), \text{ or } (T_1 = T_2) \to (b = b)$$

for any terms T_1 and T_2, since substituting a term T into a term a instead of a yields the term T, while substituting a term T into a term b instead of a changes nothing.

Third, the proof of congruence for formulas is carried out by induction on the height of the formula, which is defined recursively (in a meta-theory, for example, in PA arithmetic, in which the language of ZFC is encoded using sequence coding):

$$H[a = b] = 0; \quad H[a \in b] = 0; \quad H[\neg\varphi] = 1 + H[\varphi];$$
$$H[\varphi \wedge \psi] = 1 + \max(H[\varphi], H[\psi]); \quad H[\exists x\ \varphi[a/x]] = 1 + H[\varphi],$$

where we also consider the language of logic to be as minimal as possible, assuming that $\vee, \to, \leftrightarrow, \forall$ are expressed through the basic connectives $\neg$, $\wedge$ and $\exists$.

The proof schema then completely corresponds to the schema for proving congruence in PA, which was discussed in Theorem D3.2.

Let the height of the formula be zero, i.e., it is an atomic formula. We have only two atomic formulas:

1. $(a = b)$: here we need to check that

$$(T_1 = T_2) \to ((T_1 = b) \leftrightarrow (T_2 = b)),$$

which is true by the properties of equality;

2. $(a \in b)$: here we need to check two formulas:

$$(T_1 = T_2) \to ((T_1 \in b) \leftrightarrow (T_2 \in b)),$$
$$(T_1 = T_2) \to ((a \in T_1) \leftrightarrow (a \in T_2)),$$

which are obtained by the corresponding substitutions into the axioms of congruence and extensionality.[1]

[1] It is worth noting here that the Axiom of Extensionality should be split into two parts, represented as a conjunction of two implications. The first implication, $(A = B) \to (\forall x\ (x \in A) \leftrightarrow (x \in B))$, would then be precisely the missing axiom of congruence, preserving the substitution of equals into the second argument of the membership predicate.

Thus, for atomic formulas, congruence follows directly from the axioms. Let us introduce the induction hypothesis that for formulas of height no greater than n, congruence holds. Let the formula φ have a height of no more than n. Then by the induction hypothesis

$$(T_1 = T_2) \to (\varphi[a/T_1] \leftrightarrow \varphi[a/T_2]),$$

and the consequent of this implication is equivalent to $\neg\varphi[a/T_1] \leftrightarrow \neg\varphi[a/T_2]$ by virtue of predicate calculus, from which congruence for $\neg\varphi$ follows. The case of conjunction is considered similarly. The case of the existential quantifier follows from Lemma D2.8. This completes the induction.

Thus, in the simplest language, congruence is obtained.

Now we enrich the language with the standard symbols for operations, relations, and constants, which were defined and discussed in detail in Section C1.1. When extending the language, it is necessary to prove the extension of the congruence of equality to these new symbols.

Suppose, for example, that we have the definition for the intersection of sets

$$\forall x\ (x \in A \cap B) \overset{\text{def}}{\leftrightarrow} (x \in A \land x \in B). \qquad \text{(D4.1)}$$

It is then necessary for the following properties to hold:

$$(T_1 = T_2) \to (T_1 \in A \cap B \leftrightarrow T_2 \in A \cap B),$$
$$(T_1 = T_2) \to (A \cap B \in T_1 \leftrightarrow A \cap B \in T_2),$$
$$(T_1 = T_2) \to (a \in T_1 \cap B \leftrightarrow a \in T_2 \cap B),$$
$$(T_1 = T_2) \to (a \in A \cap T_1 \leftrightarrow a \in A \cap T_2).$$

The first formula is obtained by substituting T_1 and T_2 for x in the definition (D4.1) by the properties of equality. The second formula follows from Theorem B2.6, and the third and fourth are obtained at the inductive step when we prove the congruence of equality with respect to terms by induction on their complexity (since the term $T_1 \cap B$ is more complex than the term T_1).

By proceeding in this way for all newly introduced function and predicate symbols, we can show the congruence of equality with respect to all formulas of the extended language.

However, as was already noted when considering arithmetic, we can always consider formulas like (D4.1) as syntactic sugar or a meta-theoretic abbreviation for longer formulas of the original language. In that case, the check of congruence automatically reduces to the already proven congruence in the initial language.[2]

Theorem D4.1 (Basic properties of operations). *Let $A, B, C \in \mathcal{P}(U)$, where U is an arbitrary non-empty set. Then the following properties hold.*

1° *Commutativity:* $A \cup B = B \cup A$ *and* $A \cap B = B \cap A$.

2° *Associativity:* $(A \cup B) \cup C = A \cup (B \cup C)$ *and* $(A \cap B) \cap C = A \cap (B \cap C)$.

3° *Distributivity:* $A \cap (B \cup C) = (A \cap B) \cup (A \cap C)$ *and* $A \cup (B \cap C) = (A \cup B) \cap (A \cup C)$.

4° *Identity elements:* $A \cup \emptyset = A$ *and* $A \cap U = A$.

5° *Law of excluded middle:* $A \cup (U \setminus A) = U$.

6° *Law of non-contradiction:* $A \cap (U \setminus A) = \emptyset$.

7° *Idempotence:* $A \cup A = A$ *and* $A \cap A = A$.

8° *Absorption laws:* $A \cup (A \cap B) = A$ *and* $A \cap (A \cup B) = A$.

9° *De Morgan's laws:* $U \setminus (A \cup B) = (U \setminus A) \cap (U \setminus B)$ *and* $U \setminus (A \cap B) = (U \setminus A) \cup (U \setminus B)$.

10° *Set difference:* $A \setminus B = A \cap (U \setminus B)$.

11° *Symmetric difference:* $A \triangle B = (A \setminus B) \cup (B \setminus A) = (A \cap (U \setminus B)) \cup (B \cap (U \setminus A))$.

12° $\langle \mathcal{P}(U), \triangle \rangle$ *is an abelian group, i.e.,*

- 12.1° *associativity:* $(A \triangle B) \triangle C = A \triangle (B \triangle C)$;
- 12.2° *commutativity:* $A \triangle B = B \triangle A$;
- 12.3° *identity element:* $A \triangle \emptyset = A$;
- 12.4° *inverse element:* $A \triangle A = \emptyset$;

13° *Distributivity:* $A \cap (B \triangle C) = (A \cap B) \triangle (A \cap C)$.

All these properties are, in fact, direct consequences of logical operations, and we discussed this connection in detail in Section A2.4.

---

[2] Furthermore, as we have already mentioned, congruence for all terms and formulas is often assumed to be given axiomatically and does not require verification. But since we have included the elementary axiom of congruence here, and also to foster a deeper confidence in the internal consistency of our formalism, at least a brief proof of its extrapolation to the entire language turns out to be an extremely useful step.

To use the complement of a set to a universe, which corresponds to logical negation, without any problems, we have considered all operations on a universe U. In this case, the result of the operation is always a set, and moreover, it is contained in the same U. In principle, we could have taken the class of all sets, defined by the formula $x = x$, as U, and all these properties would have remained true, given that the class symbol here only appears on the right side of the relation $\in$. But such assumptions are not required, since in each specific case, one can find a sufficiently large set U within which everything happens. Therefore, for the sake of brevity, we can forget about the set U altogether and write $\setminus A$ instead of $U \setminus A$, calling this set the complement of A. Furthermore, it is convenient to write the intersection of sets as multiplication, i.e., omitting the operation $\cap$ altogether. Given these abbreviations, the notation $A \setminus B$ can be understood as $A \cap (U \setminus B)$, and this equality, as is easy to see, is indeed true provided that $A, B \in \mathcal{P}(U)$.

Let's look at the proof of the properties of the given theorem using the example of proving the associativity of the symmetric difference operation:

$$\begin{aligned}
[x \in (A \triangle B) \triangle C] &\leftrightarrow [x \in (A \triangle B) \setminus C] \vee [x \in C \setminus (A \triangle B)] \\
&\leftrightarrow [((x \in A \setminus B) \vee (x \in B \setminus A)) \wedge (x \notin C)] \vee \\
&\qquad [(x \in C) \wedge \neg((x \in A \setminus B) \vee (x \in B \setminus A))] \\
&\leftrightarrow [(\mathcal{A}\neg\mathcal{B} \vee \mathcal{B}\neg\mathcal{A})\neg\mathcal{C}] \vee [\mathcal{C}\neg(\mathcal{A}\neg\mathcal{B} \vee \mathcal{B}\neg\mathcal{A})] \\
&\equiv [(\mathcal{A}\neg\mathcal{B} \vee \mathcal{B}\neg\mathcal{A})\neg\mathcal{C}] \vee [\mathcal{C}(\neg\mathcal{A} \vee \mathcal{B})(\neg\mathcal{B} \vee \mathcal{A})] \\
&\equiv \mathcal{A}\neg\mathcal{B}\neg\mathcal{C} \vee \neg\mathcal{A}\mathcal{B}\neg\mathcal{C} \vee \neg\mathcal{A}\neg\mathcal{B}\mathcal{C} \vee \mathcal{A}\mathcal{B}\mathcal{C},
\end{aligned}$$

where we have used the notations $(x \in A) \overset{\text{def}}{\leftrightarrow} \mathcal{A}$, $(x \in B) \overset{\text{def}}{\leftrightarrow} \mathcal{B}$, $(x \in C) \overset{\text{def}}{\leftrightarrow} \mathcal{C}$, and switched to calculations in propositional logic, also writing conjunction as multiplication.

On the other hand,

$$\begin{gathered}
[x \in A \triangle (B \triangle C)] \leftrightarrow [x \in A \setminus (B \triangle C)] \vee [x \in (B \triangle C) \setminus A] \leftrightarrow \\
[\mathcal{A}\neg(\mathcal{B}\neg\mathcal{C} \vee \mathcal{C}\neg\mathcal{B})] \vee [\neg\mathcal{A}(\mathcal{B}\neg\mathcal{C} \vee \mathcal{C}\neg\mathcal{B})] \equiv \\
[\mathcal{A}(\neg\mathcal{B} \vee \mathcal{C})(\mathcal{B} \vee \neg\mathcal{C})] \vee [\neg\mathcal{A}\mathcal{B}\neg\mathcal{C} \vee \neg\mathcal{A}\neg\mathcal{B}\mathcal{C}] \equiv \\
\mathcal{A}\neg\mathcal{B}\neg\mathcal{C} \vee \mathcal{A}\mathcal{B}\mathcal{C} \vee \neg\mathcal{A}\mathcal{B}\neg\mathcal{C} \vee \neg\mathcal{A}\neg\mathcal{B}\mathcal{C}.
\end{gathered}$$

As we can see, in both cases, the same formula is obtained up to the commutativity of disjunction. Thus, property 12.1 is proven.

With these re-designations, most of the listed properties are exact copies of the tautologies that were listed in Section D2.1. And in fact, most complex identities with sets are much easier to compute in pure propositional logic.

Here we must introduce the necessary definitions from universal algebra. Let a structure $\langle M; 0, 1, \neg, \vee, \wedge \rangle$ be given, in which the operations and constants are connected by the following axioms:

BA1 associativity of $\vee, \wedge$;

BA2 commutativity of $\vee, \wedge$;

BA3 mutual distributivity of $\vee, \wedge$;

BA4 idempotence of $\vee, \wedge$;

BA5 absorption laws for $\vee, \wedge$;

BA6 rules for constants: $a \wedge 0 = 0$, $a \wedge 1 = a$, $a \vee 0 = a$, $a \vee 1 = 1$;

BA7 $a \vee \neg a = 1$, $a \wedge \neg a = 0$.

Such a structure is called a **Boolean algebra**.

As we can see, the powerset $M = \mathcal{P}(U)$ satisfies this definition if the constant 0 is the empty set, 1 is the set M, $\neg$ is the complement, $\wedge$ is the intersection, and $\vee$ is the union. In this connection, there is the designation *algebra of sets*, which refers to a Boolean algebra constructed in the specified way on the powerset of some non-empty set.

An inequality predicate is also usually introduced on a Boolean algebra, defined by the rule

$$a \leqslant b \overset{\text{def}}{\leftrightarrow} a \wedge b = a, \tag{D4.2}$$

which in an algebra of sets corresponds to the predicate $A \subseteq B$.

A Boolean algebra is generalized by a broader concept—a *bounded lattice*. We say that a structure $\langle M; \leqslant \rangle$ is called a **lattice** if

L1 the relation $\leqslant$ is a non-strict partial order;

L2 for any $a, b \in M$, there exist $\sup\{a, b\}$ and $\inf\{a, b\}$,

where $\sup\{a, b\} = \min\{x \mid a \leqslant x \wedge b \leqslant x\}$, $\inf\{a, b\} = \max\{x \mid x \leqslant a \wedge x \leqslant b\}$. A lattice is called *bounded* if there exist elements in it, denoted by $\top, \bot$ (or $1, 0$), such that $\top = \max M$, $\bot = \min M$.

It is not difficult to show that if an order relation is introduced on a Boolean algebra by rule (D4.2), then it is a partial order and

$$a \wedge b = \inf\{a, b\}, \quad a \vee b = \sup\{a, b\}, \tag{D4.3}$$

and furthermore, $1 = \max M$, $0 = \min M$ with respect to this order, so that every Boolean algebra is a bounded lattice.

Moreover, if we have an arbitrary lattice, we can define the operations $\vee, \wedge$ by rule (D4.3), and it will turn out that they have properties BA1, BA2, BA4, BA5. If the laws of distributivity also hold for these operations, then such a lattice is called *distributive*. If a distributive lattice is a bounded lattice, then its constants satisfy property BA6. In other words, a bounded distributive lattice is almost a Boolean algebra.

For a bounded distributive lattice to become a Boolean algebra, one must define the negation operation and add conditions BA7 for it. Negation is most often introduced through another operation, which is denoted by $\to$ (like implication) and must satisfy the rule

$$a \leqslant (b \to c) \overset{\text{def}}{\leftrightarrow} (a \wedge b) \leqslant c. \tag{D4.4}$$

The operation $\to$ is called a *relative pseudocomplement* or *implication*, and a bounded lattice with such an operation is called a **Heyting algebra**. Note that if we perceive the operation $\to$ as a logical implication, and the relation $\leqslant$ as logical derivability, then condition (D4.4) can be read as the Deduction Theorem: $\mathcal{A} \vdash (\mathcal{B} \to C) \Leftrightarrow (\mathcal{A} \wedge \mathcal{B}) \vdash C$. With this interpretation, one can say that Heyting algebras are bounded lattices with a "deduction theorem".

It is interesting to note that a Heyting algebra is necessarily distributive. This fact can be proven "head-on", or one can use the fact that Heyting algebras are precisely the models of intuitionistic propositional logic,[3] from which distributivity follows due to its presence in propositional logic.

If a relative pseudocomplement exists for a given bounded lattice, then it is not difficult to define in it a unary operation of *absolute pseudocomplement* (negation), by setting $\neg a \overset{\text{def}}{\rightleftharpoons} a \to 0$, or, forgetting about $\to$, negation is introduced by the axiom $(a \leqslant \neg b) \overset{\text{def}}{\leftrightarrow} (a \wedge b = 0)$. The absolute pseudocomplement automatically satisfies the law of non-contradiction[4] $a \wedge \neg a = 0$, but is not required to satisfy the law of the excluded middle $a \vee \neg a = 1$ (which is equivalent to the identity $\neg\neg a = a$).

---

[3] In intuitionistic propositional logic, the list of axioms consists of axioms A1–A8 (i.e., without the law of the excluded middle), given in Section D2.1, and the rules of inference are exactly the same: substitution and Modus Ponens.

[4] Indeed, from the definition of $\neg$ and the commutativity of $\wedge$, it follows that $(a \leqslant \neg b) \leftrightarrow (b \leqslant \neg a)$, from which, by substituting $\neg a$ for b, we get $(a \leqslant \neg\neg a) \leftrightarrow (\neg a \leqslant \neg a)$, i.e., by virtue of the reflexivity of the inequality, we always have $a \leqslant \neg\neg a$, and this is equivalent to $a \wedge \neg a = 0$ by the definition of $\neg$.

Finally, if the law of the excluded middle holds in a Heyting algebra, that is, condition BA7 is fully satisfied, then such an algebra is a Boolean algebra.

Let us note the main property of a Boolean algebra, called the **principle of duality**: if in any true identity, written in the signature of a Boolean algebra, we replace 0 with 1 and vice versa, and also $\wedge$ with $\vee$ and vice versa, then the resulting identity will also be true. This property follows directly from the axioms of a Boolean algebra, which are invariant with respect to such a replacement. Therefore, the pairs of symbols 0 and 1, $\wedge$ and $\vee$ are called dual. The negation $\neg$ is called a self-dual operation. Note also that if an order relation $\leqslant$ is introduced on a Boolean algebra by formula (D4.2), then its dual relation will be the inverse relation, i.e., $\geqslant$.

Despite its name, a Boolean algebra is not an algebra in the sense of the definition of an algebra over a ring. Indeed, it is precisely because of the duality property of the operations that a Boolean algebra cannot be a ring, since in a ring, the operations of addition and multiplication are not dual to each other.

Nevertheless, there is a way to find in a Boolean algebra such operations that will already satisfy the axioms of a ring. Namely, let us define the operations of addition and multiplication by the formulas

$$a + b \overset{\text{def}}{\rightleftharpoons} (a \wedge \neg b) \vee (\neg a \wedge b); \quad a \cdot b \overset{\text{def}}{\rightleftharpoons} a \wedge b, \tag{D4.5}$$

i.e., addition is defined exactly as the symmetric difference is defined in an algebra of sets.

Then the structure $\langle M, 0, 1, +, \cdot \rangle$ (with the same constants) will be a ring. Moreover, it will be a commutative ring with unity, in which multiplication is idempotent, i.e., $a \cdot a = a$ for all elements of the ring.

Rings with unity in which multiplication is idempotent are called **Boolean rings**. The idempotence of multiplication provides the ring with two remarkable properties at once, namely: **a)** characteristic 2, i.e., $a+a = 0$ for all $a \in M$; **b)** commutativity of multiplication, i.e., $ab = ba$ for all $a, b \in M$.

Indeed, $a+a = (a+a)(a+a) = a \cdot a+a \cdot a+a \cdot a+a \cdot a = (a+a)+(a+a)$, from which it follows that $a + a = 0$, or $a = -a$ (each element is its own additive inverse). Furthermore, $a + b = (a + b)(a + b) = a \cdot a + a \cdot b + b \cdot a + b \cdot b = (a + b) + (ab + ba)$, from which $ab + ba = 0$, and since $ba = -(ba)$, then $ab = ba$.

The set $\mathcal{P}(U)$ with the operations $\triangle, \cap$ is naturally an example of a Boolean ring by virtue of Theorem D4.1.

A Boolean ring and a Boolean algebra are connected in a two-way manner (one might say they are “dual” to each other) in the sense that just as a Boolean ring is obtained from a Boolean algebra without any additional conditions, so too is a Boolean algebra obtained from a Boolean ring without additional conditions. Indeed, let a Boolean ring $\langle M, 0, 1, +, \cdot \rangle$ be given. Then the operations $\wedge, \neg, \vee$ of a Boolean algebra are defined as follows:

$$a \wedge b \overset{\text{def}}{\rightleftharpoons} a \cdot b, \quad \neg a \overset{\text{def}}{\rightleftharpoons} 1 + a, \quad a \vee b \overset{\text{def}}{\rightleftharpoons} a + b + a \cdot b.$$

Here is how, for example, the law of the excluded middle is checked for these operations: $a \vee \neg a = a + \neg a + a \cdot \neg a = a + (1 + a) + a \cdot (1 + a) = 1$, since $a \cdot (1 + a) = a + a \cdot a = a + a = 0$.

D4.2 Finite and Infinite Sets

Having studied the algebra of sets, i.e., set-theoretic operations, we now move directly to studying the properties of sets themselves.

In Section C1.1, we examined in detail how the Axiom of Infinity introduces inductive sets into set theory, and in particular, the smallest inductive set, which is denoted by ω. We have also repeatedly used the notion of the *cardinality of a set* without giving it a concrete definition. In this section, we will delve into this and related concepts in detail and bring order to the terminology used previously.

D4.2.1 Comparison of Cardinalities

First, recall that two sets A and B are called **equinumerous** (notation: $A \simeq B$) if there exists a bijection between them. For each set A, there is a proper class of sets equinumerous to it, which we denote by $\mathrm{Card}(A)$. The difficulty with this concept is that we are introducing into use an entire proper class of sets that is not an object of the theory ZF. Consequently, we can describe exactly as many cardinalities as we can write formulas for them. Moreover, comparing and operating on proper classes in a theory without classes becomes quite inconvenient when we want to consider the arithmetic and topology of cardinalities, i.e., to study them by algebraic methods.

Therefore, by the cardinality of a set, we will understand specially chosen reference sets for this purpose, whose existence, generally speaking, is based on the axiom of choice, but for a certain class of sets, we can do without it, using the set ω.

Second, let us give the following definitions. A set is called **finite** if it is equinumerous with some natural number $n \in \omega$. For example, a set of three elements is equinumerous with the natural number $3 = \{0, 1, 2\}$. If $A \simeq n$, we write $|A| = n$, and the number n is called the **cardinality** of the set A. **Infinite** sets are all the others.

Among infinite sets, countable sets hold a special place, as there is a standard measure for them—the set ω. A set A is said to be **countable** if $A \simeq \omega$. In this case, ω is called the cardinality of A, which is written as $|A| = \omega$. This definition is a natural extension of the definition of a finite set.

Finite and countable sets together are called *at most countable*, and all others are *uncountable*.[5]

How many uncountable sets of different cardinalities are there? One of the fundamental theorems of set theory partially helps to answer this question.

Theorem D4.2 (Cantor). *No set is equinumerous with its powerset:*

$$A \not\simeq \mathcal{P}(A).$$

Proof. Suppose that $A \simeq \mathcal{P}(A)$, i.e., there exists a bijection $f : A \leftrightarrow \mathcal{P}(A)$. Now, since $f(a)$ is a subset of A, we are entitled to expect that for some a, it may happen that $a \in f(a)$, and for others, not. Using the Axiom of Separation, we can define the set

$$X = \{a \in A \mid a \notin f(a)\}.$$

Since $X \in \mathcal{P}(A)$ and the function f is a surjection, there exists an $x \in A$ such that $f(x) = X$. The question is: how are x and X related? Could it be that $x \in X$, since $x \in A$ and $X \subseteq A$?

Suppose that $x \in X$. Then x satisfies the formula defining X, i.e., $x \notin f(x)$. But $f(x) = X$. We get that $x \notin X$. On the other hand, if $x \notin X$, then x satisfies the negation of the formula defining X, i.e., $x \in f(x)$, which means $x \in X$.

We have obtained a situation analogous to Russell's paradox: $(x \in X) \leftrightarrow (x \notin X)$, i.e., we have derived a contradiction. Fortunately, in this case, we have a premise that we have thereby refuted, namely the assumption $A \simeq \mathcal{P}(A)$.

Thus, $A \not\simeq \mathcal{P}(A)$. □

[5] Sometimes in textbooks, one can find the designation "countable" used for both countably infinite and finite sets.

Note that although the sets A and $\mathcal{P}(A)$ are not equinumerous, we can judge the relationship between their cardinalities. Let's say that a set A *has cardinality less than or equal to* B (notation: $A \preccurlyeq B$) if there is a subset in B equinumerous with A, i.e., if there exists an injection from A to B. Note that the relation $\preccurlyeq$ is a preorder on the class of all sets (it is reflexive and transitive). One can also say that a set A has a strictly smaller cardinality than B if $(A \preccurlyeq B) \wedge (A \not\simeq B)$ (notation: $A \prec B$), moving to a strict preorder.

In the case of A and $\mathcal{P}(A)$, the inequality $A \preccurlyeq \mathcal{P}(A)$ is obvious, since there is a natural injection (embedding) $f : A \to \mathcal{P}(A)$, defined by the rule $f(a) = \{a\}$ (precisely the case where $a \in f(a)$ for all a). And considering Cantor's theorem, we note that every set A has a strictly smaller cardinality than $\mathcal{P}(A)$, i.e., $A \prec \mathcal{P}(A)$. The following theorem establishes the precise relationship between $\preccurlyeq$ and $\simeq$.

Lemma D4.1 (Knaster–Tarski theorem). *Let $\langle L, \leqslant\rangle$ be a complete lattice (a lattice in which every non-empty subset has a* sup *and an* inf*) and let the function $f : L \to L$ be monotone, i.e., $f(x) \leqslant f(y)$ whenever $x \leqslant y$ for all $x, y \in L$. Then the set of fixed points ($f(x) = x$) of the function f is non-empty and its supremum and infimum are fixed points.*

Proof. Consider the subset $S = \{x \in L \mid x \leqslant f(x)\}$. It is non-empty, since $\min L \in S$. Let $x_0 = \sup S$ (its existence follows from the completeness of the lattice). We need to show that $x_0 \in S$.

For any $s \in S$, firstly, $s \leqslant f(s)$, and secondly, since $s \leqslant x_0$, then $f(s) \leqslant f(x_0)$ due to the monotonicity of f, therefore, $s \leqslant f(x_0)$. This means that $f(x_0)$ is an upper bound of S. Then by the definition of supremum, $x_0 \leqslant f(x_0)$, from which it follows that $x_0 \in S$, i.e., $x_0 = \max S$.

If we assume that $x_0 < f(x_0)$, then $f(x_0) \leqslant f(f(x_0))$, but then $f(x_0) \in S$ in contradiction to the fact that $x_0 = \max S$. Consequently, $x_0 = f(x_0)$, i.e., it is a fixed point of f.

Moreover, the point x_0 is the greatest fixed point. Indeed, all fixed points are in S, and $x_0 = \max S$.

Similarly, by considering the set $S' = \{x \in L \mid x \geqslant f(x)\}$, essentially just reversing the relation in all calculations, replacing sup with inf and max with min, it is proven that $x_1 = \min S'$ exists and is the least fixed point of the function f. □

Theorem D4.3 (Cantor–Bernstein–Schröder). *If $A \preccurlyeq B$ and $B \preccurlyeq A$, then $A \simeq B$.*

Proof. By hypothesis, there exist injections $f : A \longrightarrow B$ and $g : B \longrightarrow A$, and we need to produce a bijection $h : A \leftrightarrow B$. Let's define the following function on the powerset $\mathcal{P}(A)$:

$$H(X) = A \setminus g[B \setminus f[X]], \quad X \in \mathcal{P}(A),$$

where the square brackets indicate taking the *image* of the set with respect to the function.

Suppose that the equation $H(X) = X$ has a solution, i.e., such a set X exists. Then f and g^{-1} act on X and $A \setminus X$ "in parallel," mapping them, respectively, to $f[X]$ and $B \setminus f[X]$, i.e., the function

$$h(a) = \begin{cases} f(a), & a \in X, \\ g^{-1}(a), & a \in A \setminus X, \end{cases}$$

is the desired bijection.

How to find such an X? First, let's note the monotonicity property of the function H: if $X \subseteq Y$, then $H(X) \subseteq H(Y)$, which follows easily from the definition of H. Thus, H is a monotone function on the lattice of sets $\mathcal{P}(A)$.

Furthermore, the powerset is a *complete lattice* ($\sup X = \bigcup X$, $\inf X = \bigcap X$). By the Knaster–Tarski theorem, if a monotone function is defined on a complete lattice, then the set of fixed points of this function is non-empty and, moreover, its infimum and supremum are fixed points. So for our case, the existence of a root of the equation $H(X) = X$ is a direct consequence of this theorem. □

The Cantor–Bernstein–Schroeder theorem allows for relatively easy establishment of certain equinumerosities. For example, it is easy to construct injections from $\mathbb{N}$ to $\mathbb{Z}$ and in the reverse direction. Furthermore, the theorem can be generalized as follows: If a cycle $A_1 \preccurlyeq A_2 \preccurlyeq \ldots \preccurlyeq A_n \preccurlyeq A_1$ exists, then all sets $A_1, \ldots, A_n$ are pairwise equinumerous. For example, we can show that $\mathbb{Z} \preccurlyeq \mathbb{Q} \preccurlyeq \mathbb{Z}\times\mathbb{Z} \preccurlyeq \mathbb{Z}$, from which it follows that all three sets are equinumerous.

The equinumerosity of the interval $[0; 1]$ and the square $[0; 1]^2$ is proven similarly. The injection from the interval to the square is trivial, and to construct an injection from the square to the interval, one can use the "interleaving method" of decimal expansions: if $(a, b) \in [0; 1]^2$, then

$a = 0.a_0a_1a_2\ldots$, $b = 0.b_0b_1b_2\ldots$ (where a_i, b_i are decimal digits), then from these two numbers, the number $c = 0.a_0b_0a_1b_1a_2b_2\ldots$, which belongs to $[0; 1]$, is constructed.

By iterating the powerset operation an arbitrary number of times, we can come up with a whole countable sequence of infinite sets with pairwise distinct cardinalities:

$$\omega \prec \mathcal{P}(\omega) \prec \mathcal{P}(\mathcal{P}(\omega)) \prec \ldots \prec \mathcal{P}^n(\omega) \prec \ldots$$

It is clear that all sets here, except for the first, are uncountable and not equinumerous.

Note one important point: as much as we would like to write that a set A is infinite if and only if $\omega \preccurlyeq A$, we cannot assert this in pure ZF theory, because we cannot prove in it that if A is infinite (not equinumerous with any $n \in \omega$), then there exists an injection from ω to A. We will have this ability only by involving the countable axiom of choice AC_ω (see below).

By the way, the set $\mathcal{P}(\omega)$ is equinumerous with the real interval $[0; 1]$, so the cardinality of $\mathcal{P}(\omega)$ is called the **continuum**. A special representative of the class $\mathrm{Card}(\mathcal{P}(\omega))$, denoted by $\mathfrak{c}$, which will henceforth be established as the concept of the cardinality of the continuum, is also called the continuum.

Georg Cantor unsuccessfully tried to find a set of intermediate cardinality—between countable and continual, as a result of which he constructed a set of beautiful examples of very "thin" subsets of $[0; 1]$, but eventually formulated his famous **Continuum Hypothesis** CH: $\neg\exists A : \omega \prec A \prec \mathfrak{c}$.

The Generalized Continuum Hypothesis GCH: if X is infinite, then $\neg\exists A : X \prec A \prec \mathcal{P}(X)$ (in the case of finite sets, as is easy to see, this is not true).

As we now know, CH and GCH are independent of the axioms of ZF. Moreover, CH is independent of ZFC, and from GCH follows AC, with GCH being strictly stronger than AC.

D4.2.2 Finite Sets

By reducing the assessment of the cardinality of finite and countable sets to specific objects—the set ω and its elements—we have thereby constructed a scale for them within the theory ZF. To measure uncountable cardinalities, we will be able to extend this scale beyond ω only with the help of a stronger axiom than the axiom of infinity, namely, with the help of the axiom of choice.

But first, we need to deal with the structure of the set ω and some properties of finite sets.

Theorem D4.4. *The set ω is (strictly) well-ordered by the relation $\in$.*

Proof. First, let's recall the notation $S(a) \overset{\text{def}}{\rightleftharpoons} a \cup \{a\}$ for the successor of a set a.

1) Let us prove that the principle of arithmetic induction in the form of the language of second-order arithmetic works on the set ω:

$$\forall A\ ((\emptyset \in A) \wedge (\forall x \in A : S(x) \in A) \to \omega \subseteq A). \tag{D4.6}$$

Indeed, every set A that satisfies the antecedent of this implication is inductive, and therefore contains ω as the minimal inductive set.

From this also follows the principle of induction with an arbitrary formula φ of the language **ZF**:

$$[\varphi(\emptyset) \wedge \forall x\ (\varphi(x) \to \varphi(S(x)))] \to \forall x \in \omega : \varphi(x).$$

This is easily derived from the previous statement by setting $A = \{x \mid (x \in \omega) \wedge \varphi(x)\}$.

2) Let us first prove several useful lemmas. Recall the definition: a set a is called *transitive* if $\forall x\ (x \in a) \to (x \subseteq a)$.

Lemma D4.2 (on transitivity). The set ω and all its elements are transitive.

Proof. To prove the transitivity of ω, we need to show that $\forall n\ (n \in \omega) \to (n \subseteq \omega)$. Consider the set $A = \{x \in \omega \mid x \subseteq \omega\}$.[6]

Base case of induction ($\emptyset \in A$): this is obvious, since $\emptyset \in \omega$ (due to the inductiveness of ω) and $\emptyset \subseteq \omega$.

Inductive step: let us introduce the hypothesis $(x \in \omega) \wedge (x \subseteq \omega)$. Then $S(x) \in \omega$ (due to the inductiveness of ω) and $S(x) = x \cup \{x\} \subseteq \omega$, since $x \subseteq \omega$ by hypothesis, and $\{x\} \subseteq \omega$, since $x \in \omega$ by hypothesis. Thus, if $x \in A$, then $S(x) \in A$, consequently, $\omega \subseteq A$, i.e., $\forall x \in \omega : (x \in \omega) \wedge (x \subseteq \omega)$. If we now denote $\mathcal{A} \overset{\text{def}}{\leftrightarrow} (x \in \omega)$ and $\mathcal{B} \overset{\text{def}}{\leftrightarrow} (x \subseteq \omega)$, then in the last formula under the quantifier is the implication $\mathcal{A} \to (\mathcal{A} \wedge \mathcal{B})$, from which in propositional logic the formula $\mathcal{A} \to \mathcal{B}$ is derived, which corresponds to our target formula.

[6] Effectively, this means that we are going to conduct induction for the formula $(x \in \omega) \wedge (x \subseteq \omega)$, and not at all for the original formula.

Let us prove that the elements of ω are also transitive, i.e., that for any $n \in \omega$, it holds that $\forall y \in n : (y \subseteq n)$. This proof is also carried out by induction on n.

Base case of induction ($n = \emptyset$): the statement $\forall y \in \emptyset : (y \subseteq \emptyset)$ is vacuously true, since the empty set has no elements.

Inductive step: suppose that $n \in \omega$ is a transitive set. Let us prove that $S(n)$ is also transitive. Let $y \in S(n)$. This means that $y \in n$ or $y = n$.

Case 1: $y \in n$. By the induction hypothesis, n is transitive, so $y \subseteq n$. Since $n \subseteq S(n)$, then $y \subseteq S(n)$.

Case 2: $y = n$. We need to show that $n \subseteq S(n)$. This follows directly from the definition of $S(n)$. Thus, $S(n)$ is a transitive set. The induction is complete. □

It is worth noting that from the transitivity of a set, the transitivity of its elements does not generally follow, even if this set is inductive. For example, the von Neumann universe V_ω is transitive and inductive, but far from all its elements are transitive.

Lemma D4.3. *For any $n \in \omega$, if $(n \neq \emptyset)$, then $(\emptyset \in n)$.*

Proof. We prove by induction. For $n = \emptyset$, the statement is true due to the false antecedent. Next, let it be true for n, i.e., $n = \emptyset \vee \emptyset \in n$. Then $\emptyset \in S(n) = n \cup \{n\}$. The induction is complete. □

Lemma D4.4. *For any $n, k \in \omega$, if $n \in k$, then $S(n) = k$ or $S(n) \in k$.*

Proof. We use induction on k. Base case of induction ($k = \emptyset$): The premise $n \in \emptyset$ is always false, so the implication is vacuously true.

Inductive step: suppose the lemma is true for k. Let us prove that it is also true for $S(k)$.

Let $n \in S(k)$. This means that $n = k$ or $n \in k$. Let's consider both cases:

Case $n = k$: we need to prove that $S(k) = S(k)$ or $S(k) \in S(k)$. The first statement is a tautology. Thus, the lemma holds for this case.

Case $n \in k$: We need to prove that $S(n) = S(k)$ or $S(n) \in S(k)$.

By the induction hypothesis, since $n \in k$, then $S(n) = k$ or $S(n) \in k$. If $S(n) = k$, then since $k \in S(k)$, we get $S(n) \in S(k)$. If $S(n) \in k$, then since $k \subseteq S(k)$, then $S(n) \in S(k)$.

In both subcases, we arrived at the conclusion $S(n) \in S(k)$, which proves the lemma. □

3) Let us prove that $\in$ is a strict linear order on ω.

Transitivity of $\in$: if $x \in y$ and $y \in z$ (where $z \in \omega$), then, since z is a transitive set (Lemma D4.2), from $y \in z$ it follows that $y \subseteq z$. And since $x \in y$, then $x \in z$.

Irreflexivity ($x \notin x$): Let us prove by induction that $\forall n \in \omega : (n \notin n)$.

Base case ($n = \emptyset$): $\emptyset \notin \emptyset$ is true by the definition of the empty set.

Inductive step: suppose $n \notin n$. Let us prove $S(n) \notin S(n)$. If $S(n) \in S(n) = n \cup \{n\}$, then $S(n) \in n$ or $S(n) = n$. $S(n) = n$ is impossible, since $n \in S(n)$, but $n \notin n$. $S(n) \in n$ is impossible, since n is transitive, and from this it would follow that $S(n) \subseteq n$, which also contradicts $n \notin n$. Consequently, $n \notin n$ for all $n \in \omega$.

Connexity (law of trichotomy): let us prove $\forall x, y \in \omega : (x = y \vee x \in y \vee y \in x)$ by induction on x.

Base case of induction ($x = \emptyset$): We need to prove that $\forall y \in \omega : (\emptyset = y \vee \emptyset \in y \vee y \in \emptyset)$. Since $y \in \emptyset$ is false, we need to prove $\forall y \in \omega : (y = \emptyset \vee \emptyset \in y)$. And this is precisely the statement of Lemma D4.3.

Inductive step: suppose that for $n \in \omega$ the formula $\forall y \in \omega : (n = y \vee n \in y \vee y \in n)$ holds. Let us prove this for $S(n)$. Let $y \in \omega$. By the induction hypothesis for y, there are three options: **a)** $y = n$: then $y \in S(n)$, since $n \in S(n)$; **b)** $y \in n$: since $n \subseteq S(n)$, then $y \in S(n)$; **c)** $n \in y$: then by Lemma D4.4, either $S(n) = y$ or $S(n) \in y$.

Combining all cases, we see that for any pair $S(n)$ and y, one of the conditions holds: $y \in S(n)$, $y = S(n)$, or $S(n) \in y$. The induction is complete.

Thus, the relation $\in$ is a strict linear order on ω.

4) Let us prove that ω is well-ordered by the relation $\in$, i.e., that in any non-empty subset of ω, there exists an $\in$-minimal element.

Here we will follow a path analogous to what we did when establishing the equivalence of different kinds of induction in Peano arithmetic (Section D3.2), namely, we will first show a second-order version of complete induction on ω.

Lemma D4.5. *For any set A,*

$$[\forall n \in \omega : (n \subseteq A) \to (n \in A)] \to (\omega \subseteq A). \tag{D4.7}$$

Proof. Take an arbitrary set A and suppose that the premise of (D4.7) is true for it (Hypothesis 1). Let further

$$A' = \{x \in \omega \mid S(x) \subseteq A\}.$$

Let's check that the premise of induction (D4.6) holds for this set.

Base case of induction: $\emptyset \in A'$, since $\emptyset \in \omega$ (by the inductiveness of ω by definition) and $\emptyset \in A$ by Hypothesis 1, from which $S(\emptyset) = \{\emptyset\} \subseteq A$.

Inductive step: let $x \in A'$, then $x \in \omega$ and $S(x) \subseteq A$. But since $S(x) \in \omega$ (by the inductiveness of ω), then by Hypothesis 1, $S(x) \in A$, from which it follows that $S(S(x)) = S(x) \cup \{S(x)\} \subseteq A$.

Thus, A' satisfies (D4.6), from which $\omega \subseteq A'$, consequently, $\forall n \in \omega$: $S(n) \subseteq A$, from which $\forall n \in \omega : n \in A$ (since $n \in S(n)$). Thus, $\omega \subseteq A$, and formula (D4.7) is proven. □

Furthermore, formula (D4.7) in pure predicate calculus is equivalent to the formula

$$(B \neq \emptyset) \to \exists n \in \omega : (n \in B) \wedge (n \cap B = \emptyset), \tag{D4.8}$$

where $B = \omega \setminus A$. That is, in every non-empty set $B \subseteq \omega$, there is an element n such that there are no elements in B smaller than it with respect to $\in$ (there are no $x \in B$ such that $x \in n$). And this means the well-foundedness of the order $\in$ on the set ω. Together with the linearity proven in point **3**), we get that $\in$ well-orders the set ω. □

The proof could have been made shorter if we used the Axiom of Regularity, but we try not to involve this axiom without special need, to show the maximum possible generality of theorems for different axiom systems of set theory. Note also that regularity holds for all elements of ω precisely because of its $\in$-foundedness, even if we do not include the Axiom of Regularity in the list of axioms of set theory.

In view of the proven theorem, we will denote the relation $\in$ for elements of ω by the more familiar sign $<$.

Lemma D4.6. *For any $n, k \in \omega$, there exists an injection $f : n \to k$ if and only if $n \leqslant k$.*

Proof. The right-to-left direction is obvious, since $n \leqslant k$ means that $n \subseteq k$ (by the lemma on transitivity D4.2). In this case, the identity map $f(x) = x$ is the required injection.

We prove the converse statement—$\forall k$ $((f : n \to k$ is an injection$) \to (n \leqslant k))$—by induction on n.

Base case: $n = \emptyset$. The statement holds, since $\emptyset \leqslant k$ for any k by Lemma D4.3.

Inductive step: Assume the statement is true for n, and let us prove it for $S(n)$. Let there be an injection $f : S(n) \to m$. We need to show that $S(n) \leqslant m$.

Since $S(n)$ is non-empty, m is also non-empty, hence $m > 0$ and its predecessor $m - 1 = \cup m$ exists.

Our goal is to construct an injection from n to $m - 1$ in order to apply the induction hypothesis for $k = m - 1$. Consider the restriction of f to n, i.e., $f \restriction n$. This is an injection from n to m. Its range is $f[n] = \{f(x) \mid x \in n\}$.

Let us consider two cases:

Case 1: $m - 1 \notin f[n]$. In this case, the entire range of $f \restriction n$ is a subset of $m-1$. Thus, $f \restriction n$ is an injection from n to $m-1$. By the induction hypothesis, this implies that $n \leqslant m - 1$.

Case 2: $m-1 \in f[n]$. This means there exists some element $s \in n$ such that $f(s) = m - 1$. Since f is an injection and $s \in n$, we have $s \neq n$, and therefore $f(s) \neq f(n)$. This means $f(n) \neq m - 1$. Since $f(n) \in m$ and $f(n) \neq m - 1$, it follows that $f(n) \in m - 1$. Let us denote $y_n = f(n)$.

Now, let us construct a new function $g : n \to m - 1$ as follows:

$$g(x) = \begin{cases} y_n, & x = s, \\ f(x), & x \in n \text{ and } x \neq s. \end{cases}$$

This function g is an injection. Indeed, its range is $(f[n] \setminus \{m - 1\}) \cup \{y_n\}$, which is a subset of $m - 1$. The injectivity of g follows from the injectivity of f on $n \setminus \{s\}$ and the fact that $y_n = f(n)$ cannot be a value of f on $n \setminus \{s\}$.

In both cases, we have constructed an injection from n to $m - 1$. By the induction hypothesis, this means that $n \leqslant m - 1$. From $n \leqslant m - 1$, it follows that $S(n) \leqslant m$ (if $n < m - 1$, then $S(n) \leqslant m - 1 < m$; if $n = m - 1$, then $S(n) = m$). The inductive step is proven. □

From this lemma, it follows that the cardinality of a finite set is uniquely determined. Indeed, if $|X| = n$ and $|X| = k$, then there exist bijections $f : X \leftrightarrow n$ and $g : X \leftrightarrow k$, but then the composition $f \circ g^{-1}$ will be a bijection between k and n, from which by the lemma it follows that $k \leqslant n$ and $n \leqslant k$, i.e., $n = k$.

Theorem D4.5 (Pigeonhole Principle). *Let the sets X and Y be non-empty, finite, and $|Y| < |X|$. Then:*

D1 *there is no injection from X to Y;*

D1 *there is no surjection from Y to X.*

Proof. Let $|X| = n$ and $|Y| = k$. Then $k < n$ by the condition of the lemma. If there were an injection from X to Y, then there would be an injection from

n to k (it is obtained by composing the original injection with two bijections between X and n and between Y and k). Then by Lemma D4.6, $n \leqslant k$, so it is not true that $n > k$ due to the linear ordering of ω proven earlier.

Let us prove that D1→D2. Suppose there is a surjection from Y to X. Then there is a surjection $g : k \to n$, where $0 < k < n$ ($k = 0$ is excluded, since there is no surjection from an empty set to a non-empty one). Then let's define the function $f : n \to k$ by the rule

$$f(x) = \min\{y \in k \mid g(y) = x\}.$$

This definition is correct due to the surjectivity of g and the well-foundedness principle of ω. It is obvious that $f : n \to k$ is an injection, which contradicts D1. Consequently, there is no surjection from Y to X. □

The Pigeonhole Principle in the form D1 can be reformulated differently:[7] if the sets X, Y are finite, $f : X \to Y$ and $|Y| < |X|$, then there exist two *distinct* elements $x_1, x_2 \in X$ such that $f(x_1) = f(x_2)$. The Pigeonhole Principle is one of the most powerful tools for solving problems in finite mathematics. Moreover, in weak arithmetics, it turns out to be equivalent to the principle of induction, which once again emphasizes the power and significance of this principle.

From the Pigeonhole Principle follows this property:

Corollary D4.1. *If the set X is finite and $Y \subseteq X$, then Y is finite and $|Y| \leqslant |X|$.*

Proof. Let $|X| = n$. It is obvious that a bijection between X and n maps the subset Y to an equinumerous subset of n. In that case, it is sufficient for us to show that any subset $Y \subseteq n$ is finite and has cardinality $k \leqslant n$. Let us prove this by induction on n.

For $n = 0$, the statement is trivially true, since $Y = n = \emptyset$.

Inductive step: suppose it is true for n, and prove it for $S(n)$. Let $Y \subseteq S(n)$. There are two options here: $n \notin Y$ or $n \in Y$. In the first case, $Y \subseteq n$, and by the induction hypothesis, the cardinality of Y exists and is equal to $k \leqslant n$, i.e., $k \leqslant S(n)$. In the second case, consider the set $Y' = Y \setminus \{n\}$; then $Y' \subseteq n$ and also has cardinality $k \leqslant n$, i.e., there exists a bijection $f : Y' \leftrightarrow k$. Then let's construct the function

[7] If you have more pigeons than pigeonholes, and every pigeon is in a pigeonhole, then at least one pigeonhole must contain more than one pigeon.

$$g(x) = \begin{cases} f(x), & x \in Y', \\ k, & x = n. \end{cases}$$

It is easy to see that $g : Y \leftrightarrow S(k)$. But since $k \leqslant n$, then $S(k) \leqslant S(n)$, and, consequently, in the second case, we also get that Y has a cardinality of no more than $S(n)$. The induction is complete. □

Corollary D4.2. *If X is finite and $Y \subsetneq X$, then $|Y| < |X|$.*

Proof. Let $|X| = n$. Since $Y \subsetneq X$, then by the previous corollary, $|Y| = k \leqslant n$. If $k = n$, then there exists a bijection $f : Y \leftrightarrow n$. Then let's construct the function

$$g(x) = \begin{cases} f(x), & x \in Y, \\ n, & x \in X \setminus Y, \end{cases}$$

which would be a surjection from n to $S(n)$, but this is impossible by the Pigeonhole Principle D2, since $n < S(n)$. Consequently, $k < n$. □

As we can see, a finite set has a distinctive feature—it is not equinumerous with any of its proper subsets. At the same time, if we take, for example, the set ω, it is equinumerous with, for example, the set $\omega \setminus \{\emptyset\}$, since the function $S(n)$ on ω is a bijection between these sets.

Therefore, there is an approach to defining finite and infinite sets that is not related to the concept of cardinality and natural numbers. A set A is called **Dedekind-infinite** if it is equinumerous with its proper subset. Otherwise, it is called **Dedekind-finite**.

Corollary D4.3. *If X is finite, then it is Dedekind-finite.*

Proof. Let $|X| = n$, where $n \in \omega$. If X were Dedekind-infinite, there would be a bijection $f : X' \leftrightarrow X$, where $X' \subsetneq X$. But this is impossible by the previous corollary. Hence, X is Dedekind-finite. □

From this it also follows that if a set is Dedekind-infinite, then it is infinite. For example, ω is equinumerous with its proper subset, therefore, it is Dedekind-infinite and (simply) infinite.

However, the similarity of the definitions within the theory ZF ends here. The fact is that if we want to prove their equivalence, and more precisely, to show that every Dedekind-finite set is finite, we will have to use one of the weak forms of the axiom of choice: AC_ω (see Theorem D4.11 later).

D4.3 Ordinals and Cardinals

Let us now consider another powerful set-theoretic tool, namely, the scale of ordinals and the theorems associated with it. In particular, our goal is to justify the method of transfinite definitions (recursions) and the accompanying proof method—transfinite induction. All the necessary definitions have already been introduced in Section C1.3. We will only repeat that by an **ordinal** we mean a transitive set that is *strictly* well-ordered by the membership relation.[8] The class of all ordinals is usually denoted by Ord, and the notation Ord(x) means the formula that expresses the property of the set x being an ordinal.

From Theorem D4.4 and Lemma D4.2, it follows that ω and all its elements (the von Neumann natural numbers) are ordinals.

In Section C1.3, we also agreed that we would denote ordinals by Greek letters and even defined their own quantifiers for them to make it easier to write down statements about ordinals. Furthermore, as in the case of ω, we agree to use the predicate $x < \alpha$ alongside the notation $x \in \alpha$, taking into account that $\in$ is an order on the ordinal. This duality in the notation of membership allows for a better understanding of the text of proofs about ordinals.

Ordinals are divided into three classes: one contains the zero ordinal $\emptyset$, another contains all **successor ordinals**, i.e., ordinals of the form $S(\alpha)$, and the third contains all the rest, which are called **limit ordinals**. The class of all limit ordinals is usually denoted by Lim, the class of all successors by Succ. Successors are usually denoted by Greek letters from the beginning of the alphabet, while limit ordinals are typically denoted by λ (with or without indices). Furthermore, the notations for natural numbers are also preserved for the notations of finite ordinals; in particular, the ordinal $\emptyset$ is denoted as 0.

---

[8] This choice of definition is motivated by the fact that it precisely generalizes the properties of von Neumann natural numbers ($n \in \omega$), which, as was shown, possess exactly these characteristics. Furthermore, the requirement of being strictly well-ordered by $\in$ itself excludes cycles of the form $x \in \cdots \in x$, which allows us not to appeal to the Axiom of Regularity at this stage.

D4.3.1 Recursion and Induction

Theorem D4.6. *The following theorems are provable in* ZF*:*

1° *an element of an ordinal is an ordinal;*

2° *for any ordinals* α, β*:* $\alpha \in \beta \leftrightarrow (\alpha \subsetneq \beta)$*;*

3° *for any ordinals* α, β*:* $\alpha \leqslant \beta$ *or* $\beta \leqslant \alpha$*;*

4° *the union and intersection of any set of ordinals is an ordinal;*

5° *the class of all ordinals is well-ordered by the relation* $\in$ *and is a proper class;*

6° *for any ordinal* α*:* $S(\alpha)$ *is an ordinal;*

7° *the von Neumann natural numbers and* ω *are ordinals;*

8° *induction* $[\varphi]$*:* $[\forall \alpha : (\forall x \in \alpha : \varphi(x, \vec{p})) \rightarrow \varphi(\alpha, \vec{p})] \rightarrow \forall \alpha : \varphi(\alpha, \vec{p})$*;*

9° *recursive definition* $[\varphi]$*: let* $\varphi(g, b, \vec{p})$ *define a mapping of type* $g \mapsto b$*, i.e.,* $\forall x \exists ! y : \ \varphi(x, y, \vec{p})$*, then for any ordinal* α *there exists a function* f *defined on* α *such that for any* $\beta < \alpha$*:*

$$\varphi(f \restriction \beta, f(\beta), \vec{p}). \tag{D4.9}$$

10° *both successor and limit ordinals form proper classes.*

Proof. **1)** Let α be an ordinal and $x \in \alpha$. By the transitivity of the ordinal, $x \subseteq \alpha$, and thus it inherits the strict well-ordering from α. It remains to show that x is transitive. Let $y \in x$; then since $x \subseteq \alpha$, we also have that $y \in \alpha$. Take any $z \in y$; then similarly we get that $z \in \alpha$. In total we have: $x, y, z \in \alpha$ and $z \in y \in x$, from which, by the transitivity of the relation $\in$ on the ordinal α, we conclude that $z \in x$, i.e., $y \subseteq x$, so x is transitive. Thus, any $x \in \alpha$ is an ordinal.

2) The forward implication is true by the transitivity of ordinals: $\alpha \in \beta \rightarrow \alpha \subsetneq \beta$ (equality is excluded due to the irreflexivity of $\in$ on the ordinal β). Let $\alpha \subsetneq \beta$. Then there exists $\gamma = \min(\beta \setminus \alpha)$ (since $\beta \setminus \alpha \neq \emptyset$). Let $x \in \gamma$; then $x \in \beta$ and $x \notin \beta \setminus \alpha$, since $x < \min(\beta \setminus \alpha)$, from which $x \in \alpha$. Conversely, if $x \in \alpha$, then $x \in \beta$ and $x \notin \beta \setminus \alpha$, i.e., $x < \gamma$, which means $x \in \gamma$. Consequently, $\alpha = \gamma$, from which $\alpha \in \beta$.

3) It is obvious that $\gamma = \alpha \cap \beta$ is an ordinal (transitivity and well-ordering are inherited from α and β). By the previous point, $\gamma \leqslant \alpha$ and $\gamma \leqslant \beta$. If $\gamma < \alpha$ and $\gamma < \beta$, then $S(\gamma) \subseteq \alpha \cap \beta$, i.e., $S(\gamma) \subseteq \gamma$, which is impossible, since otherwise it would be $\gamma \in \gamma$. Consequently, $\gamma = \alpha$ or $\gamma = \beta$. Then together with the previous statement, we get that $\alpha \leqslant \beta$ or $\beta \leqslant \alpha$.

4) Let A be a set of ordinals. If $A = \emptyset$, then by definition $\bigcap A = \emptyset$, which is an ordinal. If $A \neq \emptyset$, then just as in the case of the intersection of two ordinals, the property of being an ordinal for the set $\bigcap A$ is inherited from all elements of A at once. Furthermore, $\bigcap A = \min A$ if $A \neq \emptyset$, since $\bigcap A \leqslant a$ for all $a \in A$ (see point 2).

By the Axiom of Union, there exists the set $a = \bigcup A$. This set is transitive, because if $x \in a$, then there exists an $\alpha \in A$ such that $x \in \alpha$, from which, by the transitivity of the ordinal α, we have $x \subseteq \alpha$ and, consequently, $x \subseteq a$. Moreover, a is a set of ordinals (applying property 1 to $x \in \alpha$).

Let $x, y, z \in a$. First, we note that $\neg(x < x)$, since there exists an $\alpha \in A$ such that $x \in \alpha$, and in the ordinal α, membership is irreflexive. Second, let $x < y < z$; then there exists an ordinal $\alpha \in A$ such that $z \in \alpha$, from which by transitivity $y \in \alpha$, from which again by transitivity $x \in \alpha$, i.e., $x, y, z \in \alpha$, but then $x < z$ by the properties of the ordinal α. Third, for arbitrary $x, y \in a$, there exist $\alpha, \beta \in A$ such that $x \in \alpha$, $y \in \beta$. By point 3, $\alpha \leqslant \beta$ or $\beta \leqslant \alpha$. In the first case, we get that $x, y \in \beta$, and there they are related by membership, i.e., $(x < y) \vee (x = y) \vee (y < x)$; the second case is analogous. Thus, the set a is linearly ordered by the relation $\in$.

Let $X \subseteq a$ and $X \neq \emptyset$. Since X is a set of ordinals, then by what has been proven, $\bigcap X = \min X$. Consequently, a is well-ordered by the relation $\in$. Thus, $a = \bigcup A$ is an ordinal.

Furthermore, it is not difficult to see that $a = \sup A$ in the class of ordinals, since, on the one hand, $x \leqslant a$ for any $x \in A$, and on the other hand, if $x \leqslant \beta$ for all $x \in A$ (i.e., β is an upper bound of A), then $a \leqslant \beta$.

5) The irreflexivity, transitivity, and connexity of the relation $\in$ for any ordinals follow from the points proven above. Further, if $\mathcal{K}$ is some non-empty subclass of the class of ordinals, then let's take any of its elements $\alpha \in \mathcal{K}$; then $\min \mathcal{K} = \min(\mathcal{K} \cap S(\alpha))$, and the latter exists, because $\mathcal{K} \cap S(\alpha)$ is a set of ordinals.

In other words, the class Ord can be considered as the largest ordinal, containing all ordinals as its initial segments. At the same time, although the class Ord formally satisfies the definition of an ordinal, it is not one, because it is a proper class, i.e., it is not a set. And this follows from the next point.

6) $S(\alpha) = \alpha \cup \{\alpha\}$. First, $S(\alpha)$ is transitive: indeed, if $x \in S(\alpha)$, then either $x \in \alpha$ or $x = \alpha$. In the first case $x \subseteq \alpha \subseteq S(\alpha)$ since α is an ordinal; in the second case, obviously, $\alpha \subseteq S(\alpha)$. Second, to prove that $S(\alpha)$ is well-ordered by the relation $\in$, it is sufficient to note that the set α is already well-ordered, and $S(\alpha)$ is obtained by adding one maximal element α. It is obvious that any non-empty subset $A \subseteq S(\alpha)$ will have a minimal element: if $A \cap \alpha \neq \emptyset$, then it is $\min(A \cap \alpha)$; if $A \cap \alpha = \emptyset$, then $A = \{\alpha\}$ and $\min A = \alpha$.

Thus, if Ord were a set, i.e., an ordinal, then $S(\mathrm{Ord})$ would also exist as a set and would be greater than any ordinal, including itself, which is impossible due to the irreflexivity of the order on ordinals. Consequently, Ord is a proper class. This proves the second part of the statement in point 5.

7) As has already been noted, from Theorem D4.4 and Lemma D4.2, it follows that ω and all its elements (the von Neumann natural numbers) are ordinals.

8) Let the premise be true for the formula φ:

$$\forall \alpha : (\forall x < \alpha : \varphi(x)) \rightarrow \varphi(\alpha).$$

Suppose that the conclusion is not true, i.e., $\exists \alpha : \neg\varphi(\alpha)$. Let then α be the least such ordinal; then $\forall x < \alpha : \varphi(x)$, from which, from the premise, we conclude that $\varphi(\alpha)$ in contradiction to the assumption. Thus, the principle of transfinite induction is established.

9) To prove the recursion theorem, we reformulate it in a more convenient form, following the notations introduced in Section C1.3. Namely, let $\Phi_{\vec{p}}$ be a class-mapping defined by the formula $\varphi(g, b, \vec{p})$. Then there exists a class-mapping F defined on the class of ordinals and such that

$$F(\alpha) = \Phi_{\vec{p}}(F \restriction \alpha).$$

We will conduct the proof by induction on α, proving the following formula:

$$\psi(\alpha) \overset{\text{def}}{\leftrightarrow} \exists! f : S(\alpha) \rightarrow \mathfrak{V} : \forall \gamma \leqslant \alpha : f(\gamma) = \Phi_{\vec{p}}(f \restriction \gamma), \qquad \text{(D4.10)}$$

after which we will prove the consistency of all such functions f, which will give us the mapping F that is total on the class of ordinals.

Inductive step: suppose that $\psi(\beta)$ for all $\beta < \alpha$. We need to prove $\psi(\alpha)$. As usual when working with ordinals, it is necessary to consider three cases according to the class membership of the ordinal.

Let $\alpha = 0$. Then $\psi(0)$ means that there exists a function $f : S(0) \rightarrow \mathfrak{V}$, defined at zero, and its value is given by the equality $f(0) = \Phi_{\vec{p}}(f \restriction \emptyset) = \Phi_{\vec{p}}(\emptyset)$. This is essentially the base case of recursion for the class-function F.

Let $\alpha = S(\beta)$; then by the induction hypothesis, $\psi(\beta)$ is true, i.e., there exists a unique function $f : S(\beta) \rightarrow \mathfrak{V}$ such that $\forall \gamma \leqslant \beta : f(\gamma) = \Phi_{\vec{p}}(f \restriction \gamma)$, including $f(\beta) = \Phi_{\vec{p}}(f \restriction \beta)$. The task is to extend this function to the point $\{\alpha\}$. Let's define the function $f' : S(\alpha) \to \mathfrak{V}$ strictly according to the required formula:

$$f'(\gamma) \overset{\text{def}}{\rightleftharpoons} \begin{cases} \Phi_{\vec{p}}(f), & \gamma = \alpha, \\ f(\gamma), & \gamma \leqslant \beta, \end{cases}$$

where we used the fact that $f' \restriction \alpha = f \restriction S(\beta) = f$.

Finally, let $\alpha \in \mathrm{Lim}$, and by the induction hypothesis, for all $\beta < \alpha$ there exist unique functions $f_\beta : S(\beta) \to \mathfrak{V}$ such that $\forall \gamma \leqslant \beta : f_\beta(\gamma) = \Phi_{\vec{p}}(f_\beta \restriction \gamma)$. Then by the Axiom of Replacement, there exists the set $\{f_\beta \mid \beta < \alpha\}$ of all such functions. Let us set

$$f \overset{\text{def}}{\rightleftharpoons} \bigcup \{f_\beta \mid \beta < \alpha\}.$$

First, we need to check that f is a function defined on α. Let $\langle \gamma, a \rangle, \langle \gamma, b \rangle \in f$. This means that there exist $\beta_1, \beta_2 < \alpha$ such that $f_{\beta_1}(\gamma) = a$ and $f_{\beta_2}(\gamma) = b$. Suppose that $a \neq b$, and then let's choose γ as the smallest such ordinal that $f_{\beta_1}(\gamma) \neq f_{\beta_2}(\gamma)$. This means that $f_{\beta_1}(\delta) = f_{\beta_2}(\delta)$ for all $\delta < \gamma$, i.e., $f_{\beta_1} \restriction \gamma = f_{\beta_2} \restriction \gamma$. But since f_{β_1} and f_{β_2} satisfy the recursive formula, i.e., $\psi(\beta_1)$ and $\psi(\beta_2)$ are satisfied, then

$$f_{\beta_1}(\gamma) = \Phi_{\vec{p}}(f_{\beta_1} \restriction \gamma) = \Phi_{\vec{p}}(f_{\beta_2} \restriction \gamma) = f_{\beta_2}(\gamma).$$

Consequently, the assumption $a \neq b$ is not true, and thus f is functional. It is not difficult to see also that f is everywhere defined on α, i.e., f is a function defined on α.

Second, f satisfies the recursive identity, i.e., for any $\beta < \alpha$: $f(\beta) = \Phi_{\vec{p}}(f \restriction \beta)$, since $f \restriction S(\beta) = f_\beta$ is an extension of $f \restriction \beta$.

Finally, let's extend f to the point $\{\alpha\}$ just as we did in the previous case, and we get the function $f' : S(\alpha) \to \mathfrak{V}$ that satisfies the recursive identity. It is not difficult to see that f' is uniquely determined in both cases. Thus, we have proven $\psi(\alpha)$. From this, by induction, we arrive at the fact that $\psi(\alpha)$ is true for all α.

Finally, we can define the class-mapping F using the formula:

$$d = F(\alpha) \overset{\text{def}}{\leftrightarrow} \exists! f : S(\alpha) \to \mathfrak{V} : ((f(\alpha) = d) \wedge \forall \gamma \leqslant \alpha : f(\gamma) = \Phi_{\vec{p}}(f \restriction \gamma)).$$

Essentially, the mapping F is the union of all functions f whose existence and uniqueness are guaranteed by the formula $\psi(\alpha)$, from which it is easy to understand that F is everywhere defined on the class Ord and is functional, i.e., if $a = F(\alpha)$ and $b = F(\alpha)$, then $a = b$ for any ordinals α and any sets a, b.

With this, the justification of the transfinite recursive definition is complete.

We discussed the use of recursive definitions in detail in Section C1.3, in particular, the fact that, as a rule, the recursion generator $\Psi_{\vec{p}}$ is defined separately for successors and for limit ordinals. Furthermore, a total definition of recursion on the entire class Ord is not always required, so it is given for the necessary fragment of this class, and for the remaining ordinals, some default value is assumed that does not interfere with the definition. It is sufficient to discuss all these details once in order to use transfinite recursion relatively freely thereafter.

10) If the class Succ were a set, then there would exist a successor $S(\sup \mathrm{Succ})$ that would not be an element of Succ. Consequently, the class Succ is a proper class of ordinals. The matter is slightly more complicated with the class Lim. Suppose that it is a set, then let $\lambda = \sup \mathrm{Lim}$. Suppose that $\lambda \in \mathrm{Lim}$, i.e., $\lambda = \max \mathrm{Lim}$. However, for any ordinal, we can recursively construct a larger limit ordinal. Let's construct a function F using the property 9 proven above. Consider the generator

$$\Phi_\lambda(g) = \begin{cases} \lambda, & g = \emptyset; \\ S(\sup(\mathrm{ran}(g))), & g \neq \emptyset, \end{cases} \qquad F(\alpha) = \Phi_\lambda(F \upharpoonright \alpha).$$

Let $\lambda_\alpha \overset{\text{def}}{\rightleftharpoons} F(\alpha)$. Let's see how the sequence behaves: $\lambda_0 = \Phi_\lambda(\emptyset) = \lambda$, $\lambda_1 = \Phi_\lambda(F \upharpoonright 1) = S(\sup\{\lambda\}) = S(\lambda)$, $\lambda_2 = \Phi_\lambda(F \upharpoonright 2) = S(\sup\{\lambda, S(\lambda)\}) = S(S(\lambda))$, and so on, $\lambda_n = \Phi_\lambda(F \upharpoonright n) = \underbrace{S \ldots S}_{n}(\lambda)$. The further values of the sequence are not interesting to us, so in fact, we could have considered the recursive definition only on the initial segment of the class of ordinals—the ordinal ω, and defined the generator $\Phi_\lambda(x)$ not totally, but only for finite subsets of ω. But since this is the first time we are applying this theorem, we have tried to follow its conditions strictly.

So, let's consider the restriction of the sequence λ_α to the ordinal ω. By the Axiom of Replacement, the range of this restriction is a set of ordinals of the form $\underbrace{S \ldots S}_{n}(\lambda)$. Then its limit $\Lambda = \sup\{\lambda_n \mid n < \omega\}$ is a limit ordinal (if it were a successor, it would have an immediate predecessor. However, Λ as the supremum of a sequence with no maximum element has no immediate predecessor). Λ is greater than λ, i.e., $\Lambda > \max \mathrm{Lim}$ and $\Lambda \in \mathrm{Lim}$. A contradiction. Consequently, the class Lim is a proper class of ordinals. □

It is easy to see that $\alpha = S(\sup \alpha)$ if and only if α is a successor. Limit ordinals satisfy the identity $\lambda = \sup \lambda$.[9]

The proof method using (transfinite) induction, which is number 8° in the theorem, requires a case analysis for the three types of the ordinal α. Indeed, in the premise of the induction, it is required to check the transition

$$\forall \alpha : (\forall x \in \alpha : \varphi(x, \vec{p})) \to \varphi(\alpha, \vec{p}),$$

which is more conveniently done by a case analysis for $\alpha = 0$, $\alpha \in \mathrm{Succ}$, and $\alpha \in \mathrm{Lim}$. For $\alpha = 0$, the vacuous truth of the antecedent of the implication requires us to directly check $\varphi(0)$ to confirm the truth of the implication, i.e., exactly as in ordinary arithmetic induction. For $\alpha = S(\beta)$, the antecedent assumes the truth of $\varphi(x)$ for all $x \leqslant \beta$, but in fact, as a rule (though not always), it is sufficient to know that $\varphi(\beta)$ to deduce $\varphi(\alpha)$, i.e., to reproduce the step of local induction, as in arithmetic. If, however, $\lambda \in \mathrm{Lim}$, then methods related to the order topology of ordinals begin to work, since we essentially need to prove that by passing to the limit $\lambda = \sup\{x \mid x < \lambda\}$, we preserve the truth of the formula φ. This is reminiscent of the familiar continuity condition from calculus:

$$\lim_{x \to \lambda^-} \varphi(x) = \varphi(\lambda).$$

So in the limit case, we in one way or another use the entire hypothesis of the inductive step, but for an ordinal that is itself defined as a limit of smaller ones (a supremum).

This fundamental difference in the proof of the inductive step for the three types of ordinals is almost always appropriate and is used in proofs.

Another variant of induction is starting not from zero, but from some given ordinal α_0, when we prove a formula of the form $\forall \alpha \geqslant \alpha_0 : \varphi(\alpha, \vec{p})$. In this case, we have, as it were, another, "vacuous" type of ordinals, namely the ordinals $\alpha < \alpha_0$, since for them, the check of the formula of the inductive premise is trivial. Therefore, this case is usually omitted, and the remaining cases are:

---

[9] Formally, $\sup \emptyset = 0$, since any ordinal is an upper bound of the empty set, so some authors classify 0 as a limit ordinal. However, playing with the empty set leads to the fact that its greatest lower bound is also any ordinal, and as $\inf \emptyset$, one can consider the class Ord and, moreover, for every upper bound, one can find a larger lower bound of the empty set, which sounds absurd. Therefore, it is better to consider $\emptyset$ as a special object and not apply the concepts of bounds, suprema/infima, or max/min to it.

$\alpha = \alpha_0$, $\alpha \in \text{Succ} \wedge \alpha \geqslant \alpha_0$, $\alpha \in \text{Lim} \wedge \alpha \geqslant \alpha_0$, which is, of course, convenient. So when using induction in the future, we will not specifically explain why we use such a case analysis when proving the inductive step.

And one more variant of induction is a proof within some ordinal, and not on the entire class Ord.

The same considerations apply to recursion: most often, the recursion generator is defined by three different formulas for the three types of ordinals, and the recursion itself can start not necessarily from zero, but from any other ordinal. In general, induction and recursion are two sides of the same coin. Induction is a method of proof that relies on the recursive structure of objects, and recursion is a method for defining such objects. Accordingly, if we define some recursion, we will prove most of its properties by induction.

D4.3.2 The von Neumann Hierarchy and Rank

Now, having at our disposal the powerful apparatus of transfinite recursion, we can formally construct the hierarchy of universes, which was discussed in Section C1.3.

The von Neumann universe hierarchy is defined recursively over the ordinals:

$$V_0 = \emptyset, \quad V_{\alpha+1} = \mathcal{P}(V_\alpha), \quad V_\lambda = \bigcup_{\beta<\lambda} V_\beta, \tag{D4.11}$$

where λ is a limit ordinal. The class $\bigcup_{\alpha\in\text{Ord}} V_\alpha$ is called the **von Neumann universe**. We have previously shown in Section C1.3 that the mapping V_α is strictly monotone, i.e., if $\alpha < \beta$, then $V_\alpha \subsetneq V_\beta$.

The **rank** of a set A is the least ordinal α such that $A \subseteq V_\alpha$.

The Axiom of Regularity guarantees that every set belongs to the von Neumann universe, i.e., has a rank.

Theorem D4.7 (on Rank). *Every set is an element and, therefore, a subset of some universe in the von Neumann hierarchy. In other words,* $\mathfrak{V} = \bigcup_{\alpha\in\text{Ord}} V_\alpha$.

Proof. Suppose that there exists at least one set that does not have a rank; let us denote it by A. Let's construct its *transitive closure* $TC(A)$, which is defined recursively:

$$A_0 = A, \quad A_{n+1} = \bigcup A_n, \quad TC(A) = \bigcup\{A_n \mid n \in \omega\}.$$

Its existence in ZF is guaranteed by the axioms of Union and Replacement, as well as the theorem on recursive definition. It is easy to see that $TC(A)$ is transitive. Indeed, if $x \in TC(A)$, then there exists an n such that $x \in A_n$, then $x \subseteq A_{n+1}$, which means $x \subseteq TC(A)$.

Consider the set $S = TC(A) \cup \{A\}$. It is also transitive, since $A \subseteq S$. Let's separate from S a subset B consisting of all elements of S that do not have a rank:

$$B \overset{\text{def}}{\rightleftharpoons} \{x \in S \mid \forall \alpha \; \neg(x \subseteq V_\alpha)\}.$$

By our initial assumption, the set A does not have a rank, and since $A \in S$, then $A \in B$. Consequently, the set B is non-empty.

Since B is a non-empty set, by the Axiom of Regularity, there exists an $\in$-minimal element in it. Let's call it x_0. This means that:

1. $x_0 \in B$ (i.e., $\mathrm{rank}(x_0)$ is not defined).
2. $x_0 \cap B = \emptyset$.

Since $x_0 \in S$ and S is transitive, all elements of x_0 are also in S. But at the same time, the elements of x_0 do not belong to B by the minimality of x_0, which means that for all elements of x_0, a rank is defined. Let $\Lambda = \{\mathrm{rank}(y) \mid y \in x_0\}$. This is a set by the Axiom of Replacement, and thus it is a set of ordinals, and it has a supremum (union), which we will denote by λ.

Now, since $\forall y \in x_0 : \mathrm{rank}(y) \leqslant \lambda$, then by the monotonicity of the sequence of universes, $\forall y \in x_0 : y \subseteq V_\lambda$, from which it follows that $x_0 \subseteq V_{\lambda+1}$, which means that x_0 has a rank, in contradiction to the choice of $x_0 \in B$. □

D4.3.3 Ordinal Arithmetic

Lemma D4.7 (on order type). *If $\langle X, \prec \rangle$ is a strictly well-ordered set, then there exists a unique ordinal α such that $\langle X, \prec \rangle \cong \langle \alpha, \in \rangle$.*

Proof. In other words, we need to show that there exists a unique ordinal α that is order-isomorphic to X. Let's give a recursive definition:

$$F(\alpha) = \begin{cases} \min\{x \in X \mid F[\alpha] \prec x\}, & \exists x \in X : F[\alpha] \prec x, \\ X, & \text{otherwise,} \end{cases}$$

where $F[\alpha] \prec x$ means that $\forall \gamma < \alpha : F(\gamma) \prec x$. In other words, the recursion generator chooses the least *strict* upper bound of the set given as input, and if there is no such bound, it returns the entire set X.

The second case is actually not of interest to us, as it is triggered when the entire isomorphism has already been constructed. It is important to show that at some ordinal α, this will happen. Suppose that this is not so, i.e., that for any α, the image $F[\alpha]$ has a strict upper bound. This means that for all ordinals, $F(\alpha) = \min\{x \in X \mid F[\alpha] \prec x\}$. From this, it is easy to see that for $\alpha < \beta$, we have $F(\alpha) \prec F(\beta)$, i.e., F is a (monotone) injection. In that case, let's take the full image $F[\mathrm{Ord}] \subseteq X$ and consider the inverse mapping: $F^{-1} : F[\mathrm{Ord}] \leftrightarrow \mathrm{Ord}$, which will be a bijection. But since $F[\mathrm{Ord}]$ is a definable part of the set X, i.e., it is a set by the Axiom of Separation, then by virtue of the Axiom of Replacement, Ord will also be a set, which, as we have shown earlier, is not true.

Consequently, at some α, we will get that the image $F[\alpha]$ does not have a strict upper bound in X. Let's denote by α_0 the least of such ordinals. Then, obviously, $F[\alpha_0] \subseteq X$, and moreover, F is strictly monotone on α_0.

Suppose that $F[\alpha_0] \subsetneq X$, i.e., there is a $y \in X \setminus F[\alpha_0]$. Let y_0 be the least of such y. Next, let's take the initial segment $Y = \{x \in X \mid x \prec y_0\}$. It is clear that for any $x \in Y$, there is a $\delta < \alpha_0$ such that $F(\delta) = x$. Moreover, the preimage $F^{-1}[Y]$ is an ordinal (transitivity follows from the fact that if $\lambda < \delta \in F^{-1}[Y]$, then $\lambda \in F^{-1}[Y]$). Let's denote this ordinal by γ; then $F(\gamma) = \min\{x \in X \mid F[\gamma] \prec x\} = \min\{x \in X \mid Y \prec x\} = y_0$. We have arrived at a contradiction with the fact that y_0 is not a value of F. Consequently, $F[\alpha_0] = X$.

So, we have found an ordinal α_0 isomorphic to the set X.

Uniqueness: suppose that $\alpha_0 \cong \langle X, \prec\rangle \cong \alpha_1$; then $\alpha_0 \cong \alpha_1$, i.e., there exists a bijection $f : \alpha_0 \leftrightarrow \alpha_1$ that preserves the order.

Let's show that $f(\beta) \geqslant \beta$ for all $\beta < \alpha_0$. Indeed, if this is not so, then let's take β as the least such that $f(\beta) < \beta$. Let $\gamma = f(\beta)$. Since f preserves the order, then $f(\gamma) < f(\beta) = \gamma$. Thus, we have found an ordinal $\gamma < \beta$ for which $f(\gamma) < \gamma$ also holds, in contradiction to the choice of β. Consequently, $f(\beta) \geqslant \beta$ for all $\beta < \alpha_0$.

Next, we note that f^{-1} is also an isomorphism, and therefore $f^{-1}(\beta) \geqslant \beta$ for all $\beta < \alpha_1$, i.e., $f(\beta) \leqslant \beta$. Then, taking into account both inequalities, we get that $f = \mathrm{id}_{\alpha_0}$, from which it follows that $\alpha_0 = \alpha_1$. Uniqueness is proven. □

The operations on ordinals are defined as follows:

$$\alpha + 0 = \alpha, \qquad \alpha + S(\beta) = S(\alpha + \beta), \qquad \alpha + \lambda = \sup\{\alpha + \gamma \mid \gamma < \lambda\},$$
$$\alpha \cdot 0 = 0, \qquad \alpha \cdot S(\beta) = \alpha \cdot \beta + \alpha, \qquad \alpha \cdot \lambda = \sup\{\alpha \cdot \gamma \mid \gamma < \lambda\},$$
$$\alpha^0 = 1, \qquad \alpha^{S(\beta)} = \alpha^\beta \cdot \alpha, \qquad \alpha^\lambda = \sup\{\alpha^\gamma \mid 0 < \gamma < \lambda\},$$

where $\lambda \in \mathrm{Lim}$. This is a recursive definition that generalizes the analogous definition of arithmetic operations in PA. Note that in the definition of α^λ, the restriction $\gamma > 0$ has been added. This is due to the fact that for $\alpha = 0$, we get $0^{S(\beta)} = 0^\beta \cdot 0 = 0$, and therefore $0^\gamma = 0$ for all $\gamma > 0$; however, if we were to allow $\gamma = 0$, we would get $0^0 = 1$, and the supremum indicated in the definition of α^λ would already be equal to 1, which would lead to an inconsistency in the definitions.

These operations also have another, set-theoretic definition.

Let two well-ordered sets X, Y with order relations $<_X, <_Y$ be given, respectively. By their *ordered sum*, we will understand the following construction: the domain of the structure will be the set $(X \times \{0\}) \cup (Y \times \{1\})$, which for brevity we will denote as $X \sqcup Y$. Multiplication by the points $\{0\}$ and $\{1\}$ protects us from possible overlaps of X and Y, since $(X \times \{0\}) \cap (Y \times \{1\}) = \emptyset$. On the set $X \sqcup Y$, a relation $\prec$ is defined by the following rule:

$$(a, b) \prec (c, d) \overset{\text{def}}{\leftrightarrow}$$
$$(b = 0 \wedge d = 1) \vee (b = d = 0 \wedge a <_X c) \vee (b = d = 1 \wedge a <_Y c), \qquad \text{(D4.12)}$$

i.e., all elements of X are less than all elements of Y, and within each of these sets, the original comparison works. It is not difficult to show that $\prec$ will be a well-ordering: all properties of the order are inherited from X and Y. It turns out that in this case, the addition of ordinals satisfies the identity

$$ot(X \sqcup Y, \prec) = ot(X, <_X) + ot(Y, <_Y). \qquad \text{(D4.13)}$$

Note that if we swap X and Y, the order $\prec$ will, generally speaking, be different.

Replacing X with the ordinal α (by the lemma on order type) and Y with the ordinal β, let us prove (D4.13) in the form $ot(\alpha \sqcup \beta) = \alpha + \beta$ by induction on β.

Suppose that the formula is true for all $\beta < \gamma$ and prove it for γ. A case analysis on the type of the ordinal γ: $\gamma = 0$, $\gamma \in \mathrm{Succ}$, $\gamma \in \mathrm{Lim}$.

a) $\gamma = 0$, then $\alpha + 0 = \alpha$ and $ot(\alpha \sqcup \emptyset) = \alpha$; the identity holds.

b) Let $\gamma = S(\beta)$. By the induction hypothesis, $ot(\alpha \sqcup \beta) = \alpha + \beta$. This means that there is a unique order isomorphism $f : \alpha \sqcup \beta \leftrightarrow \alpha + \beta$. At the same time,

$$\alpha \sqcup \gamma = \alpha \times \{0\} \cup \gamma \times \{1\} = \alpha \times \{0\} \cup (\beta \cup \{\beta\}) \times \{1\} = \\ (\alpha \times \{0\} \cup \beta \times \{1\}) \cup \{\langle \beta, 1 \rangle\} = (\alpha \sqcup \beta) \cup \{\langle \beta, 1 \rangle\}$$

and by the recursive definition of addition,

$$\alpha + \gamma = \alpha + S(\beta) = S(\alpha + \beta) = (\alpha + \beta) \cup \{\alpha + \beta\}.$$

In this case, $\alpha + \beta = \max_{\in}(\alpha + \gamma)$ and $\langle \beta, 1 \rangle = \max_{\prec}(\alpha \sqcup \gamma)$. Then let us define the function g:

$$g(x) = \begin{cases} f(x), & x \in \alpha \times \{0\} \cup \beta \times \{1\}, \\ \alpha + \beta, & x = \langle \beta, 1 \rangle, \end{cases}$$

in other words, let's extend f with the pair $\langle \langle \beta, 1 \rangle, \alpha + \beta \rangle$ of the maxima of the corresponding sets. It is obvious that g is an order isomorphism between $\alpha \sqcup \gamma$ and $\alpha + \gamma$. From this, it follows that $ot(\alpha \sqcup \gamma) = \alpha + \gamma$.

c) The most complex case: $\gamma \in \mathrm{Lim}$. By the induction hypothesis, $ot(\alpha \sqcup \beta) = \alpha + \beta$ for all $\beta < \gamma$. This means that for each $\beta < \gamma$, there exists a unique order isomorphism $f_\beta : \alpha \sqcup \beta \leftrightarrow \alpha + \beta$. These isomorphisms form a chain by inclusion: if $\beta_1 \leqslant \beta_2$, then $f_{\beta_1} \subseteq f_{\beta_2}$. Indeed, let's consider the restriction $g = f_{\beta_2} \upharpoonright \alpha \sqcup \beta_1$. This is a strictly monotone mapping into the ordinal $\alpha + \beta_2$, which has the property: $\mathrm{ran}(g)$ is an ordinal, since $\mathrm{ran}(g)$ is an initial segment of the ordinal $\alpha + \beta_2$, and any initial segment of an ordinal is itself an ordinal. Since g is an order isomorphism between $\alpha \sqcup \beta_1$ and $\mathrm{ran}(g)$, then $ot(\alpha \sqcup \beta_1) = \mathrm{ran}(g)$. By the induction hypothesis, $ot(\alpha \sqcup \beta_1) = \alpha + \beta_1$. By the uniqueness of the order type, $\mathrm{ran}(g) = \alpha + \beta_1$. And by the uniqueness of the isomorphism between two well-ordered sets, g must coincide with f_{β_1}. Consequently, $f_{\beta_1} = f_{\beta_2} \upharpoonright \alpha \sqcup \beta_1$.

From this, it is easy to see that $f = \bigcup\{f_\beta \mid \beta < \gamma\}$ is, firstly, a function defined on $\alpha \sqcup \gamma$, and secondly, it is a strictly monotone function from $\alpha \sqcup \gamma$ to $\alpha + \gamma$ (since for any $\beta < \gamma$, we have $\alpha + \beta < \alpha + \gamma$ by the definition of addition).

It is also not difficult to see that f is a surjection, since $\forall y \in \alpha + \gamma : \exists \beta < \gamma : y \in \alpha + \beta$ by the definition of addition for a limit ordinal. Then

$\exists x \in \alpha \sqcup \beta : f(x) = f_\beta(x) = y$. Consequently, f is an order isomorphism between $\alpha \sqcup \gamma$ and $\alpha + \gamma$, i.e., $ot(\alpha \sqcup \gamma) = \alpha + \gamma$. The induction is complete.

Remark: from the construction of the isomorphism $f : \alpha \times \{0\} \cup \beta \times \{1\} \to \alpha + \beta$ in the inductive transition, one of its important properties follows: it leaves the first component of the sum in place in the sense that $\operatorname{ran}(f \restriction \alpha \times \{0\}) = \alpha$.

Now let's do the same with multiplication. Let two well-ordered sets X, Y with order relations $<_X, <_Y$ be given, respectively. By their *antilexicographical product*, we will understand the following construction: the domain of the structure will be the set $X \times Y$, on which a relation $\prec$ is defined by the following rule:

$$\langle a, b\rangle \prec \langle c, d\rangle \overset{\text{def}}{\leftrightarrow} (b <_Y d) \vee (b = d \wedge a <_X c), \tag{D4.14}$$

i.e., comparison in Y has priority over comparison in X: this can be understood as inserting a clone of X at each point of Y and comparing the result first on a large scale (by points of Y), and then within each clone of X.

It is not difficult to show that the relation $\prec$ is a well-ordering. Indeed, its irreflexivity is inherited from $<_X$ and $<_Y$. To prove transitivity, let's assume that $\langle a, b\rangle \prec \langle c, d\rangle \prec \langle e, f\rangle$. Next, we need to consider all possible options for these inequalities to hold. If all pairs belong to the same copy of X, i.e., $b = d = f$, then comparison within X is used, and it is transitive. If $b <_Y d = f$ or $b = d <_Y f$ or $b <_Y d <_Y f$, then comparison in Y is used, which is also transitive. There are no other options by virtue of the definition (D4.14). Connexity: for any two pairs $\langle a, b\rangle$ and $\langle c, d\rangle$, we first look at b and d: if $b <_Y d$ or $d <_Y b$, then the comparison for the pairs holds. If, however, $b = d$, then either $a <_X c$, or $a = c$, or $c <_X a$. In any case, the connexity of the original orders guarantees the connexity of the order of the product.[10] Finally, let $A \subseteq X \times Y$ and $A \neq \emptyset$. How to find its minimum? For this, we first project it onto Y: consider the set $A_Y = \{y \mid \exists x : \langle x, y\rangle \in A\}$. This set is not empty; let $y_0 = \min A_Y$. Next, let's project the layer of the set A corresponding to y_0 onto X, i.e., consider the set $B = \{x \mid \langle x, y_0\rangle \in A\}$. Let $x_0 = \min B$. Then, as is easy to see, $\langle x_0, y_0\rangle = \min A$.

It turns out that in this case, too, the equality holds:

$$ot(X \times Y, \prec) = ot(X, <_X) \cdot ot(Y, <_Y). \tag{D4.15}$$

[10] By the way, a linear ordering on a direct product is defined in the same way; for example, one can linearly order the plane $\mathbb{C}$ in this way, although this order will not be consistent with the algebraic operations on complex numbers.

Note that if we swap X and Y, the order $\prec$ will, generally speaking, be different.

Replacing X with the ordinal α (by the lemma on order type) and Y with the ordinal β, let us prove (D4.15) in the form $ot(\alpha \times \beta) = \alpha \cdot \beta$ by induction on β.

Suppose that the formula is true for all $\beta < \gamma$ and prove it for γ. A case analysis on the type of the ordinal γ: $\gamma = 0$, $\gamma \in \mathrm{Succ}$, $\gamma \in \mathrm{Lim}$.

a) $\gamma = 0$, then $\alpha \times \gamma = \emptyset$ and $\alpha \cdot \gamma = \alpha \cdot 0 = 0$; the identity $ot(\alpha \times \emptyset) = \alpha \cdot 0$ holds.

b) Let $\gamma = S(\beta)$. By the induction hypothesis, $ot(\alpha \times \beta) = \alpha \cdot \beta$. At the same time,

$$\alpha \times \gamma = \alpha \times S(\beta) = \alpha \times (\beta \cup \{\beta\}) = (\alpha \times \beta) \cup (\alpha \times \{\beta\})$$

and by the definition of multiplication,

$$\alpha \cdot \gamma = \alpha \cdot S(\beta) = \alpha \cdot \beta + \alpha.$$

In this case, all elements of $\alpha \times \{\beta\}$ are strictly greater in the sense of $\prec$ than all elements of $\alpha \times \beta$, and are ordered in the same way as the ordinal α. Then the set $(\alpha \times \beta) \cup (\alpha \times \{\beta\})$ can be considered as the ordered sum $(\alpha \times \beta) \sqcup (\alpha \times \{\beta\})$, and therefore, by what was proven for the ordered sum, the equality holds:

$$ot((\alpha \times \beta) \sqcup (\alpha \times \{\beta\})) = ot(\alpha \times \beta) + ot(\alpha \times \{\beta\}) = \alpha \cdot \beta + \alpha = \alpha \cdot \gamma,$$

as required.

c) Let $\gamma \in \mathrm{Lim}$. By the induction hypothesis, $ot(\alpha \times \beta) = \alpha \cdot \beta$ for all $\beta < \gamma$. This means that for each $\beta < \gamma$, there exists a unique order isomorphism $f_\beta : \alpha \times \beta \leftrightarrow \alpha \cdot \beta$. These isomorphisms form a chain by inclusion: if $\beta_1 \leqslant \beta_2$, then $f_{\beta_1} \subseteq f_{\beta_2}$, since the restriction $f_{\beta_2} \upharpoonright \alpha \times \beta_1$ coincides with f_{β_1} (here one can use the same reasoning as in point c) of the proof for addition).

From this, it is easy to see that $f = \bigcup\{f_\beta \mid \beta < \gamma\}$ is, firstly, a function defined on $\alpha \times \gamma$, and secondly, it is a strictly monotone function from $\alpha \times \gamma$ to $\alpha \cdot \gamma$ (since for any $\beta < \gamma$, we have $\alpha \cdot \beta < \alpha \cdot \gamma$ by the definition of multiplication). The surjectivity of f is proven analogously to the proof for addition. Consequently, f is an order isomorphism between $\alpha \times \gamma$ and $\alpha \cdot \gamma$, i.e., $ot(\alpha \times \gamma) = \alpha \cdot \gamma$. The induction is complete.

An equivalent definition of exponentiation also exists, although it is more complex. It is related to ordering the set of functions with *finite support*. Let two well-ordered sets X, Y with order relations $<_X, <_Y$ be given, respectively. The domain of the structure defining the power of ordinals is the set of all functions from Y to X with finite support, i.e., $F^{fin}(Y, X) = \{f : Y \to X \mid |\operatorname{supp}(f)| < \omega\}$, where $\operatorname{supp}(f) = \{y \in Y \mid f(y) \neq \min X\}$ is the support of the function f.

The *antilexicographical order* $\prec$ on this set is defined as follows: for $f, g \in F^{fin}(Y, X)$, $f \neq g$, let $\delta = \max\{y \in Y \mid f(y) \neq g(y)\}$ (the maximal point where the functions differ; it exists due to the finiteness of the supports of both functions). Then

$$f \prec g \overset{\text{def}}{\leftrightarrow} f(\delta) <_X g(\delta). \tag{D4.16}$$

Let us prove that the relation $\prec$ on $F^{fin}(Y, X)$ is a strict well-ordering.

Linearity of the order. Let $f, g \in F^{fin}(Y, X)$ and $f \neq g$. The set $D = \{y \in Y \mid f(y) \neq g(y)\}$ is non-empty and is a subset of the finite set $\operatorname{supp}(f) \cup \operatorname{supp}(g)$; therefore, D is finite and has a maximal element $\delta = \max D$. Since $<_X$ is a linear order, then either $f(\delta) <_X g(\delta)$ or $g(\delta) <_X f(\delta)$. In the first case $f \prec g$, in the second $g \prec f$. Transitivity is proven by a case analysis on the positions of the maxima, analogously to the proof for the antilexicographical product.

Well-foundedness. We will prove well-foundedness constructively, by constructing for any non-empty subset of $F^{fin}(Y, X)$ its minimal element. The algorithm consists in iteratively eliminating "larger" functions. Without loss of generality, we can assume that X and Y are ordinals; this allows for simpler notation.

Let $A \subseteq F^{fin}(Y, X)$ be a non-empty set of functions. We need to find its minimum. Let us define a decreasing sequence of sets $A_k \subseteq A$ recursively.

1) Base case of recursion: $A_0 = A$. Let, moreover, $y_0 = Y$.

2) Inductive step: given A_k and y_k, let us set

$$\begin{aligned}
y_{k+1} &= \min\{\max \operatorname{supp}(f \upharpoonright y_k) \mid f \in A_k\}, \\
B &= \{f \in A_k \mid \max \operatorname{supp}(f \upharpoonright y_k) = y_{k+1}\}, \\
z &= \min\{f(y_{k+1}) \mid f \in B\}, \\
A_{k+1} &= \{f \in B \mid f(y_{k+1}) = z\},
\end{aligned}$$

where $f \upharpoonright y_k \overset{\text{def}}{\rightleftharpoons} f \upharpoonright \{y \in Y \mid y < y_k\}$, i.e., at each step of the recursion, we first select functions whose supports lying below y_k are as close to zero as possible (the maximum of the support is minimal)—this is the set B; then we look at what values all these functions take at the point y_{k+1} and leave only those functions that take the minimal value there, collecting them into the set A_{k+1}.

In this case, it is obvious that the inequalities hold: $A_{k+1} \subseteq A_k$, and moreover, there are no elements in A_k strictly smaller than all elements of A_{k+1}, i.e., A_{k+1} contains the minimum of A_k, if it exists.

Note that we get a strictly decreasing sequence $\langle y_k \rangle$ in the well-ordered set Y, and therefore it is finite, i.e., at some k, we will not be able to construct y_{k+1} and A_{k+1}.

Why might this happen? From the construction, it is clear that this is possible only if $\mathrm{supp}(f \upharpoonright y_k) = \emptyset$ for all $f \in A_k$. This means that all supports of the functions $f \in A_k$ contain only points $\geqslant y_k$.

But in A_k, we collected only functions that have the same values at the point y_k, as well as at all previous points y_j, $0 < j \leqslant k$, i.e., there is a single function in A_k, and it is the minimum of A.

Thus, well-foundedness is proven.

Since $F^{fin}(Y, X)$ is well-ordered by the relation $\prec$, there exists a unique ordinal that is its order type. It can be proven that

$$ot(F^{fin}(Y, X), \prec) = ot(X, <_X)^{ot(Y, <_Y)}. \tag{D4.17}$$

Again, for convenience, we will prove this equality for ordinals, i.e., the equality $ot(F^{fin}(\beta, \alpha)) = \alpha^\beta$. We prove by induction on β.

1) Base case of induction: $\beta = 0$, then $F^{fin}(\emptyset, \alpha) = \{\emptyset\}$, i.e., this set contains a single function—the empty one. From this, it is obvious that $ot(F^{fin}(\emptyset, \alpha)) = 1$, which corresponds to the recursive definition $\alpha^0 = 1$.

2) Inductive step: suppose that $ot(F^{fin}(\gamma, \alpha)) = \alpha^\gamma$ for all $\gamma < \beta$.

2.1) Inductive transition for a successor: $\beta = S(\gamma)$. The set $F^{fin}(\beta, \alpha)$ consists of functions $f : S(\gamma) \longrightarrow \alpha$ with finite support. To each such function corresponds a unique pair $\langle f', \delta \rangle$, where $f' = f \upharpoonright \gamma$ and $\delta = f(\gamma)$. In other words, we reduce a function on $S(\gamma)$ to two components—its restriction to the ordinal γ and its value at the point γ. In this case, $f' \in F^{fin}(\gamma, \alpha)$, $\delta \in \alpha$, so the specified correspondence is a bijection $h : F^{fin}(\beta, \alpha) \leftrightarrow F^{fin}(\gamma, \alpha) \times \alpha$.

Now, let's introduce an antilexicographical order $\prec_1$ on the set $F^{fin}(\gamma, \alpha) \times \alpha$ by formula (D4.14), and on the sets $F^{fin}(\beta, \alpha)$ and $F^{fin}(\gamma, \alpha)$—the orders $\prec$ and $\prec'$ by formula (D4.16). Let's prove that for any $f, g \in F^{fin}(\beta, \alpha)$, the formula

$$(f \prec g) \leftrightarrow (h(f) \prec_1 h(g))$$

holds. Indeed, let δ be the deciding point for f and g; then $(f \prec g) \leftrightarrow (f(\delta) < g(\delta))$. In this case, $\delta \in S(\gamma)$, i.e., either $\delta < \gamma$ or $\delta = \gamma$.

For $\delta < \gamma$, we have, firstly, that $(f \prec g) \leftrightarrow (f' \prec' g')$, and secondly, that $f(\gamma) = g(\gamma) = x$ (by the definition of the deciding point), which is equivalent to $h(f) = \langle f', x\rangle$ and $h(g) = \langle g', x\rangle$. Thus, for $\delta < \gamma$, by the definition of the order $\prec_1$, we will have:

$$(h(f) \prec_1 h(g)) \leftrightarrow (f' \prec' g') \leftrightarrow (f \prec g).$$

For $\delta = \gamma$, we get that $(f \prec g) \leftrightarrow f(\gamma) < g(\gamma)$, and this is equivalent to the relation $\langle f', f(\gamma)\rangle \prec_1 \langle g', g(\gamma)\rangle$ by the definition of the order $\prec_1$.

So, h is not just a bijection, but also an order isomorphism between the structures $\langle F^{fin}(\beta, \alpha), \prec\rangle$ and $\langle F^{fin}(\gamma, \alpha) \times \alpha, \prec_1\rangle$, i.e., they have the same order type and, consequently, the equalities are true:

$$ot(F^{fin}(\beta, \alpha)) = ot(F^{fin}(\gamma, \alpha) \times \alpha) = ot(F^{fin}(\gamma, \alpha)) \cdot \alpha = \alpha^{\gamma} \cdot \alpha = \alpha^{\beta},$$

where the second equality is obtained from (D4.15), the third by the induction hypothesis, and the fourth by the recursive definition of the power of ordinals.

2.2) Inductive transition for a limit ordinal $\beta \in \mathrm{Lim}$. Here, we note that all functions from $F^{fin}(\beta, \alpha)$ can be considered as functions from the sets $F^{fin}(\gamma, \alpha)$, $\gamma < \beta$, by passing to their restrictions $f \mapsto f'$, which immediately gives a hint for the proof.

Therefore, taking an arbitrary ordinal $\gamma < \beta$, we can construct an isomorphic embedding of $F^{fin}(\gamma, \alpha)$ into $F^{fin}(\beta, \alpha)$ by the following rule: a function $f' \in F^{fin}(\gamma, \alpha)$ is associated with a function f that extends f' to the entire ordinal β with zero values. This correspondence will be proper, since its range will not include functions whose support has points not less than γ). Such an embedding means that

$$ot(F^{fin}(\gamma, \alpha)) < ot(F^{fin}(\beta, \alpha)),$$

so, using the induction hypothesis, we conclude that $ot(F^{fin}(\beta, \alpha)) > \alpha^{\gamma}$ for all $\gamma < \beta$, from which

$$ot(F^{fin}(\beta, \alpha)) \geqslant \alpha^{\beta},$$

where we have used the recursive definition $\alpha^{\beta} = \sup\{\alpha^{\gamma} \mid \gamma < \beta\}$.

To prove the reverse inequality, let's show that the order type of any initial segment of the set $(F^{fin}(\beta, \alpha), \prec)$ does not exceed α^{β}. Let's take an arbitrary function $f \in F^{fin}(\beta, \alpha)$ and consider the initial segment defined by it: $I_f = \{g \in F^{fin}(\beta, \alpha) \mid g \preccurlyeq f\}$. Since the support of f, $\mathrm{supp}(f)$, is a finite subset of the limit ordinal β, it is bounded above and, moreover, $\gamma = S(\max \operatorname{supp}(f)) < \beta$.

Let's prove that for any function $g \prec f$, its support $\operatorname{supp}(g)$ is also a subset of γ. Suppose the contrary: let there exist a $g \prec f$ such that $\max(\operatorname{supp}(g)) \geqslant \gamma$. Let $\delta_{max} = \max(\operatorname{supp}(g))$. Then $\delta_{max} \geqslant \gamma > \max \operatorname{supp}(f)$. At the point δ_{max}, we have $g(\delta_{max}) > 0$, while $f(\delta_{max}) = 0$, since δ_{max} does not belong to the support of f. The "deciding" point for comparing f and g is $\max\{y \mid f(y) \neq g(y)\}$. Since the functions differ at the point δ_{max}, and above it g can be non-zero while f is zero, the "deciding" point will be no less than δ_{max}. But at this point $f = 0$ and $g > 0$, which would mean $f \prec g$. This contradicts our assumption that $g \prec f$. Consequently, for any function $g \prec f$, it holds that $\operatorname{supp}(g) \subseteq \gamma$.

This means that the entire initial segment I_f is a subset of the set of functions with support in γ. This set, in turn, is isomorphic to $F^{fin}(\gamma, \alpha)$ (via the correspondence $f \mapsto f \upharpoonright \gamma$). Thus, we have an order-preserving embedding from I_f into $F^{fin}(\gamma, \alpha)$. From this, it follows that the order type of I_f cannot exceed the order type of $F^{fin}(\gamma, \alpha)$:

$$ot(I_f) \leqslant ot(F^{fin}(\gamma, \alpha)).$$

By the induction hypothesis, $ot(F^{fin}(\gamma, \alpha)) = \alpha^\gamma$. So, for any function $f \in F^{fin}(\beta, \alpha)$, the order type of the set of its predecessors does not exceed α^γ for some $\gamma < \beta$.

The order type of the entire set $F^{fin}(\beta, \alpha)$ is equal to the supremum of the order types of all its proper initial segments.

$$ot(F^{fin}(\beta, \alpha)) = \sup\{ot(I_f) \mid f \in F^{fin}(\beta, \alpha)\}.$$

And since for each f we found a $\gamma < \beta$ such that $ot(I_f) \leqslant \alpha^\gamma$, then

$$ot(F^{fin}(\beta, \alpha)) \leqslant \sup\{\alpha^\gamma \mid \gamma < \beta\}.$$

The right-hand side, by the recursive definition of the power, is equal to α^β. Thus, $ot(F^{fin}(\beta, \alpha)) \leqslant \alpha^\beta$.

Combining both inequalities, we get $ot(F^{fin}(\beta, \alpha)) = \alpha^\beta$. With this, the induction is complete.[11]

---

[11] We used a direct proof path here, although if we had laid out the theory of ordered sets a little earlier, the proof of this part would have been much shorter: we would have simply said that $ot(F^{fin}(\beta, \alpha)) = \sup\{ot(F^{fin}(\gamma, \alpha)) \mid \gamma < \beta\}$.

Thus, we have proven the following statement.

Lemma D4.8 (on the equivalence of definitions). The definitions of the sum, product, and power of ordinals via recursive definitions and by formulas (D4.12), (D4.14), (D4.16) are equivalent.

This is a very useful lemma, which in a number of cases allows for a significant simplification of the proof of the algebraic properties of ordinals.

Theorem D4.8. *Algebraic properties of ordinals:*

1° $S(\alpha) = \alpha + 1$, $0 + \alpha = \alpha$, $0 \cdot \alpha = 0$, $\alpha \cdot 1 = \alpha = 1 \cdot \alpha$.

2° *Non-commutativity:* $\omega + 2 \neq 2 + \omega$, $\omega \cdot 2 \neq 2 \cdot \omega$.

3° *Right distributivity fails:* $(1+1)\cdot\omega = 2\cdot\omega = \omega$. *But* $(1\cdot\omega)+(1\cdot\omega) = \omega+\omega$.

4° *Associativity:* $(\alpha + \beta) + \gamma = \alpha + (\beta + \gamma)$, $(\alpha \cdot \beta) \cdot \gamma = \alpha \cdot (\beta \cdot \gamma)$.

5° *Left distributivity:* $\alpha \cdot (\beta + \gamma) = (\alpha \cdot \beta) + (\alpha \cdot \gamma)$.

6° *Right strict monotonicity of addition:* $\beta < \gamma \Leftrightarrow \alpha + \beta < \alpha + \gamma$.

7° *Right strict monotonicity of multiplication: if* $\alpha > 0$, *then* $\beta < \gamma \Leftrightarrow \alpha \cdot \beta < \alpha \cdot \gamma$.

8° *Left non-strict monotonicity: if* $\alpha < \beta$, *then* $\alpha+\gamma \leqslant \beta+\gamma$ *and* $\alpha\cdot\gamma \leqslant \beta\cdot\gamma$.

9° *No zero divisors: if* $\alpha \cdot \beta = 0$, *then* $\alpha = 0$ *or* $\beta = 0$.

10° *Uniqueness of subtraction: if* $\beta \leqslant \alpha$, *then there exists a unique ordinal* γ *such that* $\beta + \gamma = \alpha$.

11° *Division with remainder: for any* α *and* $\beta > 0$, *there exists a unique pair* γ, δ *such that* $\alpha = \beta \cdot \gamma + \delta$, *where* $\delta < \beta$.

12° *Properties of exponentiation:* $1^\alpha = 1$; $\alpha^1 = \alpha$;

13° *Distributivity over exponent:* $\alpha^{\beta+\gamma} = \alpha^\beta \cdot \alpha^\gamma$;

14° *Monotonicity of exponentiation: if* $\alpha > 1$ *and* $\beta < \gamma$, *then* $\alpha^\beta < \alpha^\gamma$;

15° *Multiplicativity of exponentiation:* $(\alpha^\beta)^\gamma = \alpha^{\beta\cdot\gamma}$.

16° *Failure of distributivity over base: for example,* $(2 \cdot 2)^\omega = 4^\omega = \omega$, *but* $2^\omega \cdot 2^\omega = \omega \cdot \omega = \omega^2 > \omega$.

Proof. **1)** Recall that we denote the first ordinals as natural numbers, i.e., $0 = \emptyset$, $1 = S(\emptyset)$, $2 = S(1)$, etc. Then from the definition of ordinal addition, it is easy to see that $\alpha+1 = \alpha+S(0) = S(\alpha+0) = S(\alpha)$. Note that $0+\alpha = \alpha$ has to be proven by transfinite induction: $0+0 = 0$ by definition, $0+S(\alpha) = S(0+\alpha) = S(\alpha)$, $0 + \lambda = \sup\{0 + \beta \mid \beta < \lambda\} = \lambda$, where we use the induction hypothesis and consider two different cases of the inductive step: for a successor and for a limit ordinal λ.

Similarly, we prove that $0 \cdot \alpha = 0$: $0 \cdot 0 = 0$ by definition, $0 \cdot S(\alpha) = 0 \cdot \alpha + 0 = 0$, $0 \cdot \lambda = \sup\{0 \cdot \beta \mid \beta < \alpha\} = 0$.

By definition, $\alpha \cdot 1 = \alpha \cdot S(0) = \alpha \cdot 0 + \alpha = \alpha$. We prove $1 \cdot \alpha$ by induction: $1 \cdot 0 = 0$, $1 \cdot S(\alpha) = 1 \cdot \alpha + 1 = \alpha + 1 = S(\alpha)$, $1 \cdot \lambda = \sup\{1 \cdot \beta \mid \beta < \lambda\} = \lambda$.

2) $2 + \omega = \sup\{2 + n \mid n < \omega\} = \omega$, but $\omega + 2 = S(S(\omega)) > \omega$; $\omega \cdot 2 = \omega \cdot S(1) = \omega \cdot 1 + \omega = \omega + \omega > \omega$, but $2 \cdot \omega = \sup\{2n \mid n < \omega\} = \omega$.

3) The value of $(1 + 1) \cdot \omega$ is easier to represent using the second definition of multiplication: instead of each natural number, a pair is inserted—this is the same as first taking all even numbers, and then pairing them with the following odd numbers; the resulting order type will still be ω. On the other hand, the sum $(1 \cdot \omega) + (1 \cdot \omega)$ clearly contains a "limit point in the middle" and can in no way be equal to ω.

4) Associativity is also easier to prove using the second definition of the operations.

Addition: consider the sets $(\alpha \times \{0\} \cup \beta \times \{1\}) \times \{0\} \cup \gamma \times \{1\}$ and $\alpha \times \{0\} \cup ((\beta \times \{0\} \cup \gamma \times \{1\}) \times \{1\})$. There exists a natural order-preserving isomorphism between them: $\langle\langle a, 0\rangle, 0\rangle \mapsto \langle a, 0\rangle$, $\langle\langle b, 1\rangle, 0\rangle \mapsto \langle\langle b, 0\rangle, 1\rangle$, $\langle c, 1\rangle \mapsto \langle\langle c, 1\rangle, 1\rangle$, where $a \in \alpha, b \in \beta, c \in \gamma$.

The easiest way to understand this is if both sets are reduced to the simpler $\alpha \times \{0\} \cup \beta \times \{1\} \cup \gamma \times \{2\}$, in which pairs are compared first by the second argument $(0 < 1 < 2)$, and then, if they are equal, by the first. Such a set naturally has associativity, as it is inherited from the associativity of the operation $\cup$.

Multiplication: consider the sets $(\alpha \times \beta) \times \gamma$ and $\alpha \times (\beta \times \gamma)$. There exists a natural order-preserving isomorphism between them: $\langle\langle a, b\rangle, c\rangle \mapsto \langle a, \langle b, c\rangle\rangle$. Indeed, let $\langle\langle a, b\rangle, c\rangle \prec \langle\langle x, y\rangle, z\rangle$; then there are two options:[12]

1) $c < z$, then $\langle b, c\rangle \prec \langle y, z\rangle$, and then also $\langle a, \langle b, c\rangle\rangle \prec \langle x, \langle y, z\rangle\rangle$;

2) $c = z$, then $\langle a, b\rangle \prec \langle x, y\rangle$, and here there are again two cases:

2.1) $b < y$, then $\langle b, c\rangle \prec \langle y, z\rangle$, from which $\langle a, \langle b, c\rangle\rangle \prec \langle x, \langle y, z\rangle\rangle$;

2.2) $b = y$, then $a < x$, from which $\langle a, \langle b, c\rangle\rangle \prec \langle x, \langle y, z\rangle\rangle$.

5) To show that $\alpha \cdot (\beta + \gamma) = \alpha \cdot \beta + \alpha \cdot \gamma$, we again use the second definition of the operations: it is sufficient to show that $\alpha \times (\beta \sqcup \gamma)$ is isomorphic to $(\alpha \times \beta) \sqcup (\alpha \times \gamma)$. This is done by analogy with the previous point and is almost obvious, since $A \times (B \cup C) = (A \times B) \cup (A \times C)$.

---

[12] We do not use different notations for comparisons on ordered sums, as it is too cumbersome; we only distinguish between comparison in ordinals and in sets.

6) Let's prove the formula $(\beta < \gamma) \leftrightarrow (\alpha + \beta) < (\alpha + \gamma)$ using the second definition of the operations, but partly using the recursive definition ($S(\alpha + \beta) = \alpha + S(\beta)$).

If $\beta < \gamma$, then $S(\beta) \leqslant \gamma$ and there exists a natural strictly monotone embedding $\mathrm{id} : \alpha \sqcup S(\beta) \to \alpha \sqcup \gamma$. Since $ot(\alpha \sqcup S(\beta)) = \alpha + S(\beta)$ and $ot(\alpha \sqcup \gamma) = \alpha + \gamma$, there is a strictly monotone embedding $f : S(\alpha + \beta) \to \alpha + \gamma$, and as we have noted earlier (see the proof of the lemma on order type), $f(x) \geqslant x$, therefore, $f(\alpha + \beta) \geqslant \alpha + \beta$, and at the same time $f(\alpha + \beta) < \alpha + \gamma$, from which we get that $\alpha + \beta < \alpha + \gamma$.

Conversely, let $\alpha + \beta < \alpha + \gamma$. Suppose that $\neg(\beta < \gamma)$; then $\beta \geqslant \gamma$, so by the previous argument, $\alpha + \beta \geqslant \alpha + \gamma$. A contradiction.

Along the way, we have proven another useful fact: if there is a monotone embedding of α into β, then $\alpha \leqslant \beta$, i.e., strictly monotone embeddings are "incompressible".

7) Let $\alpha > 0$ and $\beta < \gamma$. Let's prove the formula $\alpha \cdot \beta < \alpha \cdot \gamma$ by induction on $\gamma > \beta$, assuming that it is true for ordinals less than γ. A case analysis for γ: $\gamma = S(\beta)$ (no smaller ordinals), $S(\beta) < \gamma \in \mathrm{Succ}$, $\beta < \gamma \in \mathrm{Lim}$.

1) $\gamma = S(\beta)$, then $\alpha \cdot \gamma = \alpha \cdot \beta + \alpha > \alpha \cdot \beta$, since $\alpha > 0$.

2) $\gamma = S(\delta)$, where $\delta > \beta$, then $\alpha \cdot \gamma = \alpha \cdot \delta + \alpha > \alpha \cdot \beta$, since $\alpha \cdot \delta > \alpha \cdot \beta$ by the induction hypothesis.

3) $\gamma \in \mathrm{Lim}$, then $\alpha \cdot \gamma = \sup\{\alpha \cdot \delta \mid \delta < \gamma\} = \sup\{\alpha \cdot S(\delta) \mid \delta < \gamma\}$, from which it follows that $\alpha \cdot S(\beta) \leqslant \alpha \cdot \gamma$, and since $\alpha \cdot S(\beta) = \alpha \cdot \beta + \alpha > \alpha \cdot \beta$ (since $\alpha > 0$), then $\alpha \cdot \beta < \alpha \cdot \gamma$. The induction is complete.

Conversely, let $\alpha \cdot \beta < \alpha \cdot \gamma$. Assuming that $\beta \geqslant \gamma$, we would arrive at the inequality $\alpha \cdot \beta \geqslant \alpha \cdot \gamma$ by what has just been proven. A contradiction. Thus, $\beta < \gamma$.

8) Left non-strict monotonicity is proven similarly. But in the proof, one cannot resort to the help of $S(\delta)$ to obtain a strict inequality, since we do not have a recursive rule for the left component of addition and multiplication. Therefore, the monotonicity turns out to be non-strict.

Moreover, one can easily give an example where strict monotonicity is violated: compare $1 < 2$, but $1 + \omega = 2 + \omega = \omega$, and also $1 \cdot \omega = 2 \cdot \omega = \omega$.

9) If $\alpha, \beta > 0$, then $\alpha \geqslant 1$ and $\beta \geqslant 1$, from which by the previous properties we get that $\alpha \cdot \beta \geqslant 1 \cdot 1 = 1 > 0$. So if $\alpha \cdot \beta = 0$, then $(\alpha = 0) \vee (\beta = 0)$.

10) Let $\alpha \geqslant \beta$, then let's set $\alpha \dot{-} \beta = ot(\alpha \setminus \beta)$. It is not difficult to see that $\alpha = \beta \sqcup (\alpha \setminus \beta)$ (here we don't even need to multiply by 0 and 1), so by the second definition of addition, we have $\alpha = \beta + ot(\alpha \setminus \beta)$, i.e., $\gamma = a \dot{-} \beta$ is the desired ordinal. If there were two ordinals satisfying this equality, $\gamma_1 < \gamma_2$, then by property 6 we would get $\beta + \gamma_1 < \beta + \gamma_2$, i.e., $\alpha < \alpha$, which is impossible for ordinals.

11) Let's move on to the theorem on division with a remainder. Let $\beta > 0$ and α be an arbitrary ordinal. Let's set $\gamma' = \min\{\delta \mid \beta \cdot \delta > \alpha\}$. This set is non-empty, since $\beta \cdot (\alpha + 1) > \alpha$, so the specified γ' exists. Let's show that $\gamma' \in \mathrm{Succ}$. First, we note that $\gamma' > 0$. Second, suppose that $\gamma' \in \mathrm{Lim}$. Then for any $\delta < \gamma'$, we have $\beta \cdot \delta' \leqslant \alpha$ and by the definition of multiplication, $\beta \cdot \gamma' = \sup\{\beta \cdot \delta \mid \delta < \gamma'\} \leqslant \alpha$, which contradicts the choice of γ'. Consequently, $\gamma' = \gamma + 1$ for some γ. Moreover, by the choice of γ', we get that $\alpha \geqslant \beta \cdot \gamma$.

Let further $\delta = \alpha \dot{-} (\beta \cdot \gamma)$. Then by the previous property, $\alpha = \beta \cdot \gamma + \delta$.

Suppose that $\alpha = \beta \cdot \gamma_1 + \delta_1 = \beta \cdot \gamma_2 + \delta_2$, where $\delta_1, \delta_2 < \beta$. Let $\gamma_1 < \gamma_2$. Then $\gamma_1 + 1 \leqslant \gamma_2$, so $\alpha \geqslant \beta \cdot \gamma_2 \geqslant \beta \cdot (\gamma_1 + 1) = \beta \cdot \gamma_1 + \beta > \beta \cdot \gamma_1 + \delta_1 = \alpha$. A contradiction. Similarly, it cannot be that $\gamma_1 > \gamma_2$, so $\gamma_1 = \gamma_2$. From this, by property 10, we get that $\delta_1 = \delta_2$ as well, so the pair γ, δ is uniquely determined.

12) Let's move on to proving the properties of ordinal exponentiation.

$1^\alpha = 1$ is proven by induction on α: $1^0 = 1$ by definition, then $1^{S(\alpha)} = 1^\alpha \cdot 1 = 1$ and $1^\lambda = \sup\{1^\gamma \mid \gamma < \lambda\} = 1$, where $\lambda \in \mathrm{Lim}$.

Next, $\alpha^1 = \alpha^{S(0)} = \alpha^0 \cdot \alpha = 1 \cdot \alpha = \alpha$.

13) The identity $\alpha^{\beta+\gamma} = \alpha^\beta \cdot \alpha^\gamma$ can be proven using the set-theoretic definitions of operations on ordinals. Namely, we need to prove that

$$\langle F^{fin}(\beta \sqcup \gamma, \alpha), \prec\rangle \cong \langle F^{fin}(\beta, \alpha) \times F^{fin}(\gamma, \alpha), \prec\rangle.$$

On the left, we have functions with finite support in $\beta \sqcup \gamma$. It is clear that each such function f can be represented as the union $f' \cup f''$, where $f' = f \upharpoonright \beta \times \{0\}$, $f'' = f \upharpoonright \gamma \times \{1\}$. The correspondence $f \mapsto \langle f', f''\rangle$ is one-to-one. On the right, we have functions g and h with finite support from two ordinals β and γ, which we consider independently. From these functions, it is easy to get our f' and f'' by the following rule:

$$f'(x, 0) = g(x), \quad f''(x, 1) = h(x),$$

and this correspondence is also one-to-one. Consequently, between the base sets $F^{fin}(\beta \sqcup \gamma, \alpha)$ and $F^{fin}(\beta, \alpha) \times F^{fin}(\gamma, \alpha)$, we have a bijection.

It remains to check that this is an order isomorphism. Without loss of generality, instead of the original functions g, h, we can consider their corresponding f', f'', i.e., instead of the base set $F^{fin}(\beta, \alpha) \times F^{fin}(\gamma, \alpha)$, we can immediately consider the set $F^{fin}(\beta \times \{0\}, \alpha) \times F^{fin}(\gamma \times \{1\}, \alpha)$.

Let $f, g \in F^{fin}(\beta \sqcup \gamma, \alpha)$. Let also $\delta = \max\{x \mid f(x) \neq g(x)\}$ be the "deciding" point. For it, there are two possibilities: $\delta \in \beta \times \{0\}$ or $\delta \in \gamma \times \{1\}$. Let's consider the first case; then δ will be the deciding point for f' and g', and moreover, $(f \prec g) \leftrightarrow (f' \prec g')$, and also $f'' = g''$. Thus, in the first case, $(f \prec g) \leftrightarrow \langle f', f'' \rangle \prec \langle g', g'' \rangle$, where by the symbol $\prec$ we denote different orderings on the corresponding sets, defined by formulas (D4.14) and (D4.16). In the second case, we have $f'' \prec g''$, so in the direct product, the comparison is by the second component, and again the equivalence $(f \prec g) \leftrightarrow \langle f', f'' \rangle \prec \langle g', g'' \rangle$ arises.

Thus, the specified correspondence $f \mapsto \langle f', f'' \rangle$ is an order isomorphism. From this, it follows that $\alpha^{\beta+\gamma} = \alpha^{\beta} \cdot \alpha^{\gamma}$.

14) Let's now prove the monotonicity of exponentiation: if $\alpha > 1$ and $\beta < \gamma$, then $\alpha^{\beta} < \alpha^{\gamma}$. This property is also easier to obtain from the set-theoretic definition of exponentiation. On the left of the inequality, we have the base set consisting of functions $f : \beta \to \alpha$ with finite support, and on the right—of functions $g : \gamma \to \alpha$ with finite support. From this, it is immediately clear why it is required that $\alpha > 1$: if $\alpha = 1 = \{0\}$, then all functions become identically equal to zero, i.e., the base sets consist of a single point, and we get $\alpha^{\beta} = \alpha^{\gamma}$; and if $\alpha = 0$, the inequality is not true at all, since $0^0 > 0^1$.

Let $\alpha > 1$, i.e., there are at least two elements in α. Then let's consider the function $g : \gamma \to \alpha$, given by the rule $g(\beta) = 1$ and $g(x) = 0$ for $x \neq \beta$. Such a function will be strictly greater than any function f (to compare them correctly, all f must be extended by zeros to γ). Thus, in the order defined on $F^{fin}(\gamma, \alpha)$, there will be at least one point greater than all points from $F^{fin}(\beta, \alpha)$. From this, it follows that $\alpha^{\beta} < \alpha^{\gamma}$.

15) Let's prove the identity $\alpha^{\beta \cdot \gamma} = (\alpha^{\beta})^{\gamma}$, for a change, by induction on γ.

Base case: $\alpha^{\beta \cdot 0} = \alpha^0 = 1$ and $(\alpha^{\beta})^0 = 1$.

Inductive step for a successor: $\alpha^{\beta \cdot (\gamma+1)} = \alpha^{\beta \cdot \gamma + \beta} = \alpha^{\beta \cdot \gamma} \cdot \alpha^{\beta} = (\alpha^{\beta})^{\gamma} \cdot (\alpha^{\beta})^1 = (\alpha^{\beta})^{\gamma+1}$, where we have used the previous property.

Inductive step for a limit ordinal: first, we note that if $\lambda \in \mathrm{Lim}$, then $\beta \cdot \lambda \in \mathrm{Lim}$ if $\beta > 0$. Therefore, we first check the identity for $\beta = 0$: $\alpha^{0 \cdot \lambda} = \alpha^0 = 1$ and $(\alpha^0)^{\gamma} = 1^{\gamma} = 1$.

Next, let $\beta > 0$. Since $\beta \cdot \lambda \in \mathrm{Lim}$, then $\alpha^{\beta \cdot \lambda} = \sup\{\alpha^{\delta} \mid \delta < \beta \cdot \lambda\}$, and for each $\delta < \beta \cdot \lambda$, one can find a $\xi < \lambda$ such that $\delta < \beta \cdot \xi$. How to do this? Divide δ by β with a remainder; then $\delta = \beta \cdot \sigma + \rho$, where $\rho < \beta$, and also $\sigma < \lambda$ (otherwise we would get $\delta \geqslant \beta \cdot \lambda$). Then $\sigma + 1 < \lambda$ (due to the limit nature of λ), from which $\beta \cdot (\sigma + 1) > \delta$ and $\beta \cdot (\sigma + 1) < \beta \cdot \lambda$, so the desired $\xi = \sigma + 1$.

By the monotonicity of exponentiation, we get that:

$$\sup\{\alpha^\delta \mid \delta < \beta \cdot \lambda\} = \sup\{\alpha^{\beta \cdot \xi} \mid \xi < \lambda\},$$

since, on the one hand, every $\delta < \beta \cdot \lambda$ can be bounded above by $\beta \cdot \xi < \beta \cdot \lambda$, where $\xi < \lambda$, and on the other hand, every $\beta \cdot \xi$ can be bounded above by some $\delta < \beta \cdot \lambda$ (for example, by setting $\delta = \beta \cdot \xi$).[13]

Using the induction hypothesis, we replace $\alpha^{\beta \cdot \xi}$ with $(\alpha^\beta)^\xi$, from which we get that

$$\alpha^{\beta \cdot \lambda} = \sup\{(\alpha^\beta)^\xi \mid \xi < \lambda\} = (\alpha^\beta)^\lambda,$$

as required. □

Thus, the class of all ordinals Ord behaves in the theory ZF in some sense just like ω behaves in HF, where the axiom of infinity is false: it is a well-ordered scale that defines a kind of infinity in the corresponding universe of sets. This scale, moreover, has its own arithmetic with good properties like division with a remainder (Euclidean) and no zero divisors. But there is also a difference from the natural numbers: the operations of addition and multiplication are non-commutative in the infinite domain, and there are also many equinumerous ordinals. For example, $\omega + \omega \simeq \omega$.

The situation can be attempted to be corrected if we single out special representatives among the ordinals, responsible for the uniqueness of cardinality. This is how the definition of cardinal numbers arises.

D4.3.4 Cardinal Arithmetic

Unlike ordinals, which serve as standards for all well-ordered sets, cardinals are intended to measure the "size" or "cardinality" of arbitrary sets.

An ordinal that is not equinumerous with any of its elements is called a **cardinal** (or a *cardinal number*).

An equivalent definition: a cardinal is an initial ordinal, that is, an ordinal that is the smallest among all ordinals of a given cardinality. In other words, a cardinal realizes the most "compact" order on a set.

Previously, we introduced the notion of cardinality for an arbitrary set as a class of equinumerous sets. The problem with these classes is that, first, they are not objects of the theory ZF but its formulas, and consequently, we cannot formalize statements like "for any cardinality" or "there exists a cardinality,"

[13] In this case, it is said that the set $\{\beta \cdot \xi \mid \xi < \lambda\}$ is cofinal in the ordinal $\beta \cdot \lambda$.

as this would mean quantifying over formulas. Second, without the Axiom of Choice, they cannot be linearly ordered; that is, it is not guaranteed that any two sets can be compared by their cardinality.

By introducing cardinals as a subclass of the scale of ordinals, we immediately solve these two problems automatically. Cardinals are legitimate sets in ZF, one can quantify over them (apply a quantifier), and any two cardinals can be compared. The price for this gain is that one cannot prove the existence of an equinumerous cardinal for every set, again, in the absence of the Axiom of Choice. When the Axiom of Choice is added to the theory, cardinals become universal measures of the cardinalities of sets.

If a set X is equinumerous with some cardinal τ, then it is said that τ is **the cardinality of the set** X, which is denoted as: $|X| = \tau$. In particular, all natural numbers and the ordinal ω are cardinal numbers.

Since any well-ordered set has a unique order type, it also has a cardinality. Indeed, if $ot(X) = \alpha$, then the cardinality of X will be the cardinal $\tau = \min\{\beta \mid (\beta \leqslant \alpha) \wedge (\beta \simeq \alpha)\}$.

Speaking of the arithmetic of cardinal numbers, one usually means infinite cardinals, since in the finite domain it coincides with ordinary arithmetic, and the interaction of finite cardinal numbers with infinite ones, unlike on the scale of ordinals, is not distinguished by a richness of properties.

The operations are defined as follows, as a rule, immediately for arbitrary sets. Let $|A| = \tau$ and $|B| = \chi$; then the *sum of cardinals* $\tau + \chi$ is defined as the cardinality $|A \sqcup B|$, the *product of cardinals* $\tau \cdot \chi$ is defined as the cardinality $|A \times B|$, and the *power of cardinals* τ^χ is defined as the cardinality $|A^B|$, if it exists. [14]

Remark: The operations on ordinals and cardinals are denoted identically. Considering that cardinals are ordinals, it is easy to confuse which operation is being applied to them, especially if it is not obvious from the notation. Therefore, one must always monitor the context to see which operations are being used. In cases where ordinal operations must be applied specifically to cardinals, different notations for these operations are introduced, for example, $+^o$ vs. $+^c$. This is a subtle question of conventions that is directly related to mathematical language. The attempt to make the language more convenient

[14] The well-ordering of the disjoint union and the Cartesian product can be constructed explicitly in ZF using the rules from formulas (D4.12) and (D4.14), respectively. This guarantees that their cardinalities exist if the cardinalities of the original sets exist. The situation for exponentiation is fundamentally different, since the domain for cardinal exponentiation (A^B) is the set of *all* functions, whereas for ordinal exponentiation it is the set of functions with *finite support*, and a well-ordering for A^B cannot be constructed in ZF alone.

for perception inevitably leads to the fact that each specific text has certain semantic conventions and predefined agreements.

If τ is a cardinal, then τ^+ denotes the smallest cardinal greater than τ. The cardinal τ^+ is called a *successor cardinal*. If a cardinal is not a successor and is not equal to zero, it is called a (*weakly*) *limit cardinal*. Here the first problems with cardinals begin. In ZF, it is impossible to prove that τ^+ exists for every τ, because we cannot guarantee the existence of a well-ordering of the set $\mathcal{P}(\tau)$. This means we cannot prove that cardinals form a proper class, nor can we prove that they form a set. These statements are independent of ZF. By adding the Axiom of Choice, with the help of Zermelo's theorem, we can already guarantee the well-ordering of any set, and thus guarantee the existence of a cardinality for that set, as well as the existence of arbitrarily large cardinalities. So in the theory ZFC, cardinals form a proper class.

Cardinals are usually denoted by Greek letters from the end of the alphabet, as well as by a special enumeration of "alephs".

Theorem D4.9. *Let $\mathcal{K}$ be a proper class, well-ordered by some class-relation. Then there exists a unique class-mapping F : Ord $\leftrightarrow$ $\mathcal{K}$ that is an order isomorphism.*

This theorem is a generalization of the lemma on order type to the case of proper classes, and its proof is not fundamentally different.

In particular, we can enumerate the infinite cardinals using this theorem as follows: $\aleph_0 = \omega$, $\aleph_{\alpha+1} = \aleph_\alpha^+$, $\aleph_\lambda = \sup\{\aleph_\alpha \mid \alpha < \lambda\}$, where $\lambda \in \mathrm{Lim}$.

From this, we see that limit cardinals have indices that are limit ordinals, and successor cardinals have indices that are successor ordinals. Only the limit cardinal ω has the index 0.

Having an enumeration of a class (or even a set) by ordinals, we can run induction over this class. For example, we can prove some property $\varphi(\tau)$ only for cardinals τ, checking the inductive step, as usual, for the three types of ordinal indices: 0, Succ, and Lim. In the case of alephs, this means that we will consider the inductive step in three cases: when $\tau = \omega$, when $\tau = \chi^+$, and when τ is a limit cardinal.

Unlike ordinal arithmetic, arithmetic for infinite cardinals turns out to be much simpler, which makes it convenient for measuring the "sizes" of sets.

Theorem D4.10. *Algebraic properties of cardinals:*

1° $\kappa + 0 = \kappa$; $\kappa \cdot 1 = \kappa$; $\kappa \cdot 0 = 0$;

2° *addition and multiplication of cardinals are associative, commutative, and distributive;*

3° *there are no zero divisors;*

4° *addition and multiplication are non-strictly monotone;*

5° $\tau^0 = 1$, *in particular,* $0^0 = 1$; $0^\tau = 0$ *if* $\tau > 0$;

6° $1^\tau = 1$;

7° $\tau^{\kappa+\chi} = \tau^\kappa \cdot \tau^\chi$; $(\tau^\kappa)^\chi = \tau^{\kappa\cdot\chi}$; $(\tau \cdot \kappa)^\chi = \tau^\chi \cdot \kappa^\chi$;

8° *the exponent is monotone in both arguments;*

9° $\kappa^2 = \kappa$ *if* $\kappa \geqslant \omega$ *(Hessenberg's Theorem);*

10° *if* $\max\{\tau, \chi\} \geqslant \omega$ *and* $\tau, \chi > 0$, *then* $\tau + \chi = \tau \cdot \chi = \max\{\tau, \chi\}$;

11° $|\mathcal{P}(X)| = 2^{|X|}$ *(if exists);*

12° $\tau < 2^\tau$ *(Cantor's Theorem).*

Proof. Most of the properties (1-8, 11-12) are direct consequences of the definitions of operations on cardinalities through disjoint union, Cartesian product, and the set of mappings. Their proofs reduce to constructing the corresponding bijections and are analogous to the proofs for finite sets.

The key properties (9, 10), specific to infinite cardinals, require more detailed consideration. The central result here is **Hessenberg's Theorem**, which states that $\kappa^2 = \kappa$ for any infinite cardinal κ.

Proof of Hessenberg's Theorem. The statement is equivalent to the fact that for any ordinal α, the equality $\aleph_\alpha \cdot \aleph_\alpha = \aleph_\alpha$ holds. The proof is by transfinite induction on α. The base case of induction ($\alpha = 0$) is the equality $\aleph_0 \cdot \aleph_0 = \aleph_0$, which follows from the existence of a bijection between $\omega \times \omega$ and ω.

For the inductive step, assume that for all $\beta < \alpha$, it is true that $\aleph_\beta \cdot \aleph_\beta = \aleph_\beta$. Let us prove that $\aleph_\alpha \cdot \aleph_\alpha = \aleph_\alpha$. To do this, we introduce a well-ordering relation $\prec$ on the set $\aleph_\alpha \times \aleph_\alpha$ by the rule:

$$\langle \xi_1, \eta_1 \rangle \prec \langle \xi_2, \eta_2 \rangle$$
$$\overset{\text{def}}{\leftrightarrow} \begin{cases} \max\{\xi_1, \eta_1\} < \max\{\xi_2, \eta_2\}, & \text{or} \\ \max\{\xi_1, \eta_1\} = \max\{\xi_2, \eta_2\} \text{ and } \xi_1 < \xi_2, & \text{or} \\ \max\{\xi_1, \eta_1\} = \max\{\xi_2, \eta_2\}, & \\ \xi_1 = \xi_2 \text{ and } \eta_1 < \eta_2. & \end{cases}$$

This ordering by "frames" can be visualized on the example of the set $\omega \times \omega$, where, unlike Cantor's diagonal numbering, the numbering proceeds along the top and right edges of the squares $n \times n$. It is easy to check that this is indeed a well-ordering. Irreflexivity is obvious, since the definition uses strict inequalities. Transitivity is inherited from the original order within $\aleph_\alpha$. Connexity follows from the fact that if two points lie on different "frames," they are compared by the number of this "frame," which is equal to $\max\{\xi, \eta\}$, and if they belong to the same "frame," then comparison by coordinates within $\aleph_\alpha$ is used. The minimum of a non-empty subset A is also not difficult to find: first, one looks for points $\langle \xi, \eta \rangle$ with the minimum value of $\max\{\xi, \eta\}$, then among them, with the minimum value of ξ, and finally, among the remaining, with the minimum value of η.

Let $\delta = ot(\aleph_\alpha \times \aleph_\alpha, \prec)$. Then $|\aleph_\alpha \times \aleph_\alpha| = |\delta|$. Let us prove that $|\delta| = \aleph_\alpha$. The estimate $|\delta| \geqslant \aleph_\alpha$ is obvious. For the reverse estimate $|\delta| \leqslant \aleph_\alpha$, consider an arbitrary pair $\langle \xi, \eta \rangle \in \aleph_\alpha \times \aleph_\alpha$ and the set of its predecessors $P(\xi, \eta)$.

Let $\mu = \max\{\xi, \eta\}$. Since $\xi, \eta < \aleph_\alpha$, then $\mu < \aleph_\alpha$ as well. From the definition of $\prec$, it follows that $P(\xi, \eta) \subseteq (\mu + 1) \times (\mu + 1)$. Consequently, $|P(\xi, \eta)| \leqslant |\mu + 1|^2$. Since $\mu + 1 < \aleph_\alpha$, then $|\mu + 1| = \aleph_\beta$ for some $\beta < \alpha$. By the induction hypothesis, $|\mu + 1|^2 = \aleph_\beta \cdot \aleph_\beta = \aleph_\beta < \aleph_\alpha$. Thus, every initial segment of the set $(\aleph_\alpha \times \aleph_\alpha, \prec)$ has a cardinality less than $\aleph_\alpha$. This means that its order type δ cannot exceed $\aleph_\alpha$, i.e., $\delta \leqslant \aleph_\alpha$. From the two estimates, it follows that $|\delta| = \aleph_\alpha$, from which $\aleph_\alpha^2 = \aleph_\alpha$, which proves Hessenberg's theorem.

Corollary. Now property 10 is derived immediately. Let $\kappa \geqslant \chi$ and $\kappa \geqslant \omega$, and also $\chi > 0$ (this condition is needed for multiplication). Then the following inequalities hold:

$$\kappa \leqslant \kappa + \chi \leqslant \kappa + \kappa = 2 \cdot \kappa \leqslant \kappa \cdot \kappa = \kappa^2 = \kappa,$$

$$\kappa \leqslant \kappa \cdot \chi \leqslant \kappa \cdot \kappa = \kappa^2 = \kappa.$$

From this, by the Cantor–Bernstein–Schroeder theorem, it follows that $\kappa + \chi = \kappa \cdot \chi = \kappa = \max\{\kappa, \chi\}$.

From Hessenberg's theorem, by induction on $n \in \omega$, a more general identity is easily derived: $\kappa^n = \kappa$, where $\kappa \geqslant \omega$ and $n > 0$. □

D4.4 The Axiom of Choice

Let us recall the formulation of the (general) Axiom of Choice:

$$\text{(AC)} \qquad (\emptyset \notin A) \to \exists f : A \to \bigcup A : \ (\forall x \in A : f(x) \in x),$$

as well as its countable version:

$$(\text{AC}_\omega) \qquad (|A| = \omega) \wedge (\emptyset \notin A) \to \exists f : A \to \bigcup A : \ (\forall x \in A : f(x) \in x).$$

Other forms of the Axiom of Choice also exist. First, it is worth mentioning the equivalent formulation in the form (C1.1) for indexed families of sets.

Second, one can impose restrictions not only on the cardinality of the set A itself but also on the cardinality of its elements. Thus, for example, the axiom of choice AC_ω^{fin} states that for a countable family of *finite* sets, a choice function exists. A generalization is to consider the axiom of choice AC_κ^λ, which asserts that for any family of cardinality κ of sets of cardinality less than λ, a choice function exists, where κ, λ are infinite cardinals.

D4.4.1 Countable Choice and its Consequences

Let's see how, with the help of AC_ω, one can equate the two definitions of finite and infinite sets (the classical one and Dedekind's), returning to the discussion at the end of Section D4.2.2.

Theorem D4.11. *It follows from AC_ω that every infinite set is Dedekind-infinite.*

Proof. Let A be an infinite set, i.e., it is not equinumerous with any $n \in \omega$. We will prove by induction that for any $n \in \omega$, there is a subset of A with cardinality n.

Base case: $n = 0$. Such a subset exists, $\emptyset \subseteq A$.

Let there exist a subset $B \subseteq A$ for n such that $|B| = n$. It is obvious that $A \setminus B \neq \emptyset$, since otherwise we would have $A = B$, i.e., A would be finite. Then there exists an $x \in A \setminus B$ and, consequently, there exists the set $B' = B \cup \{x\}$. It is easy to see that $|B'| = n + 1$. The induction is complete.

Now consider the sets

$$S_n = \{B \in \mathcal{P}(A) \mid |B| = n\}, \quad n \in \omega.$$

The set S_n can be understood as a "layer" in the powerset $\mathcal{P}(A)$, consisting of finite equinumerous sets. Each S_n is non-empty by what has been proven.

Then $\{S_n\}$ is a family of non-empty sets indexed by the set ω. Using AC_ω, let's take a choice function $f : \omega \to \mathcal{P}(A)$ such that $f(n) \in S_n$. We obtain a sequence of subsets $B_n = f(n)$ having cardinality n.

The resulting sequence of sets does not yet allow us to select a countable sequence of elements of A, because it may happen that B_n is contained in the union of all preceding ones, which prevents the choice of a new point. Therefore, from the obtained sequence, we will construct a new, disjoint sequence.

Let us set $C_n \overset{\text{def}}{\leftrightarrow} B_{2^n}$. Thus, we have a sequence of sets $C_0, C_1, C_2, \ldots$, where $|C_n| = 2^n$.

Now let's define a new sequence of pairwise disjoint sets D_n as follows:

$$D_0 = C_0; \quad D_{n+1} = C_{n+1} \setminus \bigcup_{k=0}^{n} C_k$$

Let's show that all sets D_n are non-empty. For D_0, this is obvious, since $|D_0| = |C_0| = 2^0 = 1$. For D_{n+1}, let's estimate the cardinality of the union $\left|\bigcup_{k=0}^{n} C_k\right|$.

$$\left|\bigcup_{k=0}^{n} C_k\right| \leqslant \sum_{k=0}^{n} |C_k| = \sum_{k=0}^{n} 2^k = 2^{n+1} - 1.$$

Since $|C_{n+1}| = 2^{n+1}$, the cardinality $|C_{n+1} \setminus \bigcup_{k=0}^{n} C_k|$ will be at least 1.

Thus, all sets D_n are non-empty. By construction, they are also pairwise disjoint. Now we have a countable family of non-empty, pairwise disjoint sets $\{D_n\}_{n\in\omega}$. By the axiom of countable choice (AC_ω), there exists a choice function g for this family, which selects exactly one element from each set D_n. Let us set $a_n = g(D_n)$. Since all D_n are disjoint, all elements a_n are pairwise distinct.

Thus, we have proven that an arbitrary infinite set A contains a countable subset $S = \{a_0, a_1, a_2, \ldots\}$.

Now let's prove that A is Dedekind-infinite. For this, it is sufficient to show that it is equinumerous with its own proper subset. Consider $A' = A \setminus \{a_0\}$, which is obviously a proper subset of A. Let's define a bijection $h : A \to A'$ as follows:

$$h(x) = \begin{cases} x, & x \in A \setminus S, \\ a_{n+1}, & x = a_n \in S. \end{cases}$$

This function maps each element not in our sequence S to itself, and each element a_n from the sequence to the next element a_{n+1}. It is obvious that

$h : A \leftrightarrow A'$ (is a bijection), which by definition means that A is Dedekind-infinite. □

From this theorem, it immediately follows that every Dedekind-finite set is finite (assuming AC_ω).

Corollary D4.4. *In* $\mathsf{ZF} + \mathsf{AC}_\omega$, *it is provable that a set is finite if and only if it is Dedekind-finite (equivalently: a set is infinite if and only if it is Dedekind-infinite).*

In terms of comparing cardinalities, this statement can be rewritten as: It follows from AC_ω that a set A is infinite if and only if $\omega \preccurlyeq A$ (i.e., there exists an injection from ω into A).

Lemma D4.9 (Equivalent definitions of countability). For a non-empty set A, the following statements are equivalent in ZF:

(1) A is at most countable (i.e., finite or equinumerous with ω).
(2) There exists a surjection $g : \omega \to A$.
(3) There exists an injection $f : A \to \omega$.

Proof. We will prove the equivalence by the cycle (1) → (2) → (3) → (1).

(1 → 2): Let A be at most countable. If A is finite, $|A| = n$, then there exists a bijection $h : n \to A$. We can define a surjection $g : \omega \to A$ as:

$$g(k) = \begin{cases} h(k), & k < n \\ h(0), & k \geqslant n, \end{cases}$$

here we use the conditions that $A \neq \emptyset$.

If A is countable, then there exists a bijection $h : \omega \to A$, which by definition is also a surjection.

(2 → 3): Let there be a surjection $g : \omega \to A$. We need to construct an injection $f : A \to \omega$. For any element $a \in A$, its preimage $g^{-1}[\{a\}] = \{k \in \omega \mid g(k) = a\}$ is a non-empty subset of ω. Since ω is a well-ordered set (Theorem D4.4), any of its non-empty subsets has a unique minimal element. This allows us to make a canonical choice without involving the axiom of choice. Let's define the function $f : A \to \omega$ by the rule:

$$f(a) = \min(g^{-1}[\{a\}]).$$

Let's show that f is an injection. Let $f(a_1) = f(a_2)$. This means that

$$\min(g^{-1}[\{a_1\}]) = \min(g^{-1}[\{a_2\}]).$$

Let's denote this minimal number as k_0. By the definition of the function f, this means that $k_0 \in g^{-1}[\{a_1\}]$ and $k_0 \in g^{-1}[\{a_2\}]$. Then by the definition of the preimage, $g(k_0) = a_1$ and $g(k_0) = a_2$. Consequently, $a_1 = a_2$. Injectivity is proven.

(3 → 1): Let there be an injection $f : A \to \omega$. Then A is equinumerous with its image $f[A]$, which is a subset of ω. It is provable in ZF that any subset of ω is either finite or equinumerous with ω (by the lemma on order type D4.7). Since $A \simeq f[A]$, then A is also either finite or equinumerous with ω. That is, A is at most countable. □

Theorem D4.12. *In ZF + AC*$_\omega$*, it is provable: the union of an at most countable family of at most countable sets is an at most countable set.*

Proof. Let $\{A_n\}$ be an at most countable family of at most countable sets. If all A_n are empty, their union is also empty and therefore at most countable. Suppose that among the A_n there are non-empty sets; then instead of $\{A_n\}$, let's consider a subsequence in which we leave all A_n except the empty ones. The result of the union will not change. Therefore, we can immediately assume that all A_n are non-empty. Finally, if the family $\{A_n\}$ is finite, we can make it countable by extending the sequence with, for example, the set A_0. Thus, we can immediately assume that $\{A_n\}$ is a countable sequence of non-empty, at most countable sets. Let $A = \bigcup_{n<\omega} A_n$. We need to prove that there exists a surjection from ω to A.

For each n, there exists a surjection $f_n : \omega \to A_n$, since $A_n \neq \emptyset$ and is at most countable. This means that for each n, the set of all surjections from ω to A_n is non-empty. By AC_ω, we can choose one such surjection for each n, obtaining the sequence $\langle f_n \mid n \in \omega \rangle$. Let's define a mapping $h : \omega \times \omega \to A$ by the rule $h(n, k) = f_n(k)$. This mapping is a surjection from $\omega \times \omega$ to A. Since there exists a bijection $p : \omega \leftrightarrow \omega \times \omega$ (Cantor's pairing function), the composition $h \circ p : \omega \to A$ is a surjection. By Lemma D4.9, if there exists a surjection from ω to a set, then that set is at most countable. □

Remark: The converse of the theorem does not hold; the union principle for at most countable sets is strictly weaker than the full Axiom of Countable Choice (AC_ω). It turns out to be equivalent to a weaker choice axiom: the Axiom of Countable Choice for families of *countable* sets. The union principle itself is also often stated for families of *countable* sets, as this version is equivalent in ZF to the one for *at most countable* sets used in the theorem.

D4.4.2 The Axiom of Choice and its Equivalents

Theorem D4.13 (Equivalents of the Axiom of Choice). *The following statements are equivalent in the theory ZF:*

(1) *the Axiom of Choice AC;*
(2) *Zermelo's theorem ZT: any set can be well-ordered;*
(3) *Zorn's lemma ZL: if in a partially ordered set every chain has an upper bound, then the partially ordered set has a maximal element;*
(4) *Hausdorff's maximal principle HMP: in any partially ordered set, there exists a maximal chain with respect to inclusion;*
(5) *the theorem on the square: any infinite set is equinumerous with its square.*

Proof. **1)**. Let's show that AC implies Zermelo's theorem. Let X be a non-empty set. Let $f(0) = x \in X$, where 0 is the least ordinal. The further enumeration of the elements of X is constructed as follows: we remove from X all previously enumerated elements, and in the remaining set, if it is non-empty, we again choose one element and number it with the least of the unused ordinals.

More precisely, let $c : \mathcal{P}(X) \setminus \{\emptyset\} \to X$ be a choice function, and let's extend it by setting $c(\emptyset) = X$. Then by transfinite recursion, we construct the function

$$f(\alpha) = c(X \setminus \{f(\beta) \mid \beta < \alpha\}).$$

It is not difficult to show that up to some ordinal γ, this function is an injection, and at the ordinal γ, it is a bijection to the set X. At subsequent points on the ordinal scale, f takes the value X. Thus, γ induces a well-ordering on X.

2). Let's show that from Zermelo's theorem follows Zorn's lemma. Let $\langle X, <\rangle$ be a non-empty p.o.set in which every chain is bounded above. By a **chain**, we mean a subset $Y \subseteq X$ such that $\langle Y, <\rangle$ is a l.o.set.

Let's introduce a well-ordering $\prec$ on X in accordance with Zermelo's theorem. Let's construct a chain in X recursively, by choosing at each step a proper upper bound of the already constructed part of the chain using the order $\prec$.

$$f(\alpha) = \min_{\prec}\{a \in X \mid \forall \beta < \alpha : f(\beta) < a\},$$

if this set is non-empty, and $f(\alpha) = X$ otherwise. Thus, $f(0) = \min_{\prec} X$. It is not difficult to show that up to some ordinal γ, this function is an injection, and at the ordinal γ, it is a bijection to some chain $C \subseteq X$. At subsequent points on the ordinal scale, f takes the value X. Thus, a chain C is constructed in X that has no proper upper bound (such a chain is called cofinal). And since all chains in X are bounded, there exists a $\max_{<} C$, which is the maximal element of X.

3). Let's show that from Zorn's lemma follows Hausdorff's maximal principle. Let $\langle X, < \rangle$ be a non-empty p.o.set. It is obvious that there is at least one one-point l.o.set in it. Then the set of all linearly ordered subsets of X is non-empty. Let's denote it by $L(X)$ and consider it as a p.o.set with the inclusion relation $\subseteq$.

Let C be some chain in $L(X)$. The **limit** of the chain (in the case of the order relation $\subseteq$) is called $\bigcup C$. It is easy to check that $\bigcup C$ is also a linearly ordered (by $<$) subset of X, i.e., $\bigcup C \in L(X)$. Furthermore, $\bigcup C$ bounds the chain C from above.

Thus, in $L(X)$, any chain is bounded, which means, by Zorn's lemma, there exists a maximal element $L \in L(X)$. And this is the maximal linearly ordered subset of X with respect to inclusion.

4). Let's show that from Hausdorff's maximal principle follows the axiom of choice. Let X be a non-empty set such that $\emptyset \notin X$. Let $y \in X$; then on the set $\{y\} \subseteq X$, there exists a choice function $f_{\{y\}}$ which assigns to the point y some of its elements.[15]

Thus, we can consider the set of all choice functions defined on subsets of X, and it will certainly be non-empty. Let's denote it by $F(X)$ and consider it as a p.o.set with the inclusion relation $\subseteq$. The inclusion of functions means that one (the larger) extends the other, and on the intersection of their domains, they coincide.

Then from Hausdorff's maximal principle, it follows that in $F(X)$ there exists a maximal linearly ordered set with respect to inclusion (the linear order is the same—inclusion of functions), which we will denote by $\mathcal{F}$. In this set of choice functions, each function contains all smaller ones, and thus it is not

[15] Note that the existence of a choice function for finite sets is derived from the axioms of ZF, since we can write the existential quantifier for an element of a non-empty set a finite number of times and, thereby, explicitly derive the existence of a finite choice function. For infinite sets, a finite formula is, alas, not enough.

difficult to show that the limit of the chain $\mathcal{F}$, which we denote by $f = \bigcup \mathcal{F}$, is a) a function, b) a choice function on a subset of X, and c) a choice function on all of X.

The latter follows from the fact that if f were not defined at some point of X, it could be extended to a wider (at least by one point) choice function, which contradicts the maximality of $\mathcal{F}$.

So, we have shown the following implications:

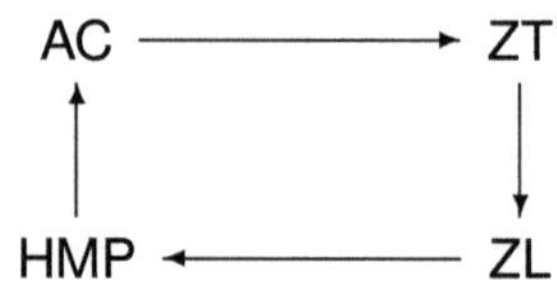

From this, it is not difficult to understand that the first four statements are equivalent.

5) Equivalence of AC and the theorem on the square.

(2) → (5): By virtue of Zermelo's theorem, we can well-order any set X and, thereby, move from sets to cardinals, and for cardinals, we have Hessenberg's theorem (Theorem D4.10).

(5) → (2): Let $X^2 \simeq X$ for any infinite set X. Our goal is to prove Zermelo's theorem, i.e., that any set can be well-ordered. It is clear that we only need to talk about infinite sets, since finite sets are equinumerous with the von Neumann natural numbers, which are well-ordered.

For convenience, let's introduce a new quantifier $\forall^{inf}$, which means quantification only over infinite sets (its legitimization in the language of ZF is analogous to the legitimization of quantifiers over ordinals).

Lemma D4.10. If $\forall^{inf} x\ (x \times x) \simeq x$, then $\forall^{inf} x, y\ (x \times y) \simeq (x \sqcup y)$.

Proof. First, we note that from the theorem on the square also follows the theorem on doubling: $\forall^{inf} x\ (x \sqcup x \simeq x)$, since[16]

$$x \preccurlyeq (x \sqcup x) \preccurlyeq (x \times x) \simeq x,$$

from which, by the Cantor–Bernstein–Schroeder theorem, we get the equality $x \simeq x \sqcup x$. Using the theorem on the square and the theorem on doubling, we find that

[16] We do not describe this embedding here, as the more general case for $x \sqcup y$ is shown a few lines below.

$$(x \sqcup y) \simeq ((x \sqcup y) \times (x \sqcup y)) \simeq ((x \times x) \sqcup (x \times y) \sqcup (x \times y) \sqcup (y \times y)) \simeq (x \sqcup y \sqcup (x \times y)),$$

consequently, $x \times y \preccurlyeq x \sqcup y$. The reverse embedding for infinite sets is obvious: x is bijectively identified with $x \times \{y_0\}$, and y is bijectively identified with $(\{x_0\} \times (y \setminus \{y_0\})) \cup \{\langle x_1, y_1 \rangle\}$, where $x_0 \neq x_1$ and $y_0 \neq y_1$ are arbitrary points of the sets x and y, respectively (they can be found due to the infinity of both sets).

From this, by the Cantor–Bernstein–Schroeder theorem, we get $x \times y \simeq x \sqcup y$, which, by the way, corresponds to point 10 of Theorem D4.10. □

Lemma D4.11. If $\forall^{inf} x, y \ (x \times y) \simeq (x \sqcup y)$, then for any set A and any well-ordered set M, the alternative holds: $A \preccurlyeq M$ or $M \preccurlyeq A$. In other words, from the theorem on the equality of sum and product, it follows that any set is comparable in cardinality with any well-ordered set.

Proof. Let's immediately skip the case where at least one of the sets is finite, since in that case, the lemma is true even without its condition. Therefore, let A and M be infinite sets.

Let $<$ be a well-ordering on M. Consider the direct product $A \times M$. Since there exists a bijection $f : A \sqcup M \leftrightarrow A \times M$, let us set $M' = f[M]$ (the image of M). Since $M' \subseteq A \times M$, two cases are possible for M': either the projection of M' onto A coincides with A, or not.

In the first case, the projection onto A, defined by the rule $\mathrm{pr}_A(a, m) = a$, maps M' onto A, i.e., the composition $\mathrm{pr}_A \circ f$ is a surjection onto A. Therefore, from it, one can construct a reverse injection $g : A \longrightarrow M$ by the rule

$$g(a) = \min_{<}(f^{-1} \circ \mathrm{pr}_A^{-1}(\{a\})), \quad a \in A,$$

which is correctly determined while the set $M' \cap (\{a\} \times M)$ is non-empty due to the surjectivity of the projection.

The existence of an injection from A to M automatically means $A \preccurlyeq M$.

In the second case, when the projection of M' does not coincide with A, there exists a subset $\{a\} \times M$ that lies entirely in the complement of M', and this complement is equinumerous with A, since it coincides with the image $f[A]$. Thus, we have a trivial injection from M to A, and thus, $M \preccurlyeq A$. □

Note that an important restriction has appeared here—the well-ordering of one of the sets. The question is—where to get it, if we are given only one infinite set A?

For this, there is the concept of Hartogs number. Let us set

$$\mathfrak{h}(A) \overset{\text{def}}{\rightleftharpoons} \min\{\alpha \mid \neg(\alpha \preccurlyeq A)\},$$

In other words, the **Hartogs number** is the minimal ordinal that cannot be injectively embedded into the set A.

Note that $\mathfrak{h}(A)$ exists for any A. First, the class of ordinals that can be injectively embedded into A is non-empty, since at least 0 can be embedded into any set. Second, this class cannot be a proper class of ordinals. Indeed, let $WO(A)$ be the set of all well-orderings of subsets of A (at least, there exist finite well-ordered subsets there). Let's factor it by the relation of order isomorphism $\cong$, obtaining the set $W = WO(A)/_{\cong}$. The set W is a definable part of the set $\mathcal{P}^2(A \times A)$ (when the ordered pair is defined according to Kuratowski). Each element of W is in one-to-one correspondence with the order type of its elements, i.e., there exists an injective mapping from W into the class Ord. If we could embed any ordinal into A, then its own element would be found in W, and then we would get a bijection between W and the class Ord, which is a proper class. And this contradicts the Axiom of Replacement. Consequently, not all ordinals can be embedded into A, which means the class of ordinals in the definition of Hartogs number is non-empty (in fact, it represents the class of all ordinals not associated with elements of W), i.e., $\mathfrak{h}(A)$ always exists.

Furthermore, it is not difficult to see that $\mathfrak{h}(A) = \sup\{ot(R) + 1 \mid R \in WO(A)\}$,[17] and also that Hartogs number is a cardinal, since $WO(A)$ is closed with respect to the equinumerosity of orders.

So, we have an infinite set A and a cardinal $\mathfrak{h}(A)$, which is well-ordered by the relation $\in$. Then, within the framework of the theorem on the square, by the two previous lemmas, we get that $A \preccurlyeq \mathfrak{h}(A)$ or $\mathfrak{h}(A) \preccurlyeq A$. However, the latter is impossible by the definition of Hartogs number; consequently, $A \preccurlyeq \mathfrak{h}(A)$ (and, moreover, $A \prec \mathfrak{h}(A)$).

From this, it follows that A can be well-ordered by transferring the order from $\mathfrak{h}(A)$.

Thus, we have shown that from the theorem on the square follows Zermelo's theorem, which, as shown above, is equivalent to AC. The theorem is proven. □

[17] The addition of $+1$ here is needed only for the case of a finite set A.

D4.5 On the Interconnection of Formal Theories

As a concluding example illustrating the deep connections between, at first glance, different areas of mathematical logic, we will consider the relationship between Peano Arithmetic (PA) and the theory of hereditarily finite sets (HF). The fundamental result is that these two theories are **mutually interpretable** and, moreover, **bi-interpretable**, which means not only the possibility of translating theorems from one theory to another, but also the provability in each theory of the fact that a sequential translation there and back preserves the original statements. This establishes an extremely close connection, showing that both theories, in essence, speak about the same thing, but in different languages.

Interpretation of PA in HF. In fact, we have already done the main work of interpreting arithmetic in set theory when we constructed arithmetic on the basis of von Neumann ordinals. We defined natural numbers and operations on them using only basic set-theoretic concepts. It is important to emphasize that the Axiom of Infinity is not required to define the set of natural numbers as the class of finite ordinals and to prove the validity of arithmetic operations on them (addition, multiplication). All the necessary constructions (the concepts of the empty set, pair, union, transitive closure) are already available in HF.

However, there is a methodological "pitfall" here concerning the induction schema. When we assert that the translation of each axiom of induction from PA is provable in HF, we must understand the nature of this statement. PA contains a *schema* of axioms, one for each arithmetic formula. In HF, we prove each such instance separately. The proof that *all* translations of the axioms are provable requires induction on the construction of the formula in our *meta-theory*. Thus, the proof of the induction that we attribute to the model of PA inside HF is, in a sense, "external". The situation changes if we consider HF not as an axiomatic system, but semantically, as the set of all true sentences about the standard model of hereditarily finite sets V_ω. In this complete theory, induction is, of course, true.

Interpretation of HF in PA. The reverse interpretation is considerably more complex and uses the technique of Gödel numbering to encode finite sets and the membership relation with natural numbers. As was shown in Section B2.3.2, we can establish a one-to-one correspondence between hereditarily finite sets and natural numbers (for example, using Ackermann's coding, $\text{Code}(S) = \sum_{s \in S} 2^{\text{Code}(s)}$). The membership relation $a \in b$ is translated into an arithmetic predicate that checks whether the bit at the position numbered

Code(a) in the binary expansion of the number Code(b) is one. This predicate is a Δ_0-formula (assuming that the exponentiation 2^x is already defined). Next, it is necessary to show that the translations of all axioms of HF are provable statements in PA. This can be done for all axioms, including the axiom of power set and the schema of separation, as the induction in PA is strong enough to prove the existence of codes for the results of these operations.

This mutual interpretability is not an isolated case. PA is also bi-interpretable with the theory Z-Inf (Zermelo's theory without the axiom of infinity). This demonstrates that Peano's axioms with the full schema of induction precisely capture the expressive power necessary to formalize the concept of a finite set and basic algorithmic operations.

The mutual interpretability of PA and HF is a classic result demonstrating that the arithmetic of natural numbers with induction is equivalent in its expressive power to the basic theory of finite sets. A detailed exposition of these constructions can be found in [37] or [39].

Note that Robinson's arithmetic, which we mentioned in Section D3.1, is mutually interpretable with the so-called weak theory of concatenation TC, developed by Grzegorczyk [29]. This theory is essentially a finite axiomatization of our model of bracket notations from Section B2.3.2, which does not contain the axiom of induction.

Speaking of the interconnection of theories, one cannot fail to mention other fundamental results concerning set theory itself.

First, **ZF and ZFC are also mutually interpretable**. The interpretation of ZFC in ZF is one of Gödel's greatest triumphs. He showed that if ZF is consistent, then its internal "constructible universe" L is a model for ZFC. The reverse interpretation is trivial, as any model of ZFC is also a model of ZF. This means that the Axiom of Choice cannot be disproven in ZF, if ZF itself is consistent.

Secondly, there is an even more surprising fact connecting the consistency of arithmetic with the existence of models for strong theories. It can be shown that the theory PA + Con(ZF), that is, PA with the addition of an axiom asserting the consistency of Zermelo-Fraenkel set theory, is capable of **interpreting the theory** ZFC + CH (i.e., with the added continuum hypothesis) [21]. This shows that adding the axiom of ZF's consistency to arithmetic results in a surprisingly powerful axiomatic system, from which quite non-trivial consequences about the structure of the universe of sets arise.

The indicated pairs of mutually interpretable theories once again emphasize the deep connection between basic mathematical objects such as arithmetic, algorithms, and sets.

Exercises

D4.1 Write the dual statement for the following identity valid in Boolean algebras: $(a \vee b) \wedge (a \vee \neg b) = a$. Verify if the dual is also true.

D4.2 Let X be an infinite set. Consider the collection $\mathcal{A} \subset \mathcal{P}(X)$ defined as:

$$\mathcal{A} = \{S \subseteq X \mid |S| < \omega \vee |X \setminus S| < \omega\}.$$

(In other words, $\mathcal{A}$ contains all finite subsets of X and their complements).

1. Prove that $\mathcal{A}$ forms a Boolean subalgebra of $\mathcal{P}(X)$ (i.e., it contains $\emptyset$, X and is closed under union, intersection, and complement).
2. Is $\mathcal{A}$ a complete Boolean algebra? (Recall that completeness requires arbitrary unions $\bigcup S_i$ to remain in the algebra).
3. An element a of a Boolean algebra is called an *atom* if a is not the zero element (here $\emptyset$), and for any b in the algebra, $b \subseteq a$ implies $b = \emptyset$ or $b = a$ (i.e., a is minimal among non-empty sets). Describe explicitly the atoms of the algebra $\mathcal{A}$. Does every non-empty element of $\mathcal{A}$ contain an atom?

D4.3 In the text, we showed how to turn a Boolean algebra into a Boolean ring. Now consider the reverse. Let $\mathcal{R} = \langle R, +, \cdot, 0, 1\rangle$ be a commutative ring with unity. Let $B = \{e \in R \mid e^2 = e\}$ be the set of all *idempotents* of the ring. Define operations on B:

$$x \wedge y \overset{\text{def}}{\rightleftharpoons} x \cdot y, \quad x \vee y \overset{\text{def}}{\rightleftharpoons} x + y - x \cdot y, \quad \neg x \overset{\text{def}}{\rightleftharpoons} 1 - x.$$

Prove that $\langle B, \vee, \wedge, \neg, 0, 1\rangle$ is a Boolean algebra. *Hint: First check that the operations are closed, i.e., result in an idempotent.*

D4.4 Recall that in a Heyting algebra, the implication $a \rightarrow b$ is defined by the fundamental adjunction property:

$$\forall x \; (x \leqslant (a \rightarrow b) \iff x \wedge a \leqslant b).$$

1. Prove that this definition implies that $a \rightarrow b$ is the greatest element of the set $S = \{x \mid x \wedge a \leqslant b\}$.
2. Using the adjunction property and the definition of the join (supremum), prove the identity:

$$(a \vee b) \to c = (a \to c) \wedge (b \to c).$$

Hint: To prove equality $A = B$, show that for any test element x, the condition $x \leqslant A$ is logically equivalent to $x \leqslant B$. Use the universal property of the join: $u \vee v \leqslant w \iff (u \leqslant w$ and $v \leqslant w)$.

D4.5 In Boolean algebras, the negation satisfies perfect symmetry (duality). In Heyting algebras, however, this symmetry is broken.

1. Prove analytically that the first De Morgan law holds: $\neg(x \vee y) = \neg x \wedge \neg y$.
2. Show that the second law $\neg(x \wedge y) = \neg x \vee \neg y$ does **not** always hold. Use the Heyting algebra of open sets of $\mathbb{R}$ (where $\neg U = \mathrm{Int}(\mathbb{R} \setminus U)$) and consider the intervals $U = (-\infty, 0)$ and $V = (0, +\infty)$ as a counterexample.

D4.6 Prove that every finite Boolean algebra $\mathcal{B}$ is isomorphic to the power set algebra $\mathcal{P}(A)$ of some finite set A. *Constructive steps:*

1. Let A be the set of all atoms of $\mathcal{B}$. Show that for any $x \in \mathcal{B}$, $x = \bigvee\{a \in A \mid a \leqslant x\}$. (Use the fact that in a finite algebra, every non-zero element contains an atom).
2. Define the map $\varphi : \mathcal{B} \to \mathcal{P}(A)$ by $\varphi(x) = \{a \in A \mid a \leqslant x\}$. Prove that φ is an isomorphism.

D4.7 Prove that the set of points on the plane $\mathbb{R} \times \mathbb{R}$ has the same cardinality as the set of points on the line $\mathbb{R}$. *Hint: Use the Cantor-Bernstein-Schroeder theorem.*

1. Construct a trivial injection $\mathbb{R} \to \mathbb{R} \times \mathbb{R}$.
2. Construct an injection $\mathbb{R} \times \mathbb{R} \to \mathbb{R}$ using decimal expansions. (Warning: Be careful with the non-uniqueness of decimal representations, e.g., $0.1999\cdots = 0.2000\ldots$. How can you avoid this issue?)

D4.8 Let R be a *finite* commutative ring with unity ($1 \neq 0$). Recall that an element $a \neq 0$ is a zero divisor if there exists $b \neq 0$ such that $ab = 0$. R is called an integral domain if it has no zero divisors. Prove that a finite integral domain is necessarily a field (i.e., every non-zero element has a multiplicative inverse). *Hint: For a fixed $a \neq 0$, consider the map $f : R \to R$ defined by $f(x) = ax$. Use the Pigeonhole Principle (specifically, that an injection on a finite set is surjective).*

D4.9 Let $\omega^{<\omega}$ denote the set of all finite sequences of natural numbers (i.e., functions from some $n \in \omega$ to ω). Prove that $|\omega^{<\omega}| = \aleph_0$. *Hint: Represent* $\omega^{<\omega}$ *as a union* $\bigcup_{n\in\omega} \omega^n$ *and use the fact that the finite product of countable sets is countable.*

D4.10 A real number is called *algebraic* if it is a root of a non-zero polynomial with rational coefficients. Otherwise, it is called *transcendental.*

1. Using the theorem on the countable union of countable sets (proven in the text), show that the set of all algebraic numbers $\mathbb{A}$ is countable.
2. Deduce that the set of transcendental numbers is uncountable.

D4.11 Use transfinite induction to prove that for any ordinal α, $1 + \alpha = \alpha$ if $\alpha \geqslant \omega$.

D4.12 Cantor Normal Form Theorem. Prove that for every non-zero ordinal α, there exist unique natural numbers $n \geqslant 1, k_1, \dots, k_n$ $(0 < k_i < \omega)$ and unique ordinals $\beta_1 > \beta_2 > \dots > \beta_n$ such that:

$$\alpha = \omega^{\beta_1} \cdot k_1 + \omega^{\beta_2} \cdot k_2 + \dots + \omega^{\beta_n} \cdot k_n.$$

Follow these steps:

1. Growth Lemma: Prove that $\gamma \leqslant \omega^\gamma$ for all ordinals γ.
2. The Degree (Logarithm): Using the Lemma, show that for any $\alpha > 0$, the set $E = \{\gamma \mid \omega^\gamma \leqslant \alpha\}$ is a set (not a proper class) and contains a maximal element β_1 (or $\sup E = \beta_1$ satisfies the condition) such that $\omega^{\beta_1} \leqslant \alpha < \omega^{\beta_1+1}$.
3. Main Proof: Use the Division Theorem to divide α by ω^{β_1} and apply transfinite induction on α to the remainder.

D4.13 Let $\alpha = \omega^\omega + \omega \cdot 2 + 1$ and $\beta = \omega^\omega + \omega$.

1. Compute $\alpha + \beta$ and $\beta + \alpha$ in Cantor Normal Form.
2. Compute $\alpha \cdot 2$ and $\beta \cdot \omega$.

D4.14 Let p_k denote the k-th prime number ($p_1 = 2, p_2 = 3, p_3 = 5, \dots$). By the Fundamental Theorem of Arithmetic, any natural number $n > 1$ can be uniquely represented as $n = p_{k_m}^{a_m} \dots p_{k_1}^{a_1}$ where $k_m > \dots > k_1 \geqslant 1$. Define a mapping $c : \mathbb{N} \setminus \{0\} \to \mathrm{Ord}$ such that $c(1) = 0$, and for $n > 1$:

$$c(n) = \omega^{k_m - 1} \cdot a_m + \dots + \omega^{k_1 - 1} \cdot a_1.$$

1. Calculate the ordinal values $c(n)$ for the following numbers (where $k, j, t \in \omega \setminus \{0\}$):
$$n_1 = 2^k, \quad n_2 = 3^k, \quad n_3 = 2^k 3^j 5^t.$$

2. Prove that the image of this mapping is exactly the ordinal ω^ω and that the mapping is an order isomorphism between $\langle \mathbb{N} \setminus \{0\}, \prec \rangle$ (where $n \prec m \iff c(n) < c(m)$) and ω^ω.

D4.15 Using the same notation for prime decomposition ($n = p_{k_m}^{a_m} \dots p_{k_1}^{a_1}$), consider the recursively defined mapping $o : \mathbb{N} \setminus \{0\} \to \mathrm{Ord}$:
$$o(1) = 0, \quad o(n) = \omega^{o(k_m)} \cdot a_m + \dots + \omega^{o(k_1)} \cdot a_1.$$

1. Calculate $o(n)$ for the following numbers ($k, j, t \geqslant 1$):
$$n_1 = 2^k, \quad n_2 = 3^k, \quad n_3 = 2^k 3^j 5^t.$$

2. Consider the sequence of numbers defined recursively: $h_0 = 1$ and $h_{n+1} = p_{h_n}$ (the h_n-th prime number). Calculate the values h_0, h_1, h_2, h_3, h_4.
3. Calculate the ordinal values $o(h_n)$ for these terms.
4. Prove that $\sup\{o(h_n) \mid n \in \omega\} = \varepsilon_0$.

D4.16 Simplify the expression $\aleph_1 + \aleph_0 \cdot \aleph_1$.

D4.17 Let κ be an infinite cardinal ($\kappa \geqslant \aleph_0$). Simplify the following cardinal expressions using the properties from D4.10:

1. $\kappa + \aleph_0$;
2. $3 \cdot \kappa$;
3. $(\kappa^2)^\kappa$.

D4.18 Recall the definition of the hierarchy of sets: $V_0 = \emptyset$, $V_{\alpha+1} = \mathcal{P}(V_\alpha)$, and $V_\lambda = \bigcup_{\xi<\lambda} V_\xi$ for limit λ.

1. Determine the cardinality of the sets V_0, V_1, V_2, V_3.
2. Prove that for any α, V_α is a transitive set.
3. Prove that if $x \in V_\alpha$, then $\mathrm{rank}(x) < \alpha$.

D4.19 In the text, the rank of a set x was defined via the Von Neumann hierarchy:
$$\mathrm{rank}(x) = \min\{\alpha \mid x \subseteq V_\alpha\}.$$

Let us define a secondary notion of rank recursively:

$$\rho(x) = \sup\{\rho(y) + 1 \mid y \in x\}, \quad \rho(\emptyset) = 0.$$

Prove that these two definitions are equivalent: for all sets x, $\text{rank}(x) = \rho(x)$. *Hint: Prove* $\rho(x) \leqslant \text{rank}(x)$ *and* $\text{rank}(x) \leqslant \rho(x)$ *separately. Recall that* $y \in V_\alpha \iff \text{rank}(y) < \alpha$.

D4.20 We defined transfinite recursion via the restriction of the function: $F(\alpha) = \Psi(F \restriction \alpha)$. Consider an alternative schema based on the image (set of previous values): $G(\alpha) = \Phi(G[\alpha])$. Prove that these two schemes are equivalent in their expressive power. That is, any function defined by the restriction schema can be obtained (possibly using auxiliary coding) using the image schema, and vice versa. *Hint:* To reduce restriction-recursion to image-recursion, consider the auxiliary function $G(\alpha) = \langle \alpha, F(\alpha) \rangle$.

D4.21 Consider the set of rational numbers $\mathbb{Q}$ with the standard order $<$.

1. Prove that the structure $\langle \mathbb{Q}, < \rangle$ is not isomorphic to any ordinal number.
2. [Cantor's Theorem on Dense Orders] Prove that any two countable, dense linear orders without endpoints are isomorphic. (A linear order is *dense* if for any x and y s.t. $x < y$, there exists z s.t. $x < z < y$. It is *without endpoints* if for any x there exist y, z s.t. $y < x < z$). Conclude that any such order is isomorphic to $\langle \mathbb{Q}, < \rangle$.

Hint for (b): Use the "back-and-forth" method to construct an isomorphism step by step.

Recommended Reading

Axiomatic set theory is the language in which almost all of modern mathematics is spoken today. It is a vast and rapidly developing world.

For anyone who wants to engage with this field seriously, there is one book, akin to "The Lord of the Rings" — the monumental work by Thomas Jech, "Set Theory" [37]. This volume, spanning over 700 pages, contains practically everything that was created in set theory during the 20th century, including a detailed exposition of the theory of cardinals, ordinals, large cardinals, and independence proofs. It is not a textbook for beginners, but rather the specialist's bible and a reference to which one returns again and again.

The central technical method of modern set theory is forcing, which allowed for the proof of the independence of the continuum hypothesis. To understand not only the technique but also the spirit of this discovery, one should turn to the primary source — the small but very dense book by its creator, Paul Cohen [13]. It is like reading the notes of a genius at the moment of creation. However, for a systematic study of the forcing method, most mathematicians prefer the smoother and more pedagogically sound path laid out in the classic textbook by Kenneth Kunen, "Set Theory: An Introduction to Independence Proofs" [44]. It is from this book that entire generations of researchers have learned this difficult craft of constructing models of set theory.

Answers to Exercises

Level A1

A1.1 The parse tree for the quadratic formula $x = \frac{-b+\sqrt{b^2-4ac}}{2a}$ consists of the following hierarchy of templates:

- Top level: Equality $x = \ldots$
- Right side: Fraction template $\frac{\sqcup}{\sqcup}$
- Numerator: Addition template $\sqcup + \sqcup$ (where the first term is unary minus $-b$, the second is a square root).
- Under the root: Subtraction $\sqcup - \sqcup$.
- Inside subtraction: Power b^2 and Multiplication $4ac$ (which is $4 \cdot a \cdot c$).
- Denominator: Multiplication $2a$.

A1.2 1) $(a \cdot b) + (c \cdot d)$; 2) $(x^2) + (2xy) + (y^2)$; 3) $(a/b)/c$; 4) $e^{(x^2+y)}$.

A1.3 Root: =. Left child: +, Right child: 0. Children of +: Left is Power (e to $i\pi$), Right is 1.

A1.4 Outermost: $\sqrt{\sqcup}$. Inside: Addition $1 + \sqcup$. Inside: $\sqrt{\sqcup}$. Inside: $1 + \sqcup$. Innermost: $\sqrt{1+x}$.

A1.5 Objects: $x, 3, x^3, 3x, 3x^2, 3$. Actions/Relations: $\frac{d}{dx}$ (operator), + (operator), = (relation).

A1.6

1. $\exists x(Fr(x) \wedge Speak(x, \text{English}))$.

N. I. Kazimirov, *Mathematics as a Foreign Language*, Mathematics in Mind,
https://doi.org/10.1007/978-3-032-30721-7

2. $\forall x(Fr(x) \to \neg Speak(x, \text{Chinese}))$.
3. $\exists x(Phys(x) \land Love(x, \text{Math}) \land \neg Love(x, \text{QM}))$.
4. $\forall x(Studies(x, \text{Logic}) \to Knows(x, \text{SetTheory}))$.
5. $\exists p(Phys(p) \land \exists t(Theory(t) \land Proposed(p,t) \land$
   $\forall m(Math(m) \to Understands(m,t))))$.
6. $\forall x \exists y(Father(y,x) \land \forall z(Father(z,x) \to z = y))$.
7. $\exists x(Glitters(x) \land \neg Gold(x))$.
8. $\forall p(Phys(p) \land \neg Knows(p, \text{Math}) \to$
   $\neg Understands(p, \text{GR}) \land \neg Understands(p, \text{QM}))$.
9. $\neg \exists t(Theory(t) \land \forall y(Explains(t,y)))$
   or $\forall t(Theory(t) \to \exists y(\neg Explains(t,y)))$.

A1.7

1. If A is contained in B and B is contained in A, then A equals B.
2. The square of any real number is non-negative.
3. For any non-zero integer n and any integer m, one can divide m by n with a quotient q and a remainder r.

A1.8

1. $(a+b)^2 = a^2 + b^2 + 2ab$.
2. $\forall x(x \cdot 0 = 0)$.
3. $\forall \varepsilon > 0 \; \exists \delta > 0$.

A1.9

1. Valid.
2. Invalid (operator + expects a right operand, but finds another operator $\cdot$).
3. Invalid (cos expects an argument).
4. Valid.

A1.10

1. Bound: k, Free: n.
2. Bound: x, Free: a.
3. Bound: i, Free: I.

A1.11 The inner variable i would shadow the outer variable i. The range of the inner sum would depend on the variable it iterates, creating ambiguity.

A1.12

1. $3(a + b)$;
2. $(a + b)^2$;
3. $\sin(a + b)$.

A1.13 1) π and 2 (numeric constants). 2) S and r (variables).

A1.14 $2^{(3^4)}$. Calculation: $(2^3)^2 = 8^2 = 64$. $2^{(3^2)} = 2^9 = 512$.

A1.15

1. Inverse function (if f is a function).
2. Multiplicative inverse $1/f(x)$ (if $f(x)$ is a number).

A1.16

1. Unary (negation).
2. Binary (subtraction).
3. Unary (factorial).
4. Binary (union).

Level A2

A2.1

- $2 + 2 = 5$: **False proposition**.
- $x + 1 = 5$: **Predicate** (depends on x).
- All prime numbers are odd: **False proposition** (counterexample: 2).
- $5 > 3$: **True proposition**.

A2.2

- Some cat is not black.
- No student likes mathematics (or: All students dislike mathematics).
- It rains and the ground is not wet.

A2.3

- **True** (elements of A are 1, 2, 3; all are < 4).
- **True** (elements of B are 2, 3, 4; 2 and 4 are even).
- **True** ($A \cap B = \{2, 3\}$; both are $\geqslant 2$).

A2.4

- Divisible by 6 $\longrightarrow$ divisible by 3: **Relevance implication**. There is a logical connection and a common variable n.
- Eiffel Tower $\longrightarrow$ snow is white: **Material implication**. The statement is true only because both parts are true; there is no semantic connection between them.
- Grandmother $\longrightarrow$ female: **Relevance implication**. The consequent is semantically derived from the definition of the antecedent.

A2.5 **Valid**. Transitivity of inclusion: $S \subseteq R$ and $R \subseteq Q$ implies $S \subseteq Q$.

A2.6 **a, b, c, d, e**. Since Faxi $\subseteq$ Krolo and Winda $\cap$ Krolo $= \emptyset$, the sets Faxi and Winda are disjoint.

A2.7 **a, c**. We know there is an $x \in$ Donas such that $x \notin$ Kalsi. Since Donas $\subseteq$ Sudr, this x is also in Sudr.

A2.8 **b, c, d**. Since Teka $\subseteq$ Fato and Teka $\cap$ Arati $= \emptyset$, any element of Teka is a Fato that is not Arati.

A2.9 **b, c**. We know there is a $z \in$ Dax such that $z \in$ Bron. Since Bron $\subseteq$ Hanto, this z is also in Hanto.

A2.10

1. A (Absorption law);
2. A (Distributivity: $A \cup (B \cap \overline{B}) = A \cup \emptyset = A$);
3. $A \cap B$ (Double complementation relative to A);
4. U (Universe) or 1 (The expression covers all possible regions).

A2.11

1. False. $LHS = (A \setminus B) \cup (A \cap C)$, while RHS includes all of C. If there are elements in $C \setminus A$, the equality fails.
2. True. Intersection distributes over symmetric difference.

A2.12

1. Not Reflexive, Not Symmetric (if y is female, she is a sister, not brother), Transitive (brother of a brother is a brother).
2. Reflexive, Symmetric, Not Transitive (e.g., $0R0.9$ and $0.9R1.8$, but $\neg(0R1.8)$).
3. Strict Partial Order (Irreflexive, Asymmetric, Transitive).
4. Partial Order (Reflexive, Antisymmetric, Transitive).

A2.13

1. Reflexive, Not Transitive: $\{(a,a),(b,b),(c,c),(a,b),(b,c)\}$ (missing (a,c)).
2. Symmetric, Not Reflexive: $\{(a,b),(b,a)\}$.
3. Strict Partial Order: $\{(a,b),(b,c),(a,c)\}$.

A2.14 Hasse Diagram (D_{12}):

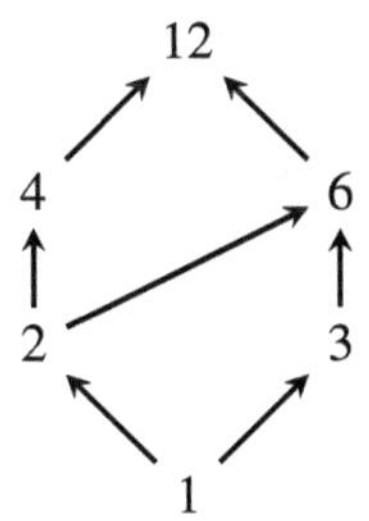

A2.15

1. Yes. $x + y$ is even $\iff$ x, y have the same parity. Classes: Evens and Odds.
2. Yes ($x = \pm y$). Classes are $\{k, -k\}$.
3. No. Not symmetric ($1 \leqslant 2$ but $2 \nleqslant 1$).

A2.16

1. Yes. Defined for all $x \in A$, unique y.
2. No. Not total (undefined for 3).
3. No. Not functional (1 maps to both 2 and 3).
4. Yes. (Constant function).

A2.17

1. A is a Knight, B is a Knave.
2. A is a Knight, B is a Knave.

A2.18

1. Invalid. (Fallacy of affirming the consequent).
2. Valid. (Syllogism / Modus Ponens).
3. Valid. (Modus Tollens / Contraposition).

A2.19

1. Inductive.
2. Deductive.
3. Deductive.

A2.20

- *Language:* "He died and was buried" vs "He was buried and died".
- *Programming:* 'if (x != 0) and (1/x > 1)' (Checks safety first) vs 'if (1/x > 1) and (x != 0)' (Causes DivisionByZero error).

A2.21

1. $\{0, 1, 2, 3, 4\}$.
2. The set of even numbers $\{0, 2, 4, \ldots\}$.
3. $\emptyset$ (Empty set).
4. $\mathbb{N}$ (The entire universe, tautology).

A2.22

1. $Grandparent(x, y) \overset{\text{def}}{\leftrightarrow} \exists z(Parent(x, z) \wedge Parent(z, y))$.
2. $GreatGrandparent(x, y) \overset{\text{def}}{\leftrightarrow} \exists z(Grandparent(x, z) \wedge Parent(z, y))$.
3. $CommonParent(x, y) \overset{\text{def}}{\leftrightarrow} \exists z(Parent(z, x) \wedge Parent(z, y))$.

Level B1

B1.1 1. Yes. 2. Yes. 3. Yes. 4. Yes (if *A* and *B* are both true, implication holds; if false, holds).

B1.2

1. $A = 0, B = 1$ (LHS=1, RHS=0).
2. $A = 0, B = 1$ (LHS=1, RHS=0).
3. $A = 0, B = 1, C = 1$ (LHS=1, RHS=0).

B1.3

1. $\neg A \wedge B$;
2. $A \wedge (\neg B \vee C)$;
3. $\exists x \forall y (\varphi(x, y) \wedge \psi(x))$ (using $\neg(\mathcal{A} \to \mathcal{B}) \equiv \mathcal{A} \wedge \neg\mathcal{B}$).

B1.4

1. $\forall x (P(x) \to Q(a))$;
2. $(\forall x P(x)) \wedge Q(a)$;
3. $\exists z (a < z \to \forall x (x = z))$;
4. $\int_0^a x^2 dx$.

B1.5

1. $\exists y\ (b < y)$;
2. $\exists y\ (S(b) < y)$;
3. Invalid because y is bound in φ. Substituting $a \to y$ would capture the variable ("exists y greater than itself" vs "greater than a").

B1.6 Use Deduction Theorem and Modus Ponens twice on $A, A \to B, B \to C$.

B1.7

1. $\mathcal{A} \to (\mathcal{A} \to \mathcal{B})$ (Hyp);
2. $\mathcal{A}$ (Hyp);
3. $\mathcal{A} \to \mathcal{B}$ (MP 1,2);
4. $\mathcal{B}$ (MP 3,2). Result $\mathcal{B}$. By Ded. Thm: $(\mathcal{A} \to (\mathcal{A} \to \mathcal{B})) \vdash \mathcal{A} \to \mathcal{B}$.

B1.8

1. $\forall x P(x)$ (Hyp).
2. $P(a)$ (Axiom $\forall$).
3. $P(a) \to \exists x P(x)$ (Axiom $\exists$).

4. $\exists x P(x)$ (MP).

B1.9 If we allow $P(a) \to \forall x P(x)$, we claim "if property holds for specific a, it holds for everyone", which is clearly false (e.g., $a = 2$, $P(x)$="x is even").

B1.10

1. $\exists x(P(x) \wedge \neg Q(x))$;
2. $\forall x \exists y(\neg R(x, y) \wedge \neg S(x))$;
3. $\exists x P(x) \wedge \exists y \neg Q(y)$.

B1.11 If x not in P, then P is constant w.r.t x. "For all x, (P is true or $Q(x)$ is true)" is equivalent to "P is true or (for all x, $Q(x)$ is true)".

B1.12 $P(x)$ means "x is even" and $Q(x)$ means "x is odd" in Ariphmetic.

B1.13

1. Universe $\{1, 2\}$, $P(1) = T, P(2) = F$.
2. Universe $\mathbb{N}$, $P(x, y) : x < y$.
3. P even, Q odd.

B1.14 Yes. Model: $\mathbb{N}$ with the order defined by divisibility (reflexive, transitive, non-linear).

B1.15

1. $(a = b) \to (P(a) \leftrightarrow P(b))$;
2. $(a = b) \to ((a < c) \leftrightarrow (b < c))$.

B1.16

1. $\exists x \exists y(x \neq y)$;
2. $\exists x \exists y \exists z(x \neq y \wedge y \neq z \wedge x \neq z)$.

B1.17

1. $\exists x \forall y(y = x)$;
2. $\exists x \exists y(x \neq y \wedge \forall z(z = x \vee z = y))$.

B1.18

1. $\forall x(x \in A \to (x \in B \vee x \in C))$;
2. $\neg \exists x(x \in A \wedge x \in B)$;
3. $\forall x((x \in A \wedge x \notin B) \to x \in C)$.

B1.19

1. True. (Every pencil is either Red or Long. p_1: Red (T), p_2: Red (T), p_3: Long (T)).
2. True. (Witness: p_1 is Red and Not Long).
3. False. (Counterexample: p_2 is Long, but it is Red, so $\neg R(p_2)$ is False).
4. True. (If x is Red (p_1, p_2), we need a Long pencil $y \neq x$. For p_1, take p_2 or p_3. For p_2, take p_3).

B1.20

1. True. (For every number x, there exists a greater number y).
2. True. (Everyone has a biological parent).
3. False. (Not every piece attacks something; e.g., a rook stuck behind pawns might attack nothing, or a King alone on the board).

Level B2

B2.1 $\forall x(x = 0 \vee \exists y(x = S(y)))$.

B2.2 $Sq(x) \overset{\text{def}}{\leftrightarrow} \exists y(x = y \cdot y)$.

B2.3

- $Even(x) \overset{\text{def}}{\leftrightarrow} \exists k(x = k + k)$ (or $x = 2 \cdot k$ if 2 is defined).
- $x \mid y \overset{\text{def}}{\leftrightarrow} \exists k(y = x \cdot k)$.
- $Prime(p) \overset{\text{def}}{\leftrightarrow} (1 < p) \wedge \forall x\ (x \mid p \to x = 1 \vee x = p)$.
- $\forall n\ (Even(n) \wedge 2 < n) \to \exists p, q\ (Prime(p) \wedge Prime(q) \wedge n = p + q))$.

B2.4 $Prime(p) \overset{\text{def}}{\leftrightarrow} \neg Inv(p) \wedge \forall d\ (d \mid p \to Inv(d) \vee (p \mid d))$. *Explanation:* A prime is a non-unit element (not equal to 1) such that any of its divisors d is either a unit (1) or is divisible by p. In the natural numbers, if $d \mid p$ and $p \mid d$, then $d = p$. Thus, the formula correctly states that the only divisors of p are 1 and p itself.

B2.5 $\forall x \exists p(Prime(p) \wedge p > x \wedge Prime(S(S(p))))$.

B2.6 $\forall x \forall y \exists z(\exists k(z = x \cdot k) \wedge \exists m(z = y \cdot m))$.

B2.7 $G(a, b, d) \overset{\text{def}}{\leftrightarrow} (d \mid a) \wedge (d \mid b) \wedge \forall k((k \mid a \wedge k \mid b) \to k \leqslant d)$.

B2.8 Let $x \mid y$ and $y \mid z$. Then $\exists k(y = x \cdot k)$ and $\exists m(z = y \cdot m)$. Substituting y: $z = (x \cdot k) \cdot m = x \cdot (k \cdot m)$ (by associativity of multiplication). Since $k \cdot m$ is a natural number, $x \mid z$.

B2.9 Base $n = 0$: $0 = 0$. Step: $S_{n+1} = S_n + (n+1) = \frac{n(n+1)}{2} + (n+1) = \frac{(n+1)(n+2)}{2}$.

B2.10 Base $n = 0$: 0 is divisible. Step: $(n+1)^3 - (n+1) = n^3 + 3n^2 + 3n + 1 - n - 1 = (n^3 - n) + 3(n^2 + n)$. Both terms divisible by 3.

B2.11 Base $n = 0$: $1 \geqslant 1$. Step: $(1+x)^{k+1} = (1+x)^k(1+x) \geqslant (1+kx)(1+x) = 1 + (k+1)x + kx^2 \geqslant 1 + (k+1)x$.

B2.12 Base $n = 1$: $1 = 1^2$. Step: $n^2 + (2(n+1) - 1) = n^2 + 2n + 1 = (n+1)^2$.

B2.13 Assume true for all $k < n$. If n is prime, done. If composite, $n = a \cdot b$. $a < n$, so a has prime factors.

B2.14 If $a^2 = 2b^2$, then a is even, $a = 2k$. $4k^2 = 2b^2 \implies 2k^2 = b^2$, so b is even. We get a smaller solution $(b/2, a/2)$, continuing infinitely.

B2.15 Root node $\{\dots\}$ with three children: leaf $\emptyset$, node $\{\emptyset\}$ (child $\emptyset$), node $\{\emptyset, \{\emptyset\}\}$ (children $\emptyset$ and $\{\emptyset\}$).

B2.16 $\text{rank}(x) = \sup\{\text{rank}(y) + 1 \mid y \in x\}$ (for finite, max $+1$).

B2.17 $(x \in A/_\sim) \overset{\text{def}}{\leftrightarrow} (x \subseteq A) \wedge (x \neq \emptyset) \wedge \forall y, z \in A\ (y \sim z) \wedge (y \in x) \to z \in x)$.

B2.18 To express that A is the ordered pair $\langle x, y\rangle = \{\{x\}, \{x, y\}\}$ fully formally (without brace notation), we must explicitly assert the existence of the constituent sets $\{x\}$ and $\{x, y\}$:

$$\exists u \exists v \Big(\underbrace{\forall t(t \in u \leftrightarrow t = x)}_{u=\{x\}} \wedge \underbrace{\forall s(s \in v \leftrightarrow (s = x \vee s = y))}_{v=\{x,y\}} \wedge \underbrace{\forall z(z \in A \leftrightarrow (z = u \vee z = v))}_{A=\{u,v\}} \Big).$$

B2.19 If $a = a' + kn$ and $b = b' + mn$, then $(a + b) = (a' + b') + n(k + m)$, so $a + b \equiv a' + b' \pmod{n}$.

B2.20 If $a = a' + kn$ and $b = b' + mn$, then $ab - a'b' = a(b - b') + b'(a - a')$, so $ac \equiv bd \pmod{n}$.

B2.21 $\text{Code}(\emptyset) = 0$; $\text{Code}(\{\emptyset\}) = 2^0 = 1$; $\text{Code}(\{\{\emptyset\}\}) = 2^1 = 2$; $\text{Code}(\{\emptyset, \{\emptyset\}\}) = 2^0 + 2^1 = 3$; $\langle\emptyset, \emptyset\rangle = \{\{\emptyset\}, \{\emptyset, \emptyset\}\} = \{\{\emptyset\}\}$, so $\text{Code}(\langle\emptyset, \emptyset\rangle) = 2$.

B2.22 a) $11 = 2^0 + 2^1 + 2^3$. Elements have codes 0, 1, 3. Since code 3 corresponds to $\{\emptyset, \{\emptyset\}\}$, the result is $\{\emptyset, \{\emptyset\}, \{\emptyset, \{\emptyset\}\}\}$. (This is the ordinal 3). b)$15 = 2^0 + 2^1 + 2^2 + 2^3$. Elements have codes 0, 1, 2, 3. The set is $\{\emptyset, \{\emptyset\}, \{\{\emptyset\}\}, \{\emptyset, \{\emptyset\}\}\}$.

Level C1

C1.1 1. $\forall x(x \in A \to x \in B)$; 2. $\neg\exists x(x \in A \land x \in B)$; 3. $\exists x\exists y(x \in A \land y \in A \land x \neq y \land \forall z(z \in A \to (z = x \lor z = y)))$; 4. $\forall x(x \in P \leftrightarrow x \subseteq A)$.

C1.2 Cantor's Theorem:

$$\neg\exists f \big((f : A \to \mathcal{P}(A)) \land \forall Y \in \mathcal{P}(A)\ \exists x \in A\ (f(x) = Y)\big).$$

(Note: The second part of the conjunction expresses that f is surjective).

Axiom of Countable Choice ($\mathbf{AC}_\omega$):

$$\forall I \forall A\big(((I : \omega \to A) \land \forall n \in \omega : \ I(n) \neq \emptyset)\big) \to \big(\exists f\ (f : \omega \to \bigcup A) \land \forall n \in \omega : \ f(n) \in I(n))\big).$$

(Here I represents the family of sets indexed by ω).

Definition of a Filter $\mathcal{F}$ on S:

$$\text{Filter}(\mathcal{F}, S) \overset{\text{def}}{\leftrightarrow} (\mathcal{F} \subseteq \mathcal{P}(S)) \land (S \in \mathcal{F}) \land (\emptyset \notin \mathcal{F}) \land [\forall A, B \in \mathcal{F} : (A \cap B \in \mathcal{F})] \land [\forall C \in \mathcal{F} : \forall D \in \mathcal{P}(S) : (C \subseteq D \to D \in \mathcal{F})].$$

Hausdorff's Maximal Principle: Let us define the auxiliary predicates for a partial order on A and a chain in A:

$$\text{Part}(A, R) \overset{\text{def}}{\leftrightarrow} \forall x, y, z \in A : \big(xRx \wedge (xRy \wedge yRz \rightarrow xRz) \wedge (xRy \wedge yRx \rightarrow x = y)\big);$$

$$\text{Chain}(C, A, R) \overset{\text{def}}{\leftrightarrow} (C \subseteq A) \wedge \forall x, y \in C : (xRy \vee yRx).$$

(Note: Since R is a partial order on A, it is already transitive and antisymmetric on any subset C, so for a chain we only need to postulate connexity).

The principle:

$$\forall A, R\big(\text{Part}(A, R) \rightarrow \forall C[\text{Chain}(C, A, R) \rightarrow \exists M(\text{Chain}(M, A, R) \wedge (C \subseteq M) \wedge \forall X(\text{Chain}(X, A, R) \wedge (M \subseteq X) \rightarrow M = X))]\big).$$

C1.3 Check the 3 properties of a filter. Empty set is not there (since S is infinite). Intersection of sets with finite complements has a finite complement. Superset of a set with finite complement has a finite complement.

C1.4 R_1: Reflexive, Symmetric, Antisymmetric, Transitive. R_2: Symmetric. R_3: Transitive.

C1.5 The function maps a pair $\langle a, b\rangle$ to $N = 2^a(2b+1)-1$. This is equivalent to $N + 1 = 2^a(2b + 1)$. By the Fundamental Theorem of Arithmetic, every positive integer M (here $M = N + 1 \geqslant 1$) can be uniquely represented as the product of a power of 2 and an odd number. Let $M = 2^k \cdot m$, where m is odd. Since m is odd, it can be uniquely represented as $2l + 1$ for some $l \in \omega$. Thus, for every $N \in \omega$, there exists a unique pair of exponents and odd factors corresponding to a unique pair $\langle a, b\rangle$ such that $N + 1 = 2^a(2b + 1)$. This guarantees both surjectivity (existence) and injectivity (uniqueness).

C1.6 Since f is surjective, the set of preimages $f^{-1}(\{y\})$ is non-empty for each y. By AC, we can choose one element x_y from each preimage. Set $g(y) = x_y$. Then $f(g(y)) = f(x_y) = y$.

C1.7 Algorithm: We proceed by recursion on the size of the set. Let i be a counter, initially 0. Let $A_{current} = A$.

► *Step:* While $A_{current} \neq \emptyset$:

1. Choose a minimal element m in $A_{current}$ (exists because the set is finite).
2. Set $f(m) = i$.

3. Remove m: $A_{current} := A_{current} \setminus \{m\}$.
4. Increment i: $i := i + 1$.

Verification: 1) *Bijectivity:* Since we remove exactly one element at each step and increment i, and the process runs n times, f maps A bijectively onto $\{0, \ldots, n-1\}$. 2) *Linearity:* The order $\preccurlyeq$ is isomorphic to the standard order on natural numbers, which is linear. 3) *Extension:* If $x < y$ in the original partial order, then y cannot be minimal as long as x is still in the set. Thus, x must be removed (assigned a value) at an earlier step than y. Consequently, $f(x) < f(y)$, which implies $x \preccurlyeq y$.

C1.8 1. Yes ($3 = \{0, 1, 2\}$). 2. No (not transitive: let $x = \{\{\emptyset\}\}$, then $x \in B$ but $x \not\subseteq B$ because $\{\emptyset\} \in x$ but $\{\emptyset\} \notin B$). 3. Yes ($\omega + 1$).

C1.9 1: 0; 2: 1; 3: 2; 4: 2.

C1.10

1. $\text{rank}(\{a\}) = \alpha + 1$. $\text{rank}(\{a, b\}) = \max(\alpha, \beta) + 1$. $\text{rank}(\langle a, b\rangle) = \text{rank}(\{\{a\}, \{a, b\}\}) = \max(\alpha + 1, \max(\alpha, \beta) + 1) + 1 =$ $= \max(\alpha, \beta) + 2$.
2. Integers $\mathbb{Z} \subseteq \mathcal{P}(\omega\times\omega)$. $\text{rank}(n) < \omega$. $\text{rank}(\langle n, m\rangle) < \omega$. $\text{rank}(\omega\times\omega) = \omega$. An equivalence class $z \in \mathbb{Z}$ is a set of pairs, so $\text{rank}(z) = \omega$. The set $\mathbb{Z}$ consists of such classes, so $\text{rank}(\mathbb{Z}) = \omega + 1$.

C1.11 1. V_1. 2. $V_{\omega+1}$. 3. $V_{\omega+2}$.

C1.12 $F(\alpha) = \alpha$.

C1.13 Fibonacci Generator.

$$\Phi(g) = \begin{cases} 0, & \text{if } \text{dom}(g) = \emptyset \text{ (step 0);} \\ 1, & \text{if } \text{dom}(g) = \{\emptyset\} \text{ (step 1);} \\ g(n-1) + g(n-2), & \text{if } \text{dom}(g) = n \geqslant 2. \end{cases}$$

Note: $g(n-1)$ is the last element of the history, $g(n-2)$ is the one before.

C1.14 Step 1: $\mathbb{R} \leftrightarrow (0, 1)$. The standard function $f(x) = \frac{1}{2} + \frac{1}{\pi}\arctan(x)$ maps $\mathbb{R}$ bijectively to the open interval $(0, 1)$.

Step 2: $(0, 1) \leftrightarrow [0, 1]$. We need to "absorb" the two missing endpoints $\{0, 1\}$ into the open interval. Choose a countable sequence of distinct points inside $(0, 1)$, for example:

$$S = \left\{\frac{1}{2}, \frac{1}{3}, \frac{1}{4}, \ldots\right\} = \left\{\frac{1}{n+2} \mid n \in \omega\right\}.$$

Let us construct a bijection $g : (0, 1) \longrightarrow [0, 1]$. We map the set S to the set $S \cup \{0, 1\}$ and leave the rest of the interval unchanged. Define $g(x)$ as follows:

$$g(x) = \begin{cases} 0, & \text{if } x = 1/2 \text{ (1st element of S)}, \\ 1, & \text{if } x = 1/3 \text{ (2nd element of S)}, \\ \frac{1}{n}, & \text{if } x = \frac{1}{n+2} \text{ for } n \geqslant 2 \text{ (shift the rest)}, \\ x, & \text{if } x \notin S. \end{cases}$$

This g is a bijection. The inverse g^{-1} moves $0 \longrightarrow 1/2$, $1 \longrightarrow 1/3$, and shifts the sequence back.

Result: The composition $g \circ f$ is the required bijection $\mathbb{R} \longrightarrow [0, 1]$.

C1.15 Let $A = (0, 1)$ and $B = (0, 1) \cup (2, 3)$. We construct a bijection $F : B \longrightarrow A$ in three steps.

Step 1: Shift and Scale. Map the two intervals of B into a single punctured interval $(0, 1) \setminus \{1/2\}$. Define $f : B \longrightarrow (0, 1) \setminus \{1/2\}$ as:

$$f(x) = \begin{cases} x/2, & \text{if } x \in (0, 1) \quad \text{(maps to } (0, 1/2)), \\ x/2 - 1/2, & \text{if } x \in (2, 3) \quad \text{(maps to } (1/2, 1)). \end{cases}$$

This function is a bijection from B to $Z = (0, 1) \setminus \{1/2\}$.

Step 2: Fill the Gap (Hilbert Hotel). We need a bijection $g : Z \longrightarrow (0, 1)$ that "absorbs" the missing point $1/2$. Choose a countable sequence inside Z: $S = \{1/3, 1/4, 1/5, \ldots\} = \{1/n \mid n \geqslant 3\}$. Define $g(y)$ to shift this sequence to include $1/2$:

$$g(y) = \begin{cases} 1/(n-1), & \text{if } y = 1/n \text{ for } n \geqslant 3 \\ & \text{(maps } 1/3 \longrightarrow 1/2, 1/4 \longrightarrow 1/3, \ldots), \\ y, & \text{if } y \notin S. \end{cases}$$

This g maps Z onto $(0, 1)$ bijectively.

Conclusion: The composition $F = g \circ f$ is the required bijection from $(0, 1) \cup (2, 3)$ to $(0, 1)$.

C1.16

1. Component-wise order ($\leqslant_{prod}$). We check $x \leqslant_{prod} y$: $2 \leqslant 3$ is True, but $100 \leqslant 1$ is False. Thus $\neg(x \leqslant_{prod} y)$. We check $y \leqslant_{prod} x$: $3 \leqslant 2$ is False. Thus $\neg(y \leqslant_{prod} x)$. *Conclusion:* x and y are incomparable. *Properties:* Since there exist incomparable elements, $\leqslant_{prod}$ is not a linear order (it is a partial order). Consequently, it is not a well-ordering (since a well-ordering must be linear).
2. Lexicographical order ($\leqslant_{lex}$). We compare the first components: $2 < 3$. According to the definition, this immediately implies $x <_{lex} y$.
   Properties:
   1. For any distinct pairs, either the first components differ (then the order is determined by them) or they are equal (then the order is determined by the second components). Thus, it is a **linear order**.
   2. Since ω is well-ordered, the lexicographical product $\omega \times \omega$ is also well-ordered. (Any non-empty subset S has a minimum: take the set of first coordinates of elements in S, find its minimum m_1, then find the minimum m_2 of the second coordinates among elements of S starting with m_1).

C1.17 If $x = \{x\}$, then $x \in x$. The set $\{x\}$ has only one element x, and $x \cap \{x\} = \{x\} \cap \{x\} = x \neq \emptyset$. This contradicts the Axiom of Regularity.

C1.18 1. ω (limit). 2. $\omega + 1$ (successor). 3. ω. 4. $\omega + \omega$.

C1.19 $(1 + 1) \cdot \omega = 2 \cdot \omega = \omega$. $1 \cdot \omega + 1 \cdot \omega = \omega + \omega = \omega \cdot 2$. Since $\omega \neq \omega \cdot 2$, right distributivity fails.

C1.20 $\omega + x = \omega \cdot 2$. Since $\omega \cdot 2 = \omega + \omega$, the solution is $x = \omega$. (Note: $x = n < \omega$ gives $\omega + n < \omega \cdot 2$, and $x = \omega + 1$ gives $\omega + (\omega + 1) = \omega \cdot 2 + 1 > \omega \cdot 2$).

C1.21 β is smaller. In ordinal arithmetic, lower terms on the right do not affect the leading power for comparison, but here the coefficients of ω matter. $3 > 2$.

C1.22 Let us analyze the structure of the order $\leqslant_{lex}$. We can group the elements of $\omega \times \omega$ by their first coordinate n. Let $R_n = \{\langle n, m\rangle \mid m \in \omega\}$ be the n-th "row".

1. Internal structure of a row: Within a fixed row R_n, the first coordinates are equal, so the order is determined by the second coordinates m. Since m ranges over ω, each row R_n is order-isomorphic to ω.

2. Ordering of rows: Since the first coordinate is compared first in $\leqslant_{lex}$, for any $n_1 < n_2$, all elements of row R_{n_1} strictly precede all elements of row R_{n_2}.

3. Total structure: The set P consists of the sequence of rows $R_0, R_1, R_2, \ldots$, where each row is a copy of ω, and they are arranged in the order type of ω.

This corresponds to the ordinal sum of ω copies of ω:

$$\underbrace{\omega}_{R_0} + \underbrace{\omega}_{R_1} + \underbrace{\omega}_{R_2} + \cdots = \omega \cdot \omega.$$

Answer: The order type is $\omega \cdot \omega$.

C1.23 The Axiom of Separation (subset axioms) leads to Russell's paradox if applied to V. Also regularity is violated (though paradox is primary).

C1.24 Complex Numbers.

1. $\langle a, b\rangle +_{\mathbb{C}} \langle c, d\rangle \overset{\text{def}}{\rightleftharpoons} \langle a +_{\mathbb{R}} c, b +_{\mathbb{R}} d\rangle$;
2. $\langle a, b\rangle \cdot_{\mathbb{C}} \langle c, d\rangle \overset{\text{def}}{\rightleftharpoons} \langle a \cdot_{\mathbb{R}} c -_{\mathbb{R}} b \cdot_{\mathbb{R}} d, a \cdot_{\mathbb{R}} d +_{\mathbb{R}} b \cdot_{\mathbb{R}} c\rangle$.

Check for $i = \langle 0, 1\rangle$: $i \cdot_{\mathbb{C}} i = \langle 0 \cdot_{\mathbb{R}} 0 -_{\mathbb{R}} 1 \cdot 1, 0 \cdot_{\mathbb{R}} 1 +_{\mathbb{R}} 1 \cdot_{\mathbb{R}} 0\rangle = \langle -1, 0\rangle$. Correct.

C1.25

Let $\mathcal{P}$ be the set of all partial orders R on A such that $\leqslant \subseteq R$. We order the set $\mathcal{P}$ by set-theoretic inclusion $\subseteq$. To apply Zorn's Lemma, we verify that every chain in $\mathcal{P}$ has an upper bound.

Let $C \subseteq \mathcal{P}$ be a non-empty chain. Define $U = \bigcup C$.

- *Reflexivity:* Since $\leqslant \subseteq U$, U is reflexive.
- *Transitivity:* Let $(x, y) \in U$ and $(y, z) \in U$. Then there exist $R_1, R_2 \in C$ such that $(x, y) \in R_1$ and $(y, z) \in R_2$. Since C is a chain, assume $R_1 \subseteq R_2$. Then both pairs are in R_2. Since R_2 is transitive, $(x, z) \in R_2 \subseteq U$.
- *Antisymmetry:* Let $(x, y) \in U$ and $(y, x) \in U$. Similarly, there exists some $R \in C$ containing both pairs. Since R is antisymmetric, $x = y$.

Thus, U is a partial order extending $\leqslant$, so $U \in \mathcal{P}$. U is obviously an upper bound for C.

By Zorn's Lemma, $\mathcal{P}$ has a maximal element L. We claim L is a linear order. Suppose not. Then there exist incomparable elements $a, b \in A$ (i.e., $\neg(aLb) \wedge \neg(bLa)$). Define a new relation L' as the transitive closure of $L \cup \{(a, b)\}$:

$$L' = L \cup \{(x, y) \mid xLa \wedge bLy\}.$$

Clearly $L \subsetneq L'$ (since $(a, b) \in L'$ but $(a, b) \notin L$). To show $L' \in \mathcal{P}$, we must check antisymmetry (transitivity and reflexivity are valid by construction). If L' is not antisymmetric, there exist x, y such that $xL'y$ and $yL'x$ with $x \neq y$. Since L is antisymmetric, at least one relation must come from the new set. This implies a cycle involving a and b which reduces to bLa, contradicting the incomparability assumption. Thus, L' is a strictly larger partial order, contradicting the maximality of L. Therefore, L must be linear.

Chapter D1

D1.1 a) π: Independent, basic (special symbol/letter); b) $\subseteq$: Dependent, basic (operator); c) $\hat{y}$: Independent, derived (letter + accent); d) ∇: Dependent, basic (operator, derived from inverted letter but considered basic operator in $\mathfrak{Math}$); e) { }: Dependent, basic (grouper).

D1.2 a) Σ — independent, basic (letter). b) $\sum$ — dependent, basic (operator, derived from letter). c) $\overline{z}$ — independent, derived (letter + accent). d) ; — dependent, basic (separator). e) $\rightarrow$ — Dependent, basic (operator).

D1.3 Rigid: a) Yes. b) Yes. c) No (compound lexeme). d) Yes.

D1.4

1. 3 args (variable, limit point, expression).
2. 2 args (base, argument).
3. 1 arg (expression).
4. 2 args (numerator function, denominator variable).

D1.5 Tree structure (from root to leaves):

- Root: Equality $\sqcup = \sqcup$ (Left: $\sigma(x)$, Right: Fraction).
- Right node (Fraction $\frac{\sqcup}{\sqcup}$): Numerator is 1, Denominator is Addition.
- Denominator (Addition $\sqcup + \sqcup$): Left is 1, Right is Exponentiation.
- Right node of Addition (Power $\sqcup^{\sqcup}$): Base is e, Exponent is Negation.
- Exponent (Negation $-\sqcup$). Operand is x.

D1.6 1) $div(a, b)$. 2) $pow(a, b)$. 3) $root(x, n)$.

D1.7 1) $\cdot + ab - cd$. 2) $ab + cd - \cdot$.

D1.8

1. $(a + \sin(\omega t)) \cdot y$;
2. $\int_0^{x^2} f(t)dt$ (Note: the variable of integration t is bound and usually not subject to substitution, the upper limit x is free);

3. $P(g(z), y) \to Q(g(z))$.

D1.9 1) $(f(u))^2$. 2) $\sin(2 \cdot f(u))$. 3) $D_{f(u)}$.

D1.10 $\sum_{k=0}^{\infty}(2k+1)$. The result is a valid lexeme.

D1.11

1. Arguments: vectors; Result: scalar (number).
2. Argument: vector; Result: scalar (non-negative number).
3. Arguments: sets; Result: logical value (boolean).

D1.12 1) $V \times V \to S$. 2) $V \times V \to V$. 3) $S \times V \to V$. 4) $V \to S$.

D1.13
1) Root: +. Children: $\sin x$ and y. (Standard convention). 2) Root: sin. Child: $x + y$.

D1.14 $(1, (2, (3, ())))$.

D1.15 $\overline{x}, x', \hat{x}$.

D1.16
1) Variable (Term). 2) Formula. 3) Connective. 4) Formula (Binding).

D1.17 $((\neg\mathcal{A}) \wedge \mathcal{B}) \to C$.

D1.18 Renaming: t: $\int_a^b t^2 dt$. Meaning does not change (dummy variable). a: $\int_a^b a^2 da$. Meaning changes! Collision with the limit of integration a.

D1.19 The lexeme is built using the following hierarchy of templates:

- Level 1 (Root): Summation template $\sum_{\sqcup}^{\sqcup} \sqcup$.
- Level 2 (Bottom placeholder): Equality template $\sqcup = \sqcup$ (combining rigid graphemes i and 1).
- Level 2 (Main placeholder): Power template $\sqcup^{\sqcup}$.
- Level 3 (Base of Power): Subscript template $\sqcup_{\sqcup}$ (combining rigid graphemes x and i).
- Leaves (Rigid Graphemes): $i, 1, n, x, 2$.

Chapter D2

D2.1 Let us define the formula $\Phi = (P \to E) \vee (L \to F)$. We expand the implications using the equivalence $A \to B \equiv \neg A \vee B$:

$$\Phi \equiv (\neg P \vee E) \vee (\neg L \vee F).$$

Using the associativity and commutativity of disjunction, we regroup the terms to match the premises:

$$\Phi \equiv (\neg P \vee F) \vee (\neg L \vee E).$$

Recall that $\neg P \vee F \equiv P \to F$ and $\neg L \vee E \equiv L \to E$. Since both premises $(P \to F)$ and $(L \to E)$ are given as True, the disjunction of them is also True.

Comment: This paradox illustrates that material implication does not require a causal connection. Even though "Paris implies England" and "London implies France" sound false individually, their disjunction is logically necessary given the premises, because at least one of the antecedent conditions (P or L) must be false, or the consequents (E or F) true, in such a way that makes the logical structure valid.[18]

D2.2

1. $\varphi(t) \to \exists x \varphi(x)$ is Axiom ($\exists$).
2. Apply MP to the hypothesis $\varphi(t)$ and the axiom.

D2.3

1. Hypothesis: $\forall x(\varphi \to \psi)$.
2. $\varphi(a) \to \psi(a)$ (Axiom $\forall$).
3. Hypothesis: $\forall x \varphi$.
4. $\varphi(a)$ (Axiom $\forall$).
5. $\psi(a)$ (MP from 2 and 4).
6. $\forall x \psi(x)$ (Gen).
7. Result by Deduction Theorem.

D2.4

[18] This derivation relies heavily on the Law of Excluded Middle. In Intuitionistic Logic, which requires constructive proofs, this statement cannot be derived because we cannot prove either of the implications individually.

1. ($\rightarrow$): Assume $\forall x(\varphi \wedge \psi)$. Then $\varphi(a) \wedge \psi(a)$. Then $\varphi(a)$ and $\psi(a)$. Generalize (Bernays' Rule R1') to $\forall x\varphi$ and $\forall x\psi$.
2. ($\leftarrow$): Assume $\forall x\varphi \wedge \forall x\psi$. Then $\varphi(a)$ and $\psi(a)$. Then $\varphi(a) \wedge \psi(a)$. Generalize.

D2.5 Similar to the previous one. Use the equivalence $\exists x\chi \leftrightarrow \neg\forall x\neg\chi$ and De Morgan's laws.

D2.6 Use case analysis on $\forall y\varphi(y) \vee \neg\forall y\varphi(y)$. Case 1: If $\forall y\varphi(y)$ is true, then for any x, $\varphi(x) \rightarrow \forall y\varphi(y)$ is true. Thus $\exists x(\dots)$ is true. Case 2: If $\neg\forall y\varphi(y)$, then $\exists x\neg\varphi(x)$. Let x_0 be such that $\neg\varphi(x_0)$. Then $\varphi(x_0) \rightarrow \dots$ is true (False $\rightarrow$ Anything). Thus $\exists x(\dots)$ is true.

D2.7

1. ($\rightarrow$): Assume $\varphi \rightarrow \forall x\psi(x)$ and φ. Then $\forall x\psi(x)$, so $\psi(a)$. Since a is arbitrary and not in φ, we get $\varphi \rightarrow \psi(a)$, then $\forall x(\varphi \rightarrow \psi(x))$.
2. ($\leftarrow$): Assume $\forall x(\varphi \rightarrow \psi(x))$ and φ. Then $\varphi \rightarrow \psi(a)$, so $\psi(a)$. Generalize to $\forall x\psi(x)$.

D2.8 Use Bernays' rule (R1): if $\Gamma \vdash \varphi \rightarrow \psi(x)$ and x is not free in φ (and Γ), then exists a finite $\Delta \subseteq \Gamma$ s.t. $\Delta \vdash \varphi \rightarrow \psi(x)$, thus using γ as a conjunction of all formulas from Δ, we get by D.T. that $\vdash \gamma \wedge \varphi \rightarrow \psi(x)$, then by Bernays's Rule (R1) it follows that $\vdash \gamma \wedge \varphi \rightarrow \forall x\psi(x)$, so $\Delta \vdash \varphi \rightarrow \forall x\psi(x)$ and, consequently $\Gamma \vdash \varphi \rightarrow \forall x\psi(x)$.

D2.9 Let $\exists x\forall y\varphi(x, y)$ is True in some model. So take such element a that $\forall y\varphi(a, y)$ is True. It means that for all b in this model $\varphi(a, b)$ is True. So, for all b there exists x (actually $x = a$) s.t. $\varphi(x, b)$ is True, i.e. $\exists x\varphi(x, b)$. And, generally, $\forall y\exists x\varphi(x, y)$ in this model. We proved that $\exists x\forall y\varphi(x, y) \Vdash \forall y\exists x\varphi(x, y)$, and by Gödel completeness theorem we get $\exists x\forall y\varphi(x, y) \vdash \forall y\exists x\varphi(x, y)$.

D2.10 $\forall x\forall y(x < y \rightarrow \exists z(x < z \wedge z < y))$ (Density). True in $\mathbb{Q}$, false in $\mathbb{Z}$.

D2.11

1. $\varphi(x, y) \equiv \exists z(z \neq 0 \wedge x + z = y)$. In $\mathbb{N}$, if $y > x$, the difference $y - x$ exists and is a natural number.
2. Consider the function $f : \mathbb{Z} \rightarrow \mathbb{Z}$ defined by $f(x) = -x$. Since $-(a + b) = (-a) + (-b)$, f is an automorphism of the additive group. However, f does not preserve order: $1 < 2$ but $-1 > -2$. Thus, $<$ cannot be defined by a formula involving only $+$.

D2.12

1. $\varphi(x) \equiv \exists y(y \cdot y = x)$. In $\mathbb{R}$, a number is a square iff it is non-negative.
2. $x \leqslant y \iff \exists z(z \cdot z = y - x)$. (Since $y - x \geqslant 0$).

D2.13

1. No (automorphisms of $\langle \mathbb{R}, < \rangle$ act transitively, e.g., $x \mapsto x + c$).
2. No (same reason).
3. No.

(Note: In a pure order structure without constants, no individual element is definable).

D2.14 It follows from $\aleph_0$-categoricity of DLO_∞.

D2.15

1. The field of algebraic numbers $\mathbb{A}$ (the set of all roots of polynomials with rational coefficients). It is countable and algebraically closed.
2. By the Upward Löwenheim-Skolem Theorem. Since PA has an infinite model $\mathbb{N}$, it must have models of any infinite cardinality $\kappa > \aleph_0$.

D2.16 The standard model $\mathbb{N}$ and the model $\mathbb{N} + \mathbb{Z}$ (naturals followed by a copy of integers) are both countable but not isomorphic (in the second, some elements have predecessors but are not reachable from 0). Thus, the theory is not categorical in cardinality $\aleph_0$.

D2.17 $a < b$: $\mathsf{DLO} \vdash \exists x(a < x \wedge x < b) \leftrightarrow (a < b)$ (density and transitivity).

D2.18 $b \leqslant a$: $\mathsf{DLO} \vdash \forall x(a < x \rightarrow b < x) \leftrightarrow (b \leqslant a)$ (connexity and transitivity).

D2.19 $\top$ (True). Since the domain is infinite, we can always find an x distinct from any two fixed elements a and b.

D2.20 The condition is $(a = 0 \wedge b > 0) \vee (a \neq 0)$. If $a = 0, b > 0$, $0x + b > 0$ is $b > 0$ (True for all x, so $\exists$ is True). If $a = 0, b \leqslant 0$, $0 > 0$ (False). If $a \neq 0$, the line $y = ax + b$ crosses zero, so there is a region where it is positive. Simplified quantifier-free form: $(a \neq 0) \vee (b > 0)$.

D2.21 $(a = 0 \wedge b = 0 \wedge c = 0) \vee (a \neq 0 \wedge b^2 - 4ac \geqslant 0) \vee (a = 0 \wedge b \neq 0)$. (Essentially: either identity $0 = 0$, or quadratic eq has real roots, or linear eq has a root).

D2.22 Let us verify the conditions of the Łoś–Vaught test:

1. *No finite models:* The axioms require infinite classes, so the domain must be infinite.
2. $\aleph_0$-*categoricity:* Let $\mathcal{M}$ be any countable model of T. Its domain M is partitioned into two classes, C_1 and C_2. Since $\mathcal{M}$ is countable and C_1, C_2 are infinite subsets, $|C_1| = \aleph_0$ and $|C_2| = \aleph_0$. Any other countable model $\mathcal{M}'$ will similarly have classes C_1', C_2' of cardinality $\aleph_0$. We can construct a bijection $f : M \to M'$ that maps C_1 to C_1' and C_2 to C_2'. Since the structure is defined solely by the equivalence relation, such a bijection is an isomorphism.

Conclusion: T is consistent, has no finite models, and is categorical in cardinality $\aleph_0$. Thus, T is complete.

Chapter D3

D3.1 1. $a + S(0) = S(a + 0)$ (PA5). 2. $a + 0 = a$ (PA5). 3. $S(a + 0) = S(a)$ (Congruence). 4. $a + S(0) = S(a)$ (Transitivity).

D3.2 $x \cdot S(0) = x \cdot 0 + x = 0 + x = x$ (PA6).

D3.3 $a < b \iff \exists x(b = a+Sx) \iff \exists x(b = S(a+x))$. By Lemma D3.1, $x = 0 \vee \exists y(x = Sy)$. If $x = 0$, $b = S(a) \implies S(a) \leqslant b$. If $x = Sy$, $b = S(a) + Sy \implies S(a) < b$. In both cases, $S(a) \leqslant b$.

D3.4 We prove $S(a) + b = S(a + b)$ by induction on b.

- *Base case* ($b = 0$): We check the equality $S(a) + 0 = S(a + 0)$.
  - LHS: Consider the term $S(a) + 0$. By axiom PA5 ($\forall x(x + 0 = x)$), substituting the term $S(a)$ for x, we obtain $S(a) + 0 = S(a)$.
  - RHS: Consider the term $S(a + 0)$. By axiom PA5, we have $a + 0 = a$. By axiom PA2 ($x = y \to S(x) = S(y)$), applying S to both sides yields $S(a + 0) = S(a)$.

  Since both sides reduce to $S(a)$, the equality holds.
- *Inductive step* ($b \to Sb$): Assume the hypothesis $S(a) + b = S(a + b)$. We must prove $S(a) + Sb = S(a + Sb)$.

- LHS: Consider $S(a) + Sb$. By axiom PA5 ($\forall x, y(x + Sy = S(x + y))$) with $x = S(a), y = b$, we have $S(a) + Sb = S(S(a) + b)$. Using the induction hypothesis ($S(a) + b = S(a + b)$) and axiom PA2, we replace the inner term: $S(S(a) + b) = S(S(a + b))$.
- RHS: Consider $S(a + Sb)$. By axiom PA5 ($a + Sb = S(a + b)$) and applying PA2 to this equality, we obtain $S(a + Sb) = S(S(a + b))$.

Both sides are equal to $S(S(a + b))$.

D3.5 Case $b < a$: $a \dot{-} b = c \implies b + c = a$. $c \leqslant a$. Case $b \geqslant a$: $a \dot{-} b = 0 \leqslant a$. For the equivalence: if $a \dot{-} b = a$, then in case $b < a$, $b + a = a \implies b = 0$. In case $b \geqslant a$, $0 = a$, so $a = 0$.

D3.6 $(a \dot{-} b) + b = a$ holds if and only if we are in the case $b < a$ (where $b + c = a$) or the case where result is 0 and $a = b$. If $b > a$, $0 + b \neq a$. Thus $b \leqslant a$.

D3.7 $x = (a + b) \dot{-} a$. Since $a + b \geqslant a$, x is the unique number such that $a + x = a + b$, which is b.

D3.8 Case $b + c \geqslant a$: LHS 0. RHS $(a \dot{-} b) \dot{-} c$: if $b \geqslant a$ then $0 \dot{-} c = 0$. If $b < a$, then let $k = a \dot{-} b$, $b + k = a$. $b + c \geqslant b + k \implies c \geqslant k$. Thus $k \dot{-} c = 0$. Case $b + c < a$: LHS is x s.t. $b + c + x = a$. RHS: $b < a$, let $k = a \dot{-} b$ ($b + k = a$). Since $b + c < b + k$, $c < k$. Then $(a \dot{-} b) \dot{-} c = k \dot{-} c = y$ s.t. $c + y = k$. Subst: $b + (c + y) = a \implies (b + c) + y = a$. Unique.

D3.9 Case $c = 0$: $0 = 0$. Case $a \leqslant b$: LHS $c \cdot 0 = 0$. RHS $ca \leqslant cb \implies 0$. Case $a > b$: let $a = b + k$. LHS ck. RHS $c(b + k) \dot{-} cb = (cb + ck) \dot{-} cb = ck$.

D3.10 Let $a \leqslant b$. We compare $a \dot{-} c$ and $b \dot{-} c$. Case $c \geqslant b$: both 0. Case $a \leqslant c < b$: LHS 0, RHS > 0. Case $c < a$: $a = c + x, b = c + y$. $a \leqslant b \implies c + x \leqslant c + y \implies x \leqslant y$.

D3.11 Let $a \leqslant b$. Compare $c \dot{-} b$ and $c \dot{-} a$. Case $c \leqslant a$: both 0. Case $a < c \leqslant b$: LHS 0, RHS > 0. Case $c > b$: $c = b + x, c = a + y$. $b + x = a + y$. Since $a \leqslant b$, $y \geqslant x$.

D3.12 $R(a, b, r) \overset{\text{def}}{\leftrightarrow} r < b \land \exists q(a = b \cdot q + r)$.

D3.13 If $a = 0$, then $a \mid b \implies b = 0$, so $a = b$. If $a \neq 0$, then $b \neq 0$ (since $b \mid a$). $a = kb, b = ma \implies a = kma$. Cancel a: $km = 1 \implies k = 1, m = 1$. Thus $a = b$.

D3.14 We need $12s - 18t = \gcd(12, 18) = 6$. Solution $s = 2, t = 1$ $(24 - 18 = 6)$.

D3.15 1) $m_0 = 1 + 1 \cdot 3 = 4$, $\beta = \text{rem}(10, 4) = 2$; 2) $m_1 = 1 + 2 \cdot 3 = 7$, $\beta = \text{rem}(10, 7) = 3$; 3) $m_2 = 1 + 3 \cdot 3 = 10$. $\beta = \text{rem}(10, 10) = 0$.

D3.16 1) $m_0 = 1+(0+1)2 = 3$, $\beta = \text{rem}(13, 3) = 1$; 2) $m_1 = 1+(1+1)2 = 5$, $\beta = \text{rem}(13, 5) = 3$.

The sequence is $\langle 1, 3\rangle$.

D3.17 Need $c \equiv 1 \pmod{1 + d}$, $c \equiv 2 \pmod{1 + 2d}$. Try $d = 1$: mods 2, 3. $c \equiv 1 \pmod 2, c \equiv 2 \pmod 3 \implies c = 5$. Pair $(5, 1)$.

D3.18 If (c, d) works, then $(c + K \cdot \text{lcm}(m_i), d)$ also works for any K.

D3.19 Positive numbers $\langle 0, n\rangle$, negative $\langle n, 0\rangle$. $5 \to \langle 0, 5\rangle$. $-3 \to \langle 3, 0\rangle$.

D3.20 $\langle 0, 5\rangle +_{\mathbb{Z}} \langle 3, 0\rangle$. Case $a_1 = b_2 = 0$: result is $\langle 0, a_2 \dot{-} b_1\rangle = \langle 0, 5 \dot{-} 3\rangle = \langle 0, 2\rangle$ (which corresponds to 2).

D3.21 $\langle 0, 2\rangle +_{\mathbb{Z}} \langle 5, 0\rangle$. Case $a_1 = b_2 = 0$: result is $\langle b_1 \dot{-} a_2, 0\rangle$ (since $b_1 > a_2$) $= \langle 5 \dot{-} 2, 0\rangle = \langle 3, 0\rangle$ (corresponds to -3).

D3.22 $\langle 2, 0\rangle \cdot_{\mathbb{Z}} \langle 3, 0\rangle$. Case "same signs" $(a_2 = b_2 = 0)$. Result is $\langle 0, (a_1 + a_2)(b_1 + b_2)\rangle = \langle 0, (2 + 0)(3 + 0)\rangle = \langle 0, 6\rangle$. (Positive, +6).

D3.23 $\langle 0, 3\rangle \cdot_{\mathbb{Z}} \langle 2, 0\rangle$. Case "different signs" (otherwise). Result is $\langle (a_1 + a_2)(b_1 + b_2), 0\rangle = \langle (0 + 3)(2 + 0), 0\rangle = \langle 6, 0\rangle$. (Negative, -6).

D3.24

1. Any provably false sentence, e.g., $0 = S0$.
2. Any provably true sentence, e.g., $0 = 0$.

D3.25

1. $\exists y\ (a = SSSSSSSSSS0 \cdot y + SSSSSS0)$.
2. $0 = S0$. It is false, and its code does not end with 6.
3. $0 = 0 \cdot (0 + 0)$. It is true, and its code ends with 6.

D3.26

1. $\neg\exists y\ (a = SSS0 \cdot y)$.
2. $0 + 0 = 0$. Its code is 13141, which is not divisible by 3 (the sum of digits is 10). Thus, $\varphi(\underline{13141})$ is true, and the sentence $0 + 0 = 0$ is also true.

Chapter D4

D4.1 Dual: $(a \wedge b) \vee (a \wedge \neg b) = a$. Proof: $a \wedge (b \vee \neg b) = a \wedge 1 = a$. True.

D4.2

1. *Zero/One:* $\emptyset$ is finite, X is cofinite (complement is $\emptyset$). *Complement:* If S is finite, S^c is cofinite. If S is cofinite, S^c is finite. *Union:* - Finite $\cup$ Finite = Finite. - Cofinite $\cup$ Anything = Cofinite (since the complement of the union is the intersection of complements: infinite $\cap$ finite = finite). Thus, $\mathcal{A}$ is closed under all Boolean operations.
2. No. Consider the family of singletons $\{\{x\} \mid x \in X\}$. Each is in $\mathcal{A}$. Their infinite union is X. However, let $X = \omega$ and take the union of singletons of all even numbers. The result is the set of even numbers E. E is infinite, and its complement (odds) is infinite. Thus $E \notin \mathcal{A}$.
3. The atoms are exactly the singleton sets $\{x\}$ for $x \in X$.

D4.3

First, check closure. Let $e, f \in B$. $\neg e$: $(1-e)^2 = 1 - 2e + e^2 = 1 - 2e + e = 1 - e$. (Closed). $e \wedge f$: $(ef)^2 = e^2 f^2 = ef$. (Closed). $e \vee f$: Let $s = e + f - ef$. Then $s^2 = e^2 + f^2 + e^2 f^2 + 2ef - 2e^2 f - 2ef^2 = e + f + ef + 2ef - 2ef - 2ef = e + f - ef = s$. (Closed). Now check distributivity (one of them): $x \wedge (y \vee z) = x(y + z - yz) = xy + xz - xyz$. $(x \wedge y) \vee (x \wedge z) = xy + xz - (xy)(xz) = xy + xz - x^2 yz = xy + xz - xyz$. They match. Other axioms hold similarly.

D4.4

1. We need to show two things: $(a \to b) \in S$ and $\forall y \in S, y \leqslant (a \to b)$.
   - ► Substitute $x = (a \to b)$ into the adjunction formula. Since $x \leqslant x$ is always true (reflexivity), the equivalence gives $(a \to b) \wedge a \leqslant b$. Thus, $(a \to b) \in S$.
   - ► Let $y \in S$. This means $y \wedge a \leqslant b$. Using the adjunction formula from right to left (substituting $x = y$), we immediately obtain $y \leqslant (a \to b)$.

   Thus, $a \to b$ is the maximum element of S.
2. We show that an arbitrary element x is less than the LHS if and only if it is less than the RHS.
   Analysis of LHS: $x \leqslant (a \vee b) \to c$. By the adjunction property, this is equivalent to:

$$x \wedge (a \vee b) \leqslant c.$$

Using the distributivity of meet over join:

$$(x \wedge a) \vee (x \wedge b) \leqslant c.$$

By the *universal property of the join* (supremum), a join is less than c iff both operands are less than c:

$$(x \wedge a \leqslant c) \quad \text{AND} \quad (x \wedge b \leqslant c).$$

Analysis of RHS: $x \leqslant (a \to c) \wedge (b \to c)$. By the property of the meet (infimum), x is less than an intersection iff it is less than each component:

$$x \leqslant (a \to c) \quad \text{AND} \quad x \leqslant (b \to c).$$

Apply the adjunction property to each inequality separately:

$$(x \wedge a \leqslant c) \quad \text{AND} \quad (x \wedge b \leqslant c).$$

Conclusion: Since the condition for $x \leqslant$ LHS reduces to the same logical conjunction as the condition for $x \leqslant$ RHS, the elements define the same lower cone and must be equal.

D4.5

1. Proof of $\neg(x \vee y) = \neg x \wedge \neg y$: Recall that $\neg a = a \to 0$. By the property of implication in Heyting algebras, $a \vee b \to c = (a \to c) \wedge (b \to c)$. Setting $c = 0$, we get:

$$(x \vee y) \to 0 = (x \to 0) \wedge (y \to 0).$$

   By definition of negation, this is exactly $\neg(x \vee y) = \neg x \wedge \neg y$.
2. Counterexample for $\neg(x \wedge y) = \neg x \vee \neg y$: Consider the topological space $\mathbb{R}$. The algebra of open sets is a Heyting algebra. Let $U = (-\infty, 0)$ and $V = (0, +\infty)$.

   - $U \wedge V = U \cap V = \emptyset$. Thus $\neg(U \wedge V) = \neg\emptyset = \mathbb{R}$ (the whole space).
   - $\neg U = \text{Int}(\mathbb{R} \setminus (-\infty, 0)) = \text{Int}([0, +\infty)) = (0, +\infty) = V$.
   - Similarly, $\neg V = U$.
   - The RHS is $\neg U \vee \neg V = V \cup U = (-\infty, 0) \cup (0, +\infty) = \mathbb{R} \setminus \{0\}$.

Since $\mathbb{R} \neq \mathbb{R} \setminus \{0\}$, the equality fails. The "law of excluded middle" gap at 0 prevents the union of negations from covering the whole negation of intersection.

D4.6

1. Let $x \in \mathcal{B}$. If $x = 0$, the set of atoms below it is empty, and $\bigvee \emptyset = 0$. If $x \neq 0$, let $y = \bigvee\{a \in A \mid a \leqslant x\}$. Clearly $y \leqslant x$. Consider $z = x \wedge \neg y$. If $z \neq 0$, then z must contain an atom a_0 (since the algebra is finite). Then $a_0 \leqslant x$ and $a_0 \leqslant \neg y$. Since a_0 is an atom under x, it is included in the join y, so $a_0 \leqslant y$. Thus $a_0 \leqslant y \wedge \neg y = 0$, a contradiction. Therefore, $z = 0$, which implies $x \leqslant y$. Hence $x = y$.
2. The map $\varphi(x) = \{a \in A \mid a \leqslant x\}$ preserves operations. $\varphi(x \wedge y) = \{a \mid a \leqslant x \wedge a \leqslant y\} = \varphi(x) \cap \varphi(y)$. Injectivity: If $\varphi(x) = \varphi(y)$, then the sets of atoms below them are identical. By step 1, x and y are the joins of these sets, so $x = y$. Surjectivity: For any subset $S \subseteq A$, let $x = \bigvee S$. Then $\varphi(x) = S$ (since atoms are disjoint, an atom b is under $\bigvee S$ iff $b \in S$).

D4.7

1. Injection $f : \mathbb{R} \longrightarrow \mathbb{R}^2$ is simply $x \mapsto \langle x, 0\rangle$.
2. Injection $g : \mathbb{R} \times \mathbb{R} \longrightarrow \mathbb{R}$. First, map $\mathbb{R}$ to $(0, 1)$ via a homeomorphism. It suffices to map $(0, 1) \times (0, 1)$ to $(0, 1)$. Let $x = 0.a_1a_2a_3\ldots$ and $y = 0.b_1b_2b_3\ldots$. To avoid the problem of ending nines, simply forbid them (use only representations without infinite tails of 9s). Define $h(x, y) = 0.a_1b_1a_2b_2a_3b_3\ldots$. This maps two sequences into one. It is injective because the position of digits is unique. The resulting number is in $(0, 1)$. By CBS theorem, $|\mathbb{R} \times \mathbb{R}| = |\mathbb{R}|$.

D4.8 Let R be a finite integral domain and let $a \in R$ with $a \neq 0$. Consider the function $\varphi_a : R \to R$ given by $\varphi_a(x) = ax$. First, we show φ_a is injective. Suppose $\varphi_a(x) = \varphi_a(y)$. Then $ax = ay \implies a(x - y) = 0$. Since R is an integral domain and $a \neq 0$, it must be that $x - y = 0$, so $x = y$. Since R is finite, any injective function from R to itself is surjective (by the Pigeonhole Principle). Therefore, 1 is in the image of φ_a. This means there exists some $b \in R$ such that $ab = 1$. Thus, a is invertible. Since a was arbitrary, R is a field.

D4.9 Let $S = \omega^{<\omega}$. We can write $S = \bigcup_{n\in\omega} \omega^n$, where ω^n is the set of sequences of length n. For $n = 1$, $|\omega| = \aleph_0$. For any n, if $|\omega^n| = \aleph_0$, then

$|\omega^{n+1}| = |\omega^n \times \omega| = \aleph_0 \cdot \aleph_0 = \aleph_0$ (by induction). Thus, S is a countable union of countable sets: $|S| = \sum_{n\in\omega} \aleph_0 = \aleph_0 \cdot \aleph_0 = \aleph_0$.

D4.10

1. The set of polynomials of degree n with rational coefficients can be identified with $\mathbb{Q}^{n+1} \setminus \{\bar{0}\}$. Since $\mathbb{Q}$ is countable, $\mathbb{Q}^{n+1}$ is countable. The set of all polynomials $\mathcal{P} = \bigcup_{n\in\omega} \mathbb{Q}^{n+1}$ is a countable union of countable sets, hence countable. Each polynomial $P \in \mathcal{P}$ has a finite number of roots. Thus, the set of algebraic numbers is a union of a countable family of finite sets, which implies $|\mathbb{A}| = \aleph_0$.
2. Since $\mathbb{R}$ is uncountable (Theorem of Cantor) and $\mathbb{R} = \mathbb{A} \cup (\mathbb{R} \setminus \mathbb{A})$, if the set of transcendental numbers $\mathbb{R} \setminus \mathbb{A}$ were countable, then $\mathbb{R}$ would be the union of two countable sets, which is impossible. Thus, transcendental numbers are uncountable.

D4.11 Base ω: $1+\omega = \sup_{n<\omega}(1+n) = \omega$. Step: $1+(\alpha+1) = (1+\alpha)+1 = \alpha+1$. Limit: $1+\lambda = \sup(1+\alpha) = \sup(\alpha) = \lambda$.

D4.12

1. Growth Lemma ($\gamma \leqslant \omega^\gamma$): We use transfinite induction on γ.
   - *Base:* $0 < 1 = \omega^0$.
   - *Successor:* If $\gamma \leqslant \omega^\gamma$, then $\gamma + 1 \leqslant \omega^\gamma + 1$. Since $\omega^\gamma \geqslant 1$, we have $\omega^\gamma + 1 \leqslant \omega^\gamma + \omega^\gamma = \omega^\gamma \cdot 2 < \omega^\gamma \cdot \omega = \omega^{\gamma+1}$. Thus $\gamma + 1 < \omega^{\gamma+1}$.
   - *Limit:* If λ is a limit ordinal and $\forall \xi < \lambda(\xi \leqslant \omega^\xi)$, then $\lambda = \sup_{\xi<\lambda} \xi \leqslant \sup_{\xi<\lambda} \omega^\xi$. By continuity of exponentiation, the RHS is ω^λ. Thus $\lambda \leqslant \omega^\lambda$.
2. The Degree: Let $\alpha > 0$. Consider $E = \{\gamma \in \mathrm{Ord} \mid \omega^\gamma \leqslant \alpha\}$. From the Growth Lemma, for any γ, if $\omega^\gamma \leqslant \alpha$, then $\gamma \leqslant \omega^\gamma \leqslant \alpha$. Thus $E \subseteq \alpha+1$, so E is a set. Let $\beta_1 = \sup E$. Since exponentiation is continuous and strictly increasing:
   - If $\beta_1 \in E$, then $\omega^{\beta_1} \leqslant \alpha$.
   - If $\beta_1 \notin E$ (limit case), then $\omega^{\beta_1} = \sup_{\gamma\in E} \omega^\gamma \leqslant \alpha$.

   Thus $\omega^{\beta_1} \leqslant \alpha$. However, $\beta_1 + 1 > \beta_1$, so $\beta_1 + 1 \notin E$ (otherwise $\sup E$ would be larger). Therefore, $\omega^{\beta_1+1} > \alpha$. We have found the unique β_1 such that $\omega^{\beta_1} \leqslant \alpha < \omega^{\beta_1+1}$.

3. Main Proof (Induction on α): Assume the theorem holds for all $\delta < \alpha$. Find the degree β_1 for α as in step 2. By the Division Theorem, divide α by ω^{β_1}:

$$\alpha = \omega^{\beta_1} \cdot k + \delta, \quad \text{where } \delta < \omega^{\beta_1}.$$

Analysis of k: Since $\alpha < \omega^{\beta_1+1} = \omega^{\beta_1} \cdot \omega$, the quotient k must be less than ω. Since $\alpha \geqslant \omega^{\beta_1}$, $k \geqslant 1$. *Analysis of δ:* Since $\delta < \omega^{\beta_1} \leqslant \alpha$, we have $\delta < \alpha$.

- If $\delta = 0$, $\alpha = \omega^{\beta_1} \cdot k$, and we are done.
- If $\delta > 0$, by the induction hypothesis, δ has a CNF: $\delta = \omega^{\gamma_1} c_1 + \ldots$. The leading term ω^{γ_1} must satisfy $\omega^{\gamma_1} \leqslant \delta$. Since $\delta < \omega^{\beta_1}$, we have $\omega^{\gamma_1} < \omega^{\beta_1}$, which implies $\gamma_1 < \beta_1$. Thus, appending the expansion of δ to $\omega^{\beta_1} \cdot k$ preserves the strictly decreasing order of exponents.

Uniqueness follows from the uniqueness of the degree β_1 and the uniqueness of quotient/remainder.

D4.13 1. Addition is not commutative. $\alpha+\beta = (\omega^\omega+\omega\cdot 2+1)+(\omega^\omega+\omega) = \omega^\omega + (\omega\cdot 2+1+\omega^\omega)+\omega$. Since $1+\omega^\omega = \omega^\omega$ and $\omega\cdot 2+\omega^\omega = \omega^\omega$, we get $\omega^\omega+\omega^\omega+\omega = \omega^\omega\cdot 2+\omega$. $\beta+\alpha = (\omega^\omega+\omega)+(\omega^\omega+\omega\cdot 2+1) = \omega^\omega\cdot 2+\omega\cdot 2+1$.

2. Multiplication: $\alpha\cdot 2 = (\omega^\omega+\ldots)\cdot 2 = \omega^\omega\cdot 2+\omega\cdot 2+1$. (Distributivity holds from the left, but here we multiply the order type). Actually, $(\omega^\omega+\delta)\cdot 2 = \omega^\omega\cdot 2+\delta$. $\beta\cdot\omega = (\omega^\omega+\omega)\cdot\omega = \omega^\omega\cdot\omega = \omega^{\omega+1}$. (Lower terms are absorbed).

D4.14

1. $n_1 = 2^k = p_1^k$. The index is 1. The exponent of ω is $1-1=0$.

$$c(2^k) = \omega^0 \cdot k = k.$$

$n_2 = 3^k = p_2^k$. The index is 2. The exponent of ω is $2-1=1$.

$$c(3^k) = \omega^1 \cdot k = \omega \cdot k.$$

$n_3 = 2^k 3^j 5^t = p_1^k p_2^j p_3^t$. Ordering indices descending: 3, 2, 1.

$$c(2^k 3^j 5^t) = \omega^{3-1}\cdot t + \omega^{2-1}\cdot j + \omega^{1-1}\cdot k = \omega^2\cdot t + \omega\cdot j + k.$$

2. The mapping translates the prime factorization into Cantor Normal Form where exponents are natural numbers. Since any finite set of natural

exponents and coefficients can be formed, the range covers all ordinals below ω^ω.

D4.15

1. First, recall base values: $o(1) = 0$, so for index 1 ($p_1 = 2$), exponent is $o(1) = 0$. For index 2 ($p_2 = 3$), exponent is $o(2) = \omega^0 \cdot 1 = 1$. For index 3 ($p_3 = 5$), exponent is $o(3) = \omega^{o(2)} \cdot 1 = \omega^1 = \omega$.
   $n_1 = 2^k$: index 1.
   $$o(2^k) = \omega^{o(1)} \cdot k = \omega^0 \cdot k = k.$$
   $n_2 = 3^k$: index 2.
   $$o(3^k) = \omega^{o(2)} \cdot k = \omega^1 \cdot k = \omega \cdot k.$$
   $n_3 = 2^k 3^j 5^t$: indices 1, 2, 3.
   $$o(2^k 3^j 5^t) = \omega^{o(3)} \cdot t + \omega^{o(2)} \cdot j + \omega^{o(1)} \cdot k = \omega^\omega \cdot t + \omega \cdot j + k.$$
2. Calculation of the sequence h_n: $h_0 = 1$. $h_1 = p_{h_0} = p_1 = 2$. $h_2 = p_{h_1} = p_2 = 3$. $h_3 = p_{h_2} = p_3 = 5$. $h_4 = p_{h_3} = p_5 = 11$. (Note: $h_5 = p_{11} = 31$).
3. Ordinal values: $o(h_0) = o(1) = 0$. $o(h_1) = o(2) = \omega^{o(1)} = \omega^0 = 1$. $o(h_2) = o(3) = \omega^{o(2)} = \omega^1 = \omega$. $o(h_3) = o(5) = \omega^{o(3)} = \omega^\omega$. $o(h_4) = o(11) = \omega^{o(5)} = \omega^{\omega^\omega}$. Generally, $o(h_{n+1}) = \omega^{o(h_n)}$.
4. Let $\alpha_n = o(h_n)$. We have the recurrence $\alpha_0 = 0$, $\alpha_{n+1} = \omega^{\alpha_n}$. The sequence is strictly increasing. Its limit $\lambda = \sup \alpha_n$ must satisfy $\lambda = \sup \omega^{\alpha_n}$. By continuity of exponentiation, $\sup \omega^{\alpha_n} = \omega^{\sup \alpha_n} = \omega^\lambda$. Thus, $\lambda = \omega^\lambda$. The ordinal ε_0 is defined as the *least* fixed point of the map $\xi \mapsto \omega^\xi$. Since our sequence starts at 0, it converges to the least fixed point. Thus $\lambda = \varepsilon_0$.

D4.16 $\aleph_1 + \aleph_0 \cdot \aleph_1 = \aleph_1 + \aleph_1 = \aleph_1$.

D4.17

1. κ (absorption).
2. κ (since $3 \leqslant \kappa$).
3. 2^κ (since $(\kappa^2)^\kappa = \kappa^\kappa = 2^\kappa$).

D4.18 1. $|V_0| = 0$, $|V_1| = 1$, $|V_2| = 2$, $|V_3| = 2^2 = 4$. (Generally $|V_{n+1}| = 2^{|V_n|}$ for finite n, tetration).

2. Induction. V_0 is transitive. If V_α is transitive, then for any $x \in V_{\alpha+1} = \mathcal{P}(V_\alpha)$, $x \subseteq V_\alpha$. If $y \in x$, then $y \in V_\alpha$. Since V_α is transitive, $y \subseteq V_\alpha$, so $y \in \mathcal{P}(V_\alpha) = V_{\alpha+1}$.

3. Proof by transfinite induction on α: We need to show that $\forall x(x \in V_\alpha \to \text{rank}(x) < \alpha)$.

- *Base case:* $\alpha = 0$. $V_0 = \emptyset$, so the premise $x \in V_0$ is false for any x. The implication holds vacuously.
- *Limit step:* Let $\alpha = \lambda$ be a limit ordinal. By definition, $V_\lambda = \bigcup_{\gamma<\lambda} V_\gamma$. If $x \in V_\lambda$, then there exists some $\gamma < \lambda$ such that $x \in V_\gamma$. By the induction hypothesis for γ, we have $\text{rank}(x) < \gamma$. Since $\gamma < \lambda$, it follows that $\text{rank}(x) < \lambda$.
- *Successor step:* Let $\alpha = \beta + 1$. By definition, $V_{\beta+1} = \mathcal{P}(V_\beta)$. If $x \in V_{\beta+1}$, then $x \subseteq V_\beta$. Thus, $\text{rank}(x) \leqslant \beta < \alpha$.

D4.19 *Part 1:* $\rho(x) \leqslant \text{rank}(x)$. Let $\text{rank}(x) = \alpha$. By definition, $x \subseteq V_\alpha$. This means for every $y \in x$, $y \in V_\alpha$. A property of the hierarchy is that $y \in V_\alpha \implies R(y) < \alpha$. (Proof: if $y \in V_\alpha$, then $y \in \mathcal{P}(V_\gamma)$ for some $\gamma < \alpha$ or union thereof, implying $y \subseteq V_\gamma$, so $\text{rank}(y) \leqslant \gamma < \alpha$). Thus, for all $y \in x$, $\text{rank}(y) < \alpha$. Assuming by induction that $\rho(y) = \text{rank}(y)$, we have $\rho(y) < \alpha$. Consequently, $\rho(y) + 1 \leqslant \alpha$. Taking the supremum over all $y \in x$, we get $\rho(x) \leqslant \alpha = \text{rank}(x)$.

Part 2: $\text{rank}(x) \leqslant \rho(x)$. Let $\rho(x) = \alpha$. By definition of supremum, for all $y \in x$, $\rho(y) + 1 \leqslant \alpha$, which implies $\rho(y) < \alpha$. By induction, $\text{rank}(y) < \alpha$. This implies $y \subseteq V_{\text{rank}(y)}$. Since V_ξ is cumulative, $V_{\text{rank}(y)} \subseteq V_\gamma$ for any $\gamma \geqslant \text{rank}(y)$. Also, $y \subseteq V_{\text{rank}(y)} \implies y \in V_{\text{rank}(y)+1}$. Since $\text{rank}(y) < \alpha$, we have $\text{rank}(y) + 1 \leqslant \alpha$. Thus $y \in V_{\text{rank}(y)+1} \subseteq V_\alpha$. So, for all $y \in x$, $y \in V_\alpha$. This means $x \subseteq V_\alpha$. By definition of $\text{rank}(x)$ as the minimum such ordinal, $\text{rank}(x) \leqslant \alpha = \rho(x)$.

D4.20

1. Image-based $\to$ Restriction-based. This direction is direct. Since the image $G[\alpha]$ is simply the range of the restriction $G \upharpoonright \alpha$, any dependency on the image can be expressed as a dependency on the restriction: $\Psi(h) = \Phi(\text{ran}(h))$.

2. Restriction-based $\to$ Image-based. Let a function F be defined by the generator Ψ: $F(\alpha) = \Psi(F \upharpoonright \alpha)$. We construct an auxiliary function $G(\alpha) = \langle \alpha, F(\alpha) \rangle$. Note that the image $G[\alpha] = \{\langle \beta, F(\beta) \rangle \mid \beta < \alpha\}$ is exactly the set of pairs constituting the function $f = F \upharpoonright \alpha$. From the set f, we can recover the ordinal α (as the set of first components, i.e., the domain) and the value $\Psi(f)$. Thus, we can define a generator Φ for G such that $\Phi(S) = \langle \text{dom}(S), \Psi(S) \rangle$, where S is treated as a relation. Then $G(\alpha) = \Phi(G[\alpha])$ is defined purely by image recursion. Finally, $F(\alpha)$ is obtained by taking the second projection of $G(\alpha)$. Conclusion: The schemes define the same class of functions (modulo coding).

D4.21

1. Suppose $f : \mathbb{Q} \to \alpha$ is an order isomorphism. Take any $q \in \mathbb{Q}$. Since $\mathbb{Q}$ has no minimal element (it extends to $-\infty$), there exists $r < q$. Then $f(r) < f(q)$. We can construct an infinite descending sequence $q_0 > q_1 > q_2 > \dots$ in $\mathbb{Q}$. This maps to an infinite descending sequence of ordinals $f(q_0) > f(q_1) > \dots$, which contradicts the well-foundedness of ordinals.
2. *Back-and-forth construction.* Let $A = \{a_0, a_1, \dots\}$ and $B = \{b_0, b_1, \dots\}$ be two countable dense linear orders without endpoints. We construct a sequence of partial isomorphisms $p_n : A_n \to B_n$ where A_n, B_n are finite subsets. *Stage 0:* Let $p_0 = \emptyset$. *Stage n (even):* Pick the first element $a \in A$ not in $\mathrm{dom}(p_n)$. We must map it to some $b \in B$ preserving order relative to already mapped elements. Since B is dense and has no endpoints, such a b always exists (in the corresponding interval). Define $p_{n+1} = p_n \cup \{\langle a, b\rangle\}$. *Stage n (odd):* Pick the first element $b \in B$ not in $\mathrm{ran}(p_n)$. Find a preimage $a \in A$ preserving order (possible by density/no endpoints of A). Define $p_{n+1} = p_n \cup \{\langle a, b\rangle\}$. The union $P = \bigcup p_n$ is the required isomorphism.

References

[1] Vladimir I. Arnold. *Huygens and Barrow, Newton and Hooke: Pioneers in mathematical analysis and catastrophe theory from evolvents to quasicrystals*. Basel: Birkhäuser, 1990.

[2] Lev D. Beklemishev. "Gödel's incompleteness theorems and the limits of their applicability. I". In: *Russian Mathematical Surveys* 65.5 (2010). Original paper at http://www.mi-ras.ru/~bekl/Papers/goedel-uspehi.pdf, pp. 857–899. DOI: 10.1070/RM2010v065n05ABEH004703.

[3] Guram Bezhanishvili, Leo Esakia, and David Gabelaia. "Some Results on Modal Axiomatization and Definability for Topological Spaces". In: *Studia Logica* 81.3 (2005), pp. 325–355.

[4] Guram Bezhanishvili, Ray Mines, and Patrick J. Morandi. "Scattered, Hausdorff-reducible, and hereditarily irresolvable spaces". In: *Topology and its Applications* 132 (2003), pp. 291–306.

[5] George Boole. *An Investigation of the Laws of Thought, on Which are Founded the Mathematical Theories of Logic and Probabilities*. London: Walton and Maberly, 1854.

[6] George Boolos. *The Logic of Provability*. Cambridge: Cambridge University Press, 1993.

[7] Nicolas Bourbaki. "L'architecture des mathématiques". In: *Les grands courants de la pensée mathématique*. Ed. by François Le Lionnais. Paris: Cahiers du Sud, 1948.

[8] Nicolas Bourbaki. *Théorie des ensembles*. English translation: *Theory of Sets*, Springer, 2004. Paris: Hermann, 1970.

N. I. Kazimirov, *Mathematics as a Foreign Language*, Mathematics in Mind,
https://doi.org/10.1007/978-3-032-30721-7

[9] Felix E. Browder, ed. *Mathematical Developments Arising from Hilbert Problems*. Vol. 28. Proceedings of Symposia in Pure Mathematics. American Mathematical Society, 1976.

[10] Wilfried Buchholz. "Explaining Gentzen's Consistency Proof within Infinitary Proof Theory". In: *Computational Logic and Proof Theory (KGC '97)*. Ed. by Georg Gottlob, Alexander Leitsch, and Daniele Mundici. Vol. 1289. Lecture Notes in Computer Science. Author's manuscript: https://www.mathematik.uni-muenchen.de/~buchholz/articles/kgc97.pdf. Berlin: Springer, 1997, pp. 4–17.

[11] Georg Cantor. *Contributions to the Founding of the Theory of Transfinite Numbers*. Trans. by Philip E. B. Jourdain. Chicago: Open Court, 1915.

[12] Noam Chomsky. *Syntactic Structures*. The Hague/Paris: Mouton, 1957.

[13] Paul J. Cohen. *Set Theory and the Continuum Hypothesis*. New York: W. A. Benjamin, 1966.

[14] Irving M. Copi, Carl Cohen, and Kenneth McMahon. *Introduction to Logic*. 14th. London: Pearson, 2014. ISBN: 978-1-292-02482-0.

[15] Richard Courant and Herbert Robbins. *What Is Mathematics?* Oxford University Press, 1941.

[16] Heinz-Dieter Ebbinghaus, Jörg Flum, and Wolfgang Thomas. *Mathematical Logic*. 3rd. Vol. 291. Graduate Texts in Mathematics. New York: Springer, 2021.

[17] Herbert B. Enderton. *A Mathematical Introduction to Logic*. 2nd. San Diego: Academic Press, 2001.

[18] Yuri L. Ershov. *Definability and Computability*. New York: Plenum Press, 1996.

[19] Yuri L. Ershov and Evgeniy A. Palyutin. *Mathematical Logic*. Moscow: Mir Publishers, 1984.

[20] Leo Esakia. "Intuitionistic logic and modality via topology". In: *Annals of Pure and Applied Logic* 127 (2004), pp. 155–170.

[21] Solomon Feferman. "Arithmetization of Metamathematics in a General Setting". In: *Fundamenta Mathematicae* 49.1 (1960), pp. 35–92.

[22] Solomon Feferman. *Three conceptual problems that bug me*. Lecture at 7th Scandinavian Logic Symposium, Uppsala. 1996.

[23] Gottlob Frege. *Begriffsschrift, eine der arithmetischen nachgebildete Formelsprache des reinen Denkens*. Halle: Louis Nebert, 1879.

[24] Gerhard Gentzen. "Neue Fassung des Widerspruchsfreiheitsbeweises für die reine Zahlentheorie". In: *Forschungen zur Logik und zur Grundlegung der exakten Wissenschaften*. Vol. 4. Neue Folge. Leipzig: S. Hirzel, 1938, pp. 19–44.

[25] Kurt Gödel. "Die Vollständigkeit der Axiome des logischen Funktionenkalküls". In: *Monatshefte für Mathematik und Physik* 37 (1930), pp. 349–360.

[26] Kurt Gödel. "Über formal unentscheidbare Sätze der Principia Mathematica und verwandter Systeme I". In: *Monatshefte für Mathematik und Physik* 38 (1931), pp. 173–198.

[27] Derek Goldrei. *Propositional and Predicate Calculus: A Model of Argument*. London: Springer, 2005.

[28] Ronald L. Graham, Donald E. Knuth, and Oren Patashnik. *Concrete Mathematics: A Foundation for Computer Science*. Addison-Wesley, 1989.

[29] Andrzej Grzegorczyk. "Undecidability without arithmetization". In: *Studia Logica* 79 (2005), pp. 163–230.

[30] Petr Hájek and Pavel Pudlák. *Metamathematics of First-Order Arithmetic*. Springer-Verlag, 1993.

[31] Felix Hausdorff. *Grundzüge der Mengenlehre*. Leipzig: Von Veit, 1914.

[32] Leon Henkin, J. Donald Monk, and Alfred Tarski. *Cylindric Algebras, Part I*. Amsterdam: North-Holland, 1971.

[33] James M. Henle. *An Outline of Set Theory*. Problem Books in Mathematics. New York: Springer-Verlag, 1986.

[34] David Hilbert and Paul Bernays. *Grundlagen der Mathematik*. Two volumes; Vol. II appeared in 1939. Second edition: 1968 (Vol. I), 1970 (Vol. II). Berlin: Springer, 1934.

[35] Douglas R. Hofstadter. *Gödel, Escher, Bach: an Eternal Golden Braid*. A classic exploration of formal systems, cognition, and self-reference, awarded the Pulitzer Prize. New York: Basic Books, 1979.

[36] Karel Hrbacek and Thomas Jech. *Introduction to Set Theory*. 3rd. New York: Marcel Dekker, 1999.

[37] Thomas Jech. *Set Theory*. The Third Millennium Edition. Berlin: Springer-Verlag, 2003.

[38] Thomas J. Jech. *The Axiom of Choice*. Amsterdam: North-Holland, 1973.

[39] Richard Kaye. *Models of Peano Arithmetic*. Oxford: Clarendon Press, 1991.

[40] Laurie Kirby and Jeff Paris. "Accessible independence results for Peano arithmetic". In: *Bulletin of the London Mathematical Society* 14.4 (1982), pp. 285–293.

[41] Stephen C. Kleene. *Introduction to Metamathematics*. Amsterdam: North-Holland, 1952.

[42] Stephen C. Kleene. *Mathematical Logic*. New York: John Wiley & Sons, 1967.

[43] Morris Kline. *Mathematics: The Loss of Certainty*. Oxford University Press, 1980.

[44] Kenneth Kunen. *Set Theory: An Introduction to Independence Proofs*. Amsterdam: North-Holland, 1980.

[45] Kazimierz Kuratowski and Andrzej Mostowski. *Set Theory*. Amsterdam: North-Holland, 1976.

[46] Elliott Mendelson. *Introduction to Mathematical Logic*. Princeton: D. Van Nostrand, 1964.

[47] Ieke Moerdijk and Jaap van Oosten. *Sets, Models and Proofs*. Springer Undergraduate Mathematics Series. Springer, 2018. ISBN: 978-3-319-92413-7.

[48] Charles Petzold. *Code: The Hidden Language of Computer Hardware and Software*. Microsoft Press, 1999.

[49] Helena Rasiowa and Roman Sikorski. *The Mathematics of Metamathematics*. Warsaw: PWN – Polish Scientific Publishers, 1963.

[50] Constance Reid. *Hilbert*. Springer-Verlag, 1970.

[51] Walter Rudin. *Principles of Mathematical Analysis*. 3rd. New York: McGraw-Hill, 1976.

[52] Kurt Schütte. *Proof Theory*. Berlin: Springer-Verlag, 1977.

[53] Stewart Shapiro. *Foundations without Foundationalism: A Case for Second-Order Logic*. Oxford: Clarendon Press, 1991.

[54] Joseph R. Shoenfield. *Mathematical Logic*. Reading, MA: Addison-Wesley, 1967.

[55] Peter Smith. *An Introduction to Gödel's Theorems*. 2nd. Cambridge: Cambridge University Press, 2013.

[56] Raymond M. Smullyan. *Theory of Formal Systems*. Princeton: Princeton University Press, 1961.

[57] Marshall H. Stone. "The Theory of Representations for Boolean Algebras". In: *Transactions of the American Mathematical Society* 40.1 (1936), pp. 37–111.

[58] Patrick Suppes. *Axiomatic Set Theory*. New York: Dover Publications, 1972.

[59] Alfred Tarski. "Sur quelques théorèmes qui équivalent à l'axiome du choix". In: *Fundamenta Mathematicae* 5.1 (1924), pp. 147–154.

[60] Alfred Tarski. *A decision method for elementary algebra and geometry*. Berkeley: University of California Press, 1951.

[61] Anne Sjerp Troelstra and Helmut Schwichtenberg. *Basic Proof Theory*. 2nd. Vol. 43. Cambridge Tracts in Theoretical Computer Science. Cambridge: Cambridge University Press, 2000.

[62] Nikolai K. Vereshchagin and Alexander Shen. *Computable Functions*. Vol. 19. Student Mathematical Library. American Mathematical Society, 2003.

[63] Vladimir A. Zorich. *Mathematical Analysis I, II*. Berlin: Springer-Verlag, 2015.

List of Notations

$\mathfrak{Math}$	Informal mathematical language (book)
$\neg\varphi$	Negation of φ
$\varphi \wedge \psi$	Conjunction (AND)
$\varphi \vee \psi$	Disjunction (OR)
$\varphi \rightarrow \psi$	Implication
$\varphi \leftrightarrow \psi$	Equivalence (iff)
$\varphi \equiv \psi$	Formula equivalence (same truth table)
$\vdash$	Syntactic derivability
$\Vdash$	Semantic consequence
$\models$	Truth in a model
$\top$	Tautology (truth constant)
$\bot$	Contradiction (falsity constant)
$\stackrel{\text{def}}{\rightleftharpoons}$	Definitional equality
$\stackrel{\text{def}}{\leftrightarrow}$	Definitional equivalence
$\mathrm{Th}(\mathcal{M})$	Complete theory of $\mathcal{M}$
$\mathrm{Pr}(n, m)$	Formula n has proof m (coding)
$\mathrm{Con}(T)$	Consistency of theory T
$\ulcorner\gamma\urcorner$	Gödel code of γ
PA	Peano Arithmetic

N. I. Kazimirov, *Mathematics as a Foreign Language*, Mathematics in Mind,
https://doi.org/10.1007/978-3-032-30721-7

Q	Robinson Arithmetic
ZF	Zermelo–Fraenkel set theory
ZFC	ZF with Choice
HF	Hereditarily finite sets (theory)
AC	Axiom of Choice
AC_ω	Countable Choice
AC_ω^{fin}	Countable Choice (finite pieces)
ZT	Zermelo's well-ordering theorem
ZL	Zorn's Lemma
CH	Continuum Hypothesis
GB	Gödel–Bernays set theory
$\forall x\ \varphi(x)$	Unbounded universal quantifier
$\exists x\ \varphi(x)$	Unbounded existential quantifier
$\exists! x\ \varphi(x)$	Unique existence (quantifier $\exists!$)
$\forall x \in A : \varphi(x)$	Bounded $\forall$ (subset of A)
$\exists x \in A : \varphi(x)$	Bounded $\exists$ (in A)
$\forall x < k : \varphi(x)$	Bounded $\forall$ (numeral bound)
$\exists x < k : \varphi(x)$	Bounded $\exists$ (numeral bound)
$\forall \alpha : \varphi(\alpha)$	Quantification over ordinals
$\exists \alpha : \varphi(\alpha)$	$\exists$ over ordinals
$\emptyset, \{\}$	Empty set
$a \in b$	a is an element of set b
$a \notin b$	Negation of $a \in b$
$A \subseteq B$	Set A is a subset of B
$A \subsetneq B$	Set A is a proper subset of B
$\{a, b, \dots, c\}$	The set consisting of elements $a, b, \dots, c$
$\{x \mid \varphi(x)\}$	Class/set comprehension (x satisfies φ)
$A \cup B$	Union of A and B
$A \cap B$	Intersection of sets A and B
$\bigcup A$	Union of all elements of set A
$\bigcap A$	Intersection of all elements of set A

$A \setminus B$	Difference of sets A and B
$A \triangle B$	Symmetric difference of sets A and B
$\mathcal{P}(A)$	The set of all subsets of A (powerset)
$A \sqcup B$	Disjoint union of sets A and B
$\langle a, b \rangle$	Ordered pair
$A \times B$	Direct (Cartesian) product of sets A and B
$TC(A)$	Transitive closure of set A
$\mathrm{rank}(x)$	Rank of set x
V_α	Stage α of the von Neumann cumulative hierarchy
$S(x)$	Successor operation (in PA and ZF)
$\omega, \aleph_0$	First infinite ordinal
$\lvert A \rvert,\ \#A$	Cardinality of set A
$\mathrm{Card}(A)$	Equinumerosity class of A
$\mathfrak{c}, 2^{\aleph_0}$	Cardinality of continuum
$A \simeq B$	A, B equinumerous
$A \preccurlyeq B$	$\lvert A \rvert \leqslant \lvert B \rvert$ ($\exists$ injection)
$A \prec B$	Strict $\lvert A \rvert < \lvert B \rvert$
Ord	Class of all ordinals
Succ, Lim	Successor / limit ordinals
$ot(X, <)$	Order type (well-order)
$\mathfrak{h}(A)$	Hartogs number of A
$f : A \to B$	Function f from set A to set B
$f : A \leftrightarrow B$	Bijection between sets A and B
$\hookrightarrow$	Embedding (injection), often order-preserving
$\mathrm{dom}(f)$	Domain of function f
$\mathrm{ran}(f)$	Range of function f
$f \restriction D$	Restriction of function f to set D
f^{-1}	Inverse function
R^{-1}	Inverse relation
χ_A	Indicator function of set A
$\mathrm{supp}(f)$	Support of function f

$F^{fin}(Y, X)$	The set of functions from Y to X with finite support
$\text{pr}_A(p)$	Projection onto the A component
aRb	$\langle a, b\rangle \in R$ (elements a, b are in relation R)
$[a]_R$	Equivalence class of element a under relation R
$A/_R$	Quotient of A by equivalence relation R
$\mathfrak{A} \cong \mathfrak{B}$	Structures $\mathfrak{A}$ and $\mathfrak{B}$ are isomorphic
$\mathbb{N}, \mathbb{Z}, \mathbb{Q}, \mathbb{R}$	Number systems (standard)
$a = b$	Equality
$a < b, a > b$	Strict inequality
$a \leqslant b, a \geqslant b$	Non-strict inequality
$a + b$	Addition
$a \cdot b, ab$	Multiplication
$a \dot{-} b$	Monus (truncated subtraction)
$a \mid b$	a divides b
$a \equiv b \pmod{m}$	a is congruent to b modulo m
$\underline{n}$	Numeral (S-iterates of 0)
$\beta(c, d, i)$	Gödel's beta function for sequence encoding
$\text{Code}(S)$	Function for encoding a set as a natural number
Λ	Empty string (in grammars)
$\max A, \min A$	Extremum in ordered A
$\sup A, \inf A$	Supremum / infimum

Index

A

Abduction 69
Absolute set theory 172
Ackermann's coding for HF-sets 163
Algebra of sets 206
Alphabet 6, 230
Antecedent 63
Atomic formula 99
Automorphism 272
Axiom 16
Axiom of Choice 179, 182, 202, 381
  countable 186, 382
  equivalents of 182, 386
Axiom of Determinacy 187
Axiomatics 104
Axioms of Peano Arithmetic 132, 289
Axioms of predicate logic 263
Axioms of propositional logic 256
Axioms of Set Theory 150, 154, 171, 333

B

Banach-Tarski paradox 186
Barber paradox 18
Bernays' rules 104, 264
Binding (operator) 22
Boolean algebra 259, 339
Bracket notations 161
Bézout's identity 313

C

Canonical bracket notation 163
Cantor's Theorem 343
Cantor–Bernstein–Schroeder Theorem 345
Cardinal 204, 277, 377
Cardinal arithmetic 378
Cardinality 179, 378
Cartesian product of sets 187
Categoricity of a theory 277
Chain 184
  bounded 184
Chinese Remainder Theorem 316
Choice function 180
Class of sets 174
Closed formula 102
Compactness Theorem 276
Completeness of a calculus 94, 257, 260
Completeness of a theory 120, 278
Comprehension 23
Computable numbers 328
Concept 41
Concept, definition of 41
Concept, extension of 41
Congruence 107, 170, 303, 334
Congruence axioms 108
Conjunction 49, 56
Consequent 63

N. I. Kazimirov, *Mathematics as a Foreign Language*, Mathematics in Mind,
https://doi.org/10.1007/978-3-032-30721-7

Consistency of a theory 115, 117, 268
Constant symbols 99
Continuum 346
Continuum Hypothesis 346
Contraposition 72
Convolution 23

D

Deduction 72
Deduction Theorem 95, 104, 262, 264
Definability via signature 273
Definable mapping 171, 177
Definition, direct 41
Definition, indirect 41
Definition, self-referential 146
Derivability relation 104
Derivability, interpretation in algebra 261
Designation 19, 41
Difference of sets 53
Disjunction 49
Disjunctive normal form 256
Divisible abelian groups 113, 282
Division Algorithm 310
Domain of interpretation 267

E

Elementary theory 217
Equality of sets 53, 154, 162, 170
Equinumerosity of sets 178, 342
Equivalence 49, 96, 116, 254
Equivalence relation 176
Euclid's Lemma 314
Euler-Venn diagram 67

F

Falsity 48, 254
First-order language 102
Formal theory 104
Formalization scheme 40
Formula 14, 51, 247
  satisfiable 268
  valid 268
Formula, propositional 49
Formulas of class Δ_0 137
Function 79, 176, 178
  domain of 176
  n-ary 201
  range of 176
Function symbols 99
Fundamental Theorem of Arithmetic 314, 323

G

Gauss's Lemma 314
Generalization rule 264
Gödel–Bernays set theory 175
Grapheme 6
  auxiliary 232
  basic 230
  derived 232
  elementary 229
  grouping 230
  independent 230
  operator 230
  rigid 233
  separating 230
Gödel numbering 324
Gödel's beta function 317
Gödel's completeness theorem 117, 270
Gödel's incompleteness theorem 120, 278, 326
Gödel–Maltsev compactness theorem 276
Gödelian theory 278

H

Hartogs number 390
Hausdorff's maximal principle 182
Hereditarily finite sets 154
Homomorphism 259, 272

I

Image under a function 177
Implication 49
Implication, material 63
Implication, relevance 63
Inclusion of sets 53

Independence of axioms 104
Independent formula 119
Induction 73, 134, 138, 306
  transfinite 194, 355
Induction, complete (course-of-values) 138, 306
Induction, mathematical 74
Inductive set 173
Inference rules 103
Interpretation of a language 110, 267
Interpretation of formulas 268
Intersection of sets 53
Isomorphism 216, 272

K

Kuratowski's ordered pair 159

L

Lattice 339
Law of double negation 57, 254
Law of the excluded middle 91
Lexeme 7, 233, 237
Lexeme, typed 247
Liar's paradox 18
Lindenbaum algebra 258
Linear order 176
Logic 38
Logical axioms 103
Logical connective 49
Logical consequence 116, 268
Łoś-Vaught test 280
Löwenheim-Skolem theorem
  downward 274
  upward 277

M

Mathematical concept 42, 174
Matryoshka principle 7
Maximal element 183
Maximum 183
Membership 53
Metalanguage 17
Minimal element 183
Minimum 183
Model of a language 110
Model of a theory 110, 208
Model of the integers 210
Model of the rational numbers 211
Model of the real numbers 212
Modus Ponens rule 92, 256
Monus 309
Morley's categoricity theorem 277

N

Natural numbers
  von Neumann 173
Negation 49, 57
Non-logical axioms 105
Non-terminal symbols 100
Normal model 112

O

Operation
  n-ary 201
Order in arithmetic 137
Order, linear 78
Order, partial 78
Ordered set 176
Ordinal 191, 192, 354
  limit 194, 354
  ω, definition of 173
  successor 194, 354
Ordinal arithmetic 197, 363
  addition 196
  multiplication 196

P

Parse tree 10, 239
Partial order 176
Peano Arithmetic (PA) 129, 282
Permutation group 179
Pigeonhole Principle 351
Predicate 14, 98
Predicate calculus 52, 60
Predicate logic 52, 60, 98, 263
Predicate logic formula 101
Predicate symbols 100
Preimage under a function 177

Preorder 176
Presburger Arithmetic 282
Prime number 44
Principle of duality 341
Principle of infinite descent 140, 308
Propositional formula 88
Propositional logic 50, 88, 253
Propositional variables 253
Pure predicate theory 105

Q

Quantifier 60
  bounded 137, 169
  uniqueness 169
Quantifier elimination 280
Quantifier, existential 60
Quantifier, universal 60
Quine's model 209

R

Rank of a set 191, 199, 361
Recursion generator 194
Recursive definition 194, 235, 355
Reduction 27
Reflection 17
Relation 74, 175
  antisymmetric 176
  connex 176
  definable (class) 177
  functional 176
  induced 184
  inverse 175
  irreflexive 175
  n-ary 201
  preorder 176
  reflexive 175
  symmetric 175
  total 176
  transitive 176
Relation, antisymmetric 78
Relation, equivalence 76
Relation, n-ary 75
Relation, preorder 76
Relation, reflexive 75
Relation, symmetric 75
Relation, transitive 76
Robinson Arithmetic (Q) 282, 291
Rule of substitution 91
Rule, Modus Ponens 71
Rule, Modus Tollens 72
Russell's paradox 146, 343

S

Satisfiability of a theory 268
Semantic consequence 116, 268
Semantics 13
Sentential logic 50
Sequence 188
Set
  countable 343
  Dedekind-finite 353
  Dedekind-infinite 353
  finite 178, 343
  infinite 178, 343
  transitive 192
Set exponentiation 188
Signature 101
Soundness of derivation 93, 116, 257, 269
Strict order relation 105
Strong definition of inequality 298
Structure
  universe of 202
Structure (algebraic) 202
Substitution operator 247
Supremum and infimum 183
Syllogism 67
Synonym 27
Syntax 13

T

Tarski's undefinability theorem 327
Tarski-Seidenberg theorem 281
Tautology 90, 93, 254
Template, derived 236
Template, linear 242
Template, simple 235
Template, substitution 234
Tennenbaum's theorem 276
Term 14, 51, 100, 247

Theorem 16
Theorem on automorphisms 272
Theorem on functional completeness 96, 255
Theorems of a theory 104
Theory of dense orders 121, 282
Theory of equality 106, 282
Theory of groups 112, 282
Theory of linear orders 111
Theory of real closed fields 209
Theory of semigroups 107
Theory of strict orders 105
Truth 48, 254
Truth in a model 268
Truth table 57, 254

U

Ultrafilter 259
Union of sets 53
Universal closure 61
Universe 45
Upper/lower bound 183

V

Valid formula 115
Valuation, assignment of variables 110, 254, 267
Variables 21, 246
  bound 23, 46, 99
  free 23, 46, 99
Variables, propositional 49
von Neumann universe 198, 361

W

Well-foundedness principle 139, 308
Well-ordering 176

Z

Zermelo's theorem 182
Zermelo–Fraenkel set theory 169, 282
Zorn's lemma 182

MIX
Papier aus verantwortungsvollen Quellen
Paper from responsible sources
FSC® C105338
FSC
www.fsc.org

If you have any concerns about our products,
you can contact us on
ProductSafety@springernature.com

In case Publisher is established outside the EU,
the EU authorized representative is:
Springer Nature Customer Service Center GmbH
Europaplatz 3, 69115 Heidelberg, Germany

Printed by Libri Plureos GmbH
in Hamburg, Germany